New Directions for Situated Cognition in Mathematics Education

Mathematics Education Library

VOLUME 45

The titles published in this series are listed at the end of this volume.

Anne Watson
Peter Winbourne
(Editors)

New Directions for Situated Cognition in Mathematics Education

Springer

Ann Watson
University of Oxford
Oxford
UK
Anne.watson@edstud.ox.ac.uk

Peter Winbourne
London South Bank Unviersity
London
UK
winboupc@lsbu.ac.uk

Series Editor:
Alan Bishop
Monash University
Melboure 3800
Australia
Alan.Bishop@Education.monash.edu.au

Library of Congress Control Number: 2007939044

ISBN -13: 978-0-387-71577-3 e-ISBN-13: 978-0-387-71579-7

Printed on acid-free paper.

9 8 7 6 5 4 3 2 1

springer.com

Contents

Contributing Authors vii

Introduction 1
ANNE WATSON AND PETER WINBOURNE

School Mathematics As A Developmental Activity 13
STANISLAV ŠTECH

Participating In What? Using Situated Cognition Theory To Illuminate Differences In Classroom Practices 31
MARIA MANUELA DAVID AND ANNE WATSON

Social Identities As Learners And Teachers Of Mathematics 57
MIKE ASKEW

Looking For Learning In Practice: How Can This Inform Teaching 77
PETER WINBOURNE

Are Mathematical Abstractions Situated? 101
MEHMET FATIH OZMANTAR AND JOHN MONAGHAN

'We Do It A Different Way At My School' 127
MARTIN HUGHES AND PAMELA GREENHOUGH

Situated Intuition And Activity Theory Fill The Gap 151
JULIAN WILLIAMS, LIORA LINCHEVSKI, AND BILHA KUTSCHER

The Role Of Artefacts In Mathematical Thinking: A Situated Learning Perspective 177
MADALENA PINTO DOS SANTOS AND JOÃO FILIPE MATOS

Exploring Connections Between Tacit Knowing And Situated Learning Perspectives In The Context Of Mathematics Education 203
CRISTINA FRADE AND JORGE TARCÍSIO DA ROCHA FALCÃO

Cognition And Institutional Setting 231
ERHAN BINGOLBALI AND JOHN MONAGHAN

School Practices With The Mathematical Notion Of Tangent Line 259
MÁRCIA PINTO AND VALÉRIA MOREIRA

Learning Mathematically As Social Practice In A Workplace Setting 285
BRIAN HUDSON

Analysing Concepts Of Community Of Practice 301
CLIVE KANES AND STEPHEN LERMAN

'No Way Is Can't': A Situated Account Of One Woman's Uses And Experiences Of Mathematics 327
SANDRA WILSON, PETER WINBOURNE, AND ALISON TOMLIN

Acknowledgements 351

Index of Authors 353

Index 357

Contributing Authors

Erhan Bingolbali is a lecturer in the School of Education, University of Firat, Turkey. He obtained his PhD in Mathematics Education from the University of Leeds in 2005. He researches mathematical thinking from an institutional perspective: learning and teaching at university and school levels, and service teaching. He is particularly interested in relationships amongst practice, knowledge and identity and how institutions, as communities of practice, influence individuals' becoming and knowledge development.

Maria Manuela David works in the Faculty of Education of the Federal University of Minas Gerais, Brazil. She teaches mathematics education on both undergraduate and graduate courses in this institution. Her current research interests include: contrasting the logical development of a mathematics topic and its conceptual development at school; teacher-student interactions in mathematics classrooms; participation and learning in different classroom practices.

Jorge Tarcísio da Rocha Falcão is a teacher and researcher at Universidade Federal de Pernambuco, Brazil. His doctorate was in the psychology of learning. His research interests have ranged from the learning of scientific concepts at school with ICT (mainly LOGO) to success in mathematics seen as an enculturation process. Currently, he is interested in the wide range of competences people must develop and use in situated contexts, such as professional and commercial situations.

Cristina Frade is a lecturer at Universidade Federal de Minas Gerais, Brazil. She teaches mathematics at a secondary school linked to the Faculty of Education, and mathematics education at the Graduate Programme of Education. Her research areas are: the tacit-explicit dimension of mathematics practice in and out of school contexts; situated learning and communities of practice; interdisciplinary mathematics and science school practices; psychology, culture and affect in mathematics education.

Pamela Greenhough worked as a primary school teacher for 15 years, holding a variety of positions including that of acting head teacher. More recently, she has been employed as a Research Fellow at the Universities of Exeter and Bristol, UK. She has worked on a number of projects funded by the ESRC including 'Homework and its Contribution to Learning' and 'The Home School Knowledge Exchange Project'. Currently, she is investigating learning out of school with Professor Martin Hughes for his ESRC Fellowship.

Brian Hudson is Professor in Educational Work (Pedagogiskt Arbete) for ICT and Learning at Department of Interactive Media and Learning in the Faculty of Teacher Education, Umeå University, Sweden, and a member of the Umeå Forskningscentrum för Matematikdidaktik (Mathematics Education Research Centre). He also works for part of his time as Professor of Education and is a National Teaching Fellow at Sheffield Hallam University.

Martin Hughes is Professor of Education at the University of Bristol, previously at the Universities of Exeter and Edinburgh, and at the Thomas Coram Research Unit, LIE, UK. He has researched widely young children's learning, focusing on mathematics, computers, and home-school relationships. Previous authored and co-authored books include *Children and Number* and *Numeracy and Beyond*. He has an ESRC professorial fellowship and works with Pamela Greenhough on out-of-school learning.

Clive Kanes is a Senior Lecturer in the Department of Education and Professional Studies at King's College London, UK. He uses activity theory in exploring cultural-historical and mediated forms of human activity in the field of education. His principal research focus is on theories of object orientation and his main current project is a critique of epistemological foci from this stance. He has published numerous works and is editor of a forthcoming book on developing professional practice to be published by Springer.

Bilha Kutscher works in the S. Amitzur Unit for Research in Mathematics Education, Hebrew University and in the David Yellin College of Education, Israel. Her research interests involve teaching and learning mathematics, especially with students-at-risk at both the primary and secondary levels. She enjoys being involved with research projects that develop pedagogical tools, student learning material and teacher professional development for the benefit of at-risk student populations.

Stephen Lerman was a secondary mathematics teacher in the UK and Israel before moving into research teacher education. He has been Chair of the British Society for Research in Learning Mathematics and President of the International Group for the Psychology of Mathematics Education. He is now Professor of Mathematics Education at London South Bank University, UK, and his research interests are in socio-cultural theories of learning and teaching, sociological perspectives on equity, and classroom research.

Liora Linchevski is the Director of S. Amitzur Unit for Research in Mathematics Education at the Hebrew University of Jerusalem, and also teaches at the David Yellin Teachers College, Israel. Her research interests involve the transition from arithmetic to algebra, teaching and learning mathematics in heterogeneous classes, and working with students-at-risk at both the primary and secondary levels. She works on projects including Israeli-Palestinian co-operation and with at-risk communities in Australia.

João Filipe Matos is Professor of Education at the University of Lisbon, Portugal. He researches mathematics education and ICT from a critical, social and political point of view. He has directed a number of research projects funded by national and European agencies. He has served as committee member for the International Group for the Psychology of Mathematics Education and the Mathematics Education and Society conference series, and coordinates the Competence Centre CRIE for ICT in education.

John Monaghan works in the Centre for Studies in Mathematics Education, University of Leeds, UK. He teaches on undergraduate, masters and preservice courses. He particularly enjoys doctoral supervision and his co-authors in this book are previous doctoral students. His research interests include students' understanding of calculus and of algebra, linking school mathematics with out-of-school activities and the use of technology in mathematics with particular interest in 'computer algebra' and in how teachers make use of technology.

Valéria Moreira teaches mathematics at Centro Federal de Educação Tecnológica in Januária, Minas Gerais, Brazil. She is a former secondary school teacher who taught across the 11-18 age groups. Her experience encompasses schools from public and private systems. When a student, Valéria received a grant from the Brazilan government to participate in the research project *Investigating the transition from school to university* which led to her Masters degree.

Mehmet Fatih Ozmantar is Assistant Professor in the School of Education, University of Gaziantep, Turkey. He received a PhD in Mathematics Education from the University of Leeds. His research interests involve teaching and learning in school, human interaction in mathematics learning and the issue of identity. He is particularly interested in construction and use of mathematical knowledge in one-to-one tutor-student as well as classroom environments, and enjoys working closely with teachers and students.

Márcia Maria Fusaro Pinto is a lecturer at Universidade Federal de Minas Gerais in Belo Horizonte, Brazil. Her experience includes teaching students from different vocational courses, including those studying to be mathematicians and those taking courses in initial mathematics teacher education. Her research interest is in teaching and learning at secondary school and at university levels. The focus is on technologies and on the analysis of classroom interactions within socio-cultural theoretical perspectives.

Madalena Santos is a member of the Centre for Research in Education at the University of Lisbon, Portugal, where she graduated in mathematics and received a PhD in mathematics education. She works on a national programme, CRIE, supporting schools in the use of ICT. She has also worked on two major research projects in mathematical thinking and learning, taking a situated perspective. She has published a number of articles and book chapters in international publications.

Stanislav Štech is Professor in Educational Psychology at Charles University Prague, Czech Republic. His interests have developed from sociocultural theory to identifying psychological assumptions, and then to empirical research about learning. Adopting a cultural psychological perspective he has developed an understanding of teaching/learning processes as domain-specific through studying them in history, biology, and mathematics. He has published books, articles and research papers mainly in Czech and in French.

Alison Tomlin worked for about 20 years in adult education, mainly in literacy and numeracy, in South London, UK. Following research in adult numeracy education, she worked for King's College London as a researcher on projects including numeracy in primary, adult and further education.

Anne Watson taught mathematics in comprehensive schools before becoming a researcher and teacher educator. She has published many books and articles for teachers as well as pursuing research into improving mathematics teaching and learning, particularly where underachievement is an issue of social justice. She uses socio-cultural ideas alongside those of mathematics itself. She is Reader in Mathematics Education at the University of Oxford, UK.

Julian Williams taught mathematics in schools before entering teaching and research at the University of Manchester, UK, where he is Professor of Mathematics Education. He recently helped found the BERA interest group on sociocultural and cultural-historical activity theory, interests reflected in *Children's mathematics 4-15* (with Julie Ryan). He leads a research project on participation in mathematics that involves narrative perspectives on students' identity as well as a longitudinal measurement design.

Sandra Wilson went to school in Scotland and took up adult education in mathematics in London, UK. She worked as a legal officer for an Inner London borough and is now retired. Sandra has contributed to several articles written from students' perspectives.

Peter Winbourne is Reader in Educational Development at London South Bank University, UK. He was a teacher of mathematics and advisory teacher in London schools for eighteen years before moving into higher education in the early 1990's and working and researching with teachers and students. His main research focus is the development and application of theories of situated cognition and identity. He also develops research into the impact of work-based professional development.

Chapter 1

Introduction

Anne Watson and Peter Winbourne
University of Oxford, London South Bank University

1. INTRODUCTION

In 1997, forty invited mathematics educators met Jean Lave at the University of Oxford for a day-long seminar during which the possible place of theories of situated cognition in mathematics education was thoroughly aired and explored. A follow-up seminar took place during 1998. This meeting gave rise to a collection of papers (Watson, 1998). Sadly, at the time, no commercial publisher could be found who would publish the collection as an authentic record of the work of the seminar, or within a reasonable time. During the intervening years, some of the papers (e.g. those by Adler, Lerman, Winbourne and Watson) became influential beyond what might be expected from a small print-run. In addition, concepts associated with a situated perspective are now taken-as-shared in mathematics education research. It is time to review the subfield by drawing together a collection of up-to-date work which could be said to have been influenced at some stage by 'situated cognition'. Many of the authors in this volume participated in the original Oxford seminars and contributed to the collection of papers. In all cases their thinking has moved on and this new collection represents mature, critical, organic perspectives on aspects of mathematics education, framed by political, social and mathematical concerns. In March 2006 most of the authors met at a video-conference to discuss key theoretical issues, and this was followed up with lively electronic discussion. The chapters of this book, while clearly the work of the individual authors or authoring teams, have been peer-reviewed within the team, many of whom communicated with each other throughout the final stages of writing.

As editors we have made no attempt to devise a unity of perspectives. Indeed we have encouraged people to elaborate their own position and interpretation of key ideas such as the relationship between community, identity, situation and cognition; we have invited them to consider each other's views, but not to feel that they need to adopt certain terms or forms of language. As editors we do not have unity of views: for Anne, learners' knowledge of trigonometry would be a place to start thinking; for Peter, Sandra's sense of self would be a starting point (see chapter 15). Some key words and distinctions, situation/context, learning/participation, knowledge/participation, practice/activity are examples, may be used in different ways by different authors. Indeed some authors may avoid certain words altogether. There has been vigorous debate about whether the word 'knowledge' carries with it the baggage of individualism, acquisition metaphors, and assumptions about a static nature. This introductory chapter is a pointer to some underlying current issues in the use of situated perspectives to inform mathematics education. It is not intended as a unifying overview, nor as merely an account of the contents of chapters.

2. MATHEMATICS

All the situations on which the chapters of this book are based exemplify activity we would describe as mathematical, whether they are classrooms, workplaces, homes, or the street. Carlgren's (2005) description of knowing as 'knowledge as contextualised relation' helps a little here, as it implies that we can say we *know* something when, in a particular context, we structure it by its relationships in that context (see Hudson's chapter in this volume). For those for whom everything is seen as a mathematical context, the relations they use to structure the world are mathematical – there is something tautological here. Can a situated perspective *usefully* explain why cos x and cos |x| are identically equal, whether they are explored by students H and S in Ozmantar and Monaghan's chapter, or by us during an editorial discussion, or by you here and now? To do so we either need to tell a long story about cultural-historical theory and how the community of mathematicians and individuals within it became able to communicate symbolically about a function we call 'cos' and its properties. Alternatively we can say 'that's maths' and recognise that if this equivalence were violated, then we would not be talking about maths but about something else. In editing this book we take the line that mathematics does not need to be explained only as a form of social practice (what is *not* a form of social practice?), or a kind of cognition, or a philosophical construct, or otherwise scalpelled out into different disciplinary slices – it is its own discipline. Hence to identify mathematical activity we look for the ways people talk, what they talk about, what they focus on, how they classify experience, what levels and kinds of generality occur to them, what is varied and what is fixed, what relationships they observe or construct and how they express them – in much the same way as we would recognise music, musicality and musicianship. Thus there are social practices which are mathematical, there are tools which might embody and express mathematics, and there are individual insights which are mathematical, and there are situations which embody mathematics when observed with a mathematical perspective.

A central problem in the literature on situated cognition is that some authors who use mathematics as a focus for their studies, but are not mathematicians, tend to have two views of mathematics, i.e. instrumental learning of formal techniques in school, and fluid ad hoc informal mathematics out of school. Our view of mathematics is more complex – we see a range of different formal and informal mathematical practices within classrooms. Wenger (1998, p. 60) talks about a distinction between process and product, the product being the reification of the process and says that engaging only with the product is 'illusory' – but mathematics is more than process and product. The processes and the products are all constituents of mathematics and the process of reification is itself a mathematical practice on classes of outcomes of earlier processes; the product of reification is also a mathematical object, and engaging with it is essential for engaging with mathematics at a higher level. The point of advanced mathematics is to engage with reifications – and this is therefore not illusory in Wenger's sense. Furthermore, mathematical processes and mathematical products are both legitimate foci for mathematical activity; both can also be mathematical tools, and both also constitute mathematical learning. This poses a real challenge for situated cognition and activity theory as researchers struggle to provide useful and convincing models for mathematics classrooms.

This is our view as editors; this does not imply that all the authors would agree with us, but it explains why we have not insisted that all authors define what they mean by 'mathematics'.

3. KNOWLEDGE

In this book a dynamic view of knowledge is taken, to a greater and lesser extent, by all authors. Knowledge is what is produced in learning environments, but this involves mediation between learners' activity and historical conventions or authoritarian views of meaning, and is seen by various authors as individual and/or distributed. For some authors 'knowing' is a participatory verb, others prefer to use 'knowledge' as a characteristic of a particular practice which is negotiable between practices; others again see 'knowledge' as a form of competence. As editors it seemed obvious to us that we should take a negotiable, dynamic approach to the word 'knowledge' as used among the team of authors, but also to ensure that it still stands for what is called 'mathematical knowledge' outside our team.

In the practice of teachers and learners, 'knowledge' adequately stands for mathematics which might be seen as a tool for problem-solving, or as rules of participation in mathematical activity, or as the outcome, through learning, of activity. We recognise many different kinds of knowledge in mathematics, even when restricting ourselves to conventions. Knowledge that 'angles of a triangle total 180 degrees', given the usual Euclidean definitions of triangle and the arbitrary measure of a degree, seems different from knowledge that 'integration accumulates small changes', or knowledge that different representations have different affordances – yet these are all mathematical knowledge. When we unpick common uses of the word 'knowledge' we find what theorists might variously call rules, patterns of participation, competence, attunements, aspects of identity, discourse, tools, artefacts and practices. It seems unnecessary to turn our backs on its 'normal' uses, but important to define it where it is used.

As Lave has pointed out, whereas from a cognitive perspective it is "learning" that is problematic, from a situated perspective '"knowledge" becomes a complex and problematic concept' (1993, p. 12). Trying to articulate 'knowledge' within situated perspectives illuminates its complex nature and multiple roles, although for many purposes it can be taken to be a static canon of authoritarian relations.

4. SITUATED PERSPECTIVES: POWER AND LIMITATION

Theories of situated cognition show that mathematical behaviour in, say, a supermarket, is essentially a different way of being than the mathematical behaviour required to be a nurse, except when he is shopping in a supermarket. Thus such perspectives offer promising ways to make sense of why people might not apply in one context what they had learnt in another. They go some way towards explaining this apparent failure to transfer 'abstract' knowledge by recognising differences in the language, tools, methods of participation, goals and patterns of social interaction offered in each situation. In particular, distinctions are drawn between school mathematics, everyday mathematics and workplace mathematics – each of these three being plural, since mathematics in one everyday, or school, or workplace context is different again from another everyday, school or workplace context. Participation with others in these different practices is seen to involve gradually becoming more expert, 'learning' to be a fluent practitioner. In these perspectives, individual learning is commonly defined as participation in a community of practice. In classrooms, therefore, learning may be about becoming a fluent member of the class and this may have little to do with doing mathematics. Instead it might be more about learning how to survive teacher's questioning, or learning how to cope with the behaviour of the student sitting behind you, or learning how to look very clever with minimal effort.

On its own, the idea that learning and knowledge are essentially situated fails to explain similarities in knowing across dissimilar situations, and fails to explain differences of participation within the same situation (except as on a continuum from novice to expert). It does not explain why some participants reject the normal patterns of behaviour, and set up alternative communities which either disrupt activity, or in which more learning of the conventional canons of mathematics happens without the teacher. It *does* explain that people may be able to use familiar forms of participation (knowledge) in situations which appear to be different but are nevertheless perceived to be similarly structured, and it is this reliance on structural similarity which provides clues about apparent successful transfers made by skilled users of mathematics. Those who see the world as mathematically structured are more likely to see similarity where others might see difference. So how is it that people learn different mathematics in the 'same' environment? To make this shift across the obvious observable boundaries of time, place and people we need to see the classroom through the eyes of the individual learners and ask 'why is the view different?'

This inclusion of notions of identity in situated theory is central to understanding classrooms as complex places in which forms of participation are not totally determined, nor is development unidirectional. Rather learning is intimately connected with, constrained by and afforded by social situations. Notions of identity also allow us to explain, and possibly even predict, how individuals might act similarly in different situations.

Indeed, some of the authors in this book have found that the notion of boundaries between situations is very unclear when thinking about individual mathematical trajectories. In the chapters by Hughes and Greenhough (chapter 7) and Wilson, Winbourne and Tomlin (chapter 15) 'transfer' is seen to have a deeply emotional dimension – this is not really developed in most theoretical expositions – which sees through time and space boundaries between school maths, home maths, kitchen maths, life maths and adult maths classes. All these contexts also carry similarities of imagination, strength and resourcefulness as qualities of participation. In contrast, for some other contexts institutional, organisational and systemic boundaries seem to be enough to explain differences, even when the same teacher might be involved in teaching 'the same' mathematical content, although the mechanisms by which this happens need expression (see Bingolbali and Monaghan's chapter). What seems to be more important is the learner's perception of similarity or difference in a situation, and this calls on emotional memory as well as identification of similar structures, affordances or patterns of participation.

5. WHY SITUATED PERSPECTIVES?

So having raised some of the problems of using situated perspectives, why have we stuck with them?

It is fundamental for us that education has to be seen as social – people become able to do things through intentional teaching that they would not otherwise be able to do, or would not even know of as do-able, on their own. But, as we see in the chapters which talk about learning in other contexts, school is not the only context in which mathematical learning happens. Our continued choice to see learning as situated in social, political and economic contexts is one of fundamental values, a desire to describe, explain and consider learning mathematics as a complex, human activity. It is not a claim of truth or totality.

Learning, as participation in practice, does not need intentional teaching. Indeed in some articulations of learning there is little distinction between learning and living. But there is a danger in constructing descriptions of learning which include everything we do, communities of practice which include everywhere we go, because of Lave and Wenger's claim that a community of practice is an 'intrinsic condition of the existence of knowledge'. We may end up with no useful distinctions from which to learn. However, where there *is* teaching we need finer tools than either community of practice theory or activity theory can provide to describe both learning and failure to learn, or differences in learning. How can we hone finer tools which still recognise the immense power of the social and cultural contexts of learning but also express differences between learners and differences in the nature of mathematical participation? There is a need to delineate concepts and relationships which enable teachers to plan while recognising that teaching cannot be deterministic, nor would we want it to be.

Situated cognition may have acquired predictive power both for researchers in classrooms and for teachers when planning. Some authors, ourselves included, have used it to suggest that full participation means that learners develop both a personal expertise in mathematics and also contribute to the construction of mathematics within the lesson – thus challenging the traditional role of the teacher and replacing it with interaction within a community. In 1998 we constructed the notion of 'local communities of mathematical practice' (LCMP) to describe temporary situations in which people, including teachers, seemed to be engaged in the same mathematical activity (Winbourne and Watson, 1998). Since these situations are often ephemeral, LCMP may have more in common with similarly structured events elsewhere than to the next or previous event in the same classroom. LCMP does not refer to what is planned, it describes what happens, taking a slice across the whole situation which may or may not relate to what the teacher has planned to happen. Wenger (1998 p. 125) argues against any localised use of 'community of practice' by pointing out that 'interactions and activities take place in the service of enterprises and identities whose definition is not confined to single events'. But in our exposition we did not overlook communities and their characteristics which make these events possible – rather we look for characteristics within an event which display features and possibilities for continued participation. The way LCMPs connect to proximal events is through the identities of the participants influencing the next moment. Something could happen habitually in one classroom as part of expected and ongoing practice, where a similar event in a neighbouring classroom might be unnoticed and unrepeated. What happens next, or again, influences and is influenced by the modes of participation.

So far, our gentle critique of situated cognition zooms in onto identity and specifically mathematical practices, but some authors in this volume present broader critiques, zooming out to question the notions of 'situation' and 'practices'. There seems to be general agreement that a practice is not merely an action or an event, but includes values, meanings and purposes. If so, then these meanings and goals suggest an activity system which consists of goal-orientated practices, rules, power relationships and different ways of participating. Practice can be seen in a collective sense of activity with shared motives. Thus some uses of the word 'practice' imply a community of practice within which meanings and identities are negotiated, while others use it to mean something which can exist without a community, but around which a community could accrete.

In macro terms, mathematics education certainly concerns itself with institutions, curricula, assessment regimes, and so on, but the authors in this book seem to be addressing individual actions, learning and knowledge, rather than systemic issues or changes. Thus, while several use or allude to activity theory to frame their thinking, they all grapple with the essential educational conundrums of people's learning, knowledge, what it means to develop this and how this happens, and why some end up with different cultural capital than others, and what effect their different initial capital has.

Finally, what is specifically mathematical about the developments from situated perspectives described in this book? Early uses of situativity to inform mathematics education performed a great service in highlighting essential differences in mathematical-ness according to situation and purpose. No longer do informed educators expect the ad hoc, informal, economically functional mathematics of work and out-of-school context to relate to formal school mathematics by transfer in or out. Application of mathematics is now seen more clearly as a highly complex, socially developed process necessarily mediated through use of physical, symbolic and discursive tools. These findings were urgent, necessary and immensely powerful in influencing mathematics educational thought. School mathematics continues to be a social gatekeeper, making it imperative to continue to analyse how and why some succeed and others do not, and also to critique the role itself. Mathematics holds increasingly important roles in economic and political activity, making it imperative to question the processes of learning to have access to, and control of, these uses. Also specifically mathematical are the chapters which address the practices of mathematics, and how students learn to participate in these.

6. THIS BOOK

We open this collection with a chapter by Štech which challenges the notion that what situated theory can say about learning is easily applicable to school mathematics. He identifies school mathematics with the conscious and deliberate awareness of procedure which, according to Vygotsky, enables intellectual development beyond that which can occur in informal contexts which do not have education as an intention. He reminds us that school education sets out to 'rupture' and 'intellectualise' mental functions so that levels of abstract conceptualisation can be achieved. Furthermore, it is only in formal education contexts that this can take place.

David and Watson show how three interactive classrooms afford significantly different insights into the practices of mathematical participation; a first analysis suggests they are similar classroom situations, but a closer analysis shows that, in one of them, the usual classroom norms are augmented by complex mathematical practices that might promote the kind of intellectual development to which Štech refers. When students show signs of such engagement their mathematical observations can significantly determine the direction of the lesson. So David and Watson exemplify a classroom mathematical community of practice and suggest how teachers might begin to plan for these.

Askew similarly uses classroom vignettes to illustrate practice, foregrounding the substantial impact of classroom relationships on pupils' learning, and giving some examples of quite specific pedagogic strategies that allow space to be opened up for the development of such relationships within classroom mathematics practices. Possibly, the connection between relationships and learning indicates the development of a community of practice.

In the following chapter, Winbourne asks how it is that some people learn mathematics in classrooms that do not have the good features described by Askew. He probes beyond observable classroom behaviour to uncover the importance of the experience of individual children for whom the mathematics lesson is merely one of a string of experiences in and out of school. For Winbourne, the central unit of analysis is identity; how people see themselves as participants in practices.

Ozmantar and Monaghan develop and explore the meaning of abstraction, given that all learning, from the perspective taken in this book, might be seen as situated. Detailed analysis of students' responses to a task which can prompt abstraction shows that situatedness includes such questions as who you are, what you know, who you are working with, what you respect and many others. These questions shift the reader from thinking about situations as merely temporal and geographical to thinking about the persons acting within situations where identity is a constituting element of their participation.

Hughes and Greenhough describe a situation in which attempts to reproduce school mathematics in the home lead to tensions and conflicts which initially obstruct learning. They use this story to reconceptualise ways in which parents and schools together can support children's learning. This chapter indicates that situativity might be more about power, discourse and expectations than time and place.

Another way of supporting learning might be to use intuitions normally situated in out-of-school contexts within a classroom mathematics activity. Williams, Linchevski and Kutscher show that children's activity has to reconstitute related knowledge, and that significant shifts of meaning affording access to powerful mathematical ideas can follow. Their insight comes from the application of cultural historical activity theory to an analysis of objects and activities which are positioned on the boundaries between sites of children's experience.

In the chapter by Santos and Matos, one particular newspaper seller is shown to straddle two practices, that of selling and that of distributing newspapers. Using activity theory the researchers discover this by examining the artefacts he constructs and uses in the course of his work. These artefacts structure and are structured by the organisational, economic and mathematical activities of all participants. The role of artefacts in the situated mathematical thinking these authors describe reminds us that artefacts structure the ways in which participants engage in mathematics in classrooms. The chapter by Williams, Linchevski and Kutscher also exploits this idea.

Frade and da Rocha Falcão draw on Polanyi's tacit dimension of knowing to make sense of the unspoken, unseen, changes in participation, in how participants see themselves, which characterise the shared dispositions and activity in a community of practice. By identifying the role of tacit knowledge in some out-of-school communities they deduce that it is worthwhile to describe classroom vignettes in terms of tacit dimensions.

Bingolbali and Monaghan provide a careful theoretical exposition of the influences on cognition of an institutional setting. Building on this theoretical base, they use undergraduates studying mathematics and mechanical engineering to show how different understandings and expectations of the derivative in graphic, algebraic and application forms develop as a result of teaching seen as situated in particular institutional cultures.

A similar theme is pursued by Pinto and Moreira, who observe mathematics lessons in two different vocational courses in a technical school. This exploration was prompted by the authors' realisation, when teaching at University, that students with different routes of access had different mathematical understandings and available tools to bring to bear on their understanding of the tangent line. The authors found not only that the vocational teachers taught mathematics which would be appropriate in the expected employment, but that also their classrooms were sites of forms of participation which were more like work than school settings; in the Highway Engineering class, both the students were engaged in highway engineering practices.

Hudson's chapter starts with a report on a study which took place over ten years before this publication. In it several workers talk about how the mathematics which they use in their work seems easy and meaningful, particularly when compared to school mathematics. The factory had linked to the school in the hope of influencing the school curriculum to be more vocationally orientated. Hudson follows this report with more recent reflections, including revisiting the research sites. The lack of progress may be seen by some as due to short-sighted curriculum and assessment practices. However, as we can see from many other chapters, the hoped-for relationship between work-place and school mathematics may never have been achievable, not least because of the educational purpose of school as identified by Štech.

Kanes and Lerman contribute to the debate by drawing a distinction between what they see as a conservative view of communities of practice into which participants are inducted and a dynamic view in which the community is constantly changing as participants come and go and negotiate new positions and practices. Their chapter appears to provide, to some extent, a theoretical closure on ways in which we might look at the social contexts of educational experiences.

The final chapter of the book, by Wilson, Winbourne and Tomlin, provides a powerful and disturbing biography of a person learning mathematics which challenges any temptations we may have to believe that we can tell meaningful stories which are in any way true. What we learn from this chapter is that the experience of any individual is more complex and multidimensional than can be rendered from within any particular theoretical perspective, or indeed, using any form of language.

The style of writing in the chapter by Wilson, Winbourne and Tomlin raises issues relating to an apparent incongruity in using academic approaches to articulate aspects of real lives. This makes for a necessarily complex piece of writing, where the form of the chapter must itself be taken as methodological discussion. Entwined, braided pathways run through it and the analysis is voiced so that reading the chapter requires more levels of awareness than the more usual 'excerpt-comment' genre. In narrative research intertextuality is acceptable (e.g. Curt, 1994; Boylan, 2004) and, while we are not advocating such complex styles as universally helpful, this chapter caused us to question our acceptance of the norms of academic writing for the rest of the book.

We do not believe that the values which drive us as editors, of social justice and valuing all of those engaged in the educational process, are at odds with the rarefied and specialist language we have used throughout this book. Nevertheless the question of appropriateness of modes of communication is one that should exercise our field of research lest we distance ourselves too far from practitioners and students.

REFERENCES

Boylan, M. (2004). *Questioning (in) school mathematics: Lifeworlds and ecologies of practice.* Unpublished PhD Thesis, Sheffield Hallam University.

Carlgren, I. (2005). The content of schooling – from knowledge and subject matter to knowledge formation and subject specific ways of knowing, *ECER 2005 – European Conference on Educational Research.* Dublin: University College Dublin.

Curt, B. C. (1994). *Textuality and tectonics: Troubling social and psychological science.* Buckingham: Open University Press.

Lave, J. (1993). The practice of learning. In S. Chaiklin & J. Lave (Eds.), *Understanding practice: Perspectives on activity and context* (pp. 3-32). New York: Cambridge University Press.

Watson, A. (Ed.). (1998). *Situated cognition in the learning of mathematics centre for mathematics education.* Oxford: University of Oxford, Department of Educational Studies.

Wenger, E. (1998). *Communities of practice: Learning, meaning and identity.* Cambridge: Cambridge University Press.

Winbourne, P., & Watson, A. (1998). Participation in learning mathematics through shared local practices. In A. Olivier & K. Newstead (Eds.), *Proceedings of the 22nd International Group for the Psychology of Mathematics Education* (Vol. 4, pp. 177-184). Stellenbosch, South Africa: University of Stellenbosch.

Chapter 2

School Mathematics As A Developmental Activity[1]

Stanislav Štech
Charles University, Prague

Abstract: This chapter points out that one of the purposes of having mathematics as a school subject is that it can contribute directly to learners' development of higher psychological functions, and hence to the development of their identity as mature people. It draws attention to the dangers of too narrow an interpretation of situated learning, and makes the case for mathematics in the school context being seen as having a deeper psychological effect than that of acquiring mathematical instruments to solve problems close to life. Rather, activity theory, with its different levels of operations, tasks and complex activities, is shown to enable mathematics in school to be seen as potentially contributing to the development of thinking, motivation and identity.

Key words: epistemology in mathematics education, situated learning, activity theory, 'intellectualisation', cognitive development

1. INTRODUCTION

In this chapter I use an activity theory perspective to draw attention to features of school mathematics which have exceptional developmental potential. My interest is in the role of mathematics within the school curriculum, and this viewpoint highlights some limitations of a situated approach to mathematics learning.

---

[1] This paper is related to a plenary address given at the 30th Conference of the International Group for the Psychology of Mathematics Education, Charles University, Prague, July 2006 and a version of it appears in the proceedings of that conference, volume 1, pp. 35-48.

The Czech Republic has recently approved a new A-level examination (Baccalaureate) in which mathematics has been dropped as a compulsory subject for the first time. This was the result of emotive resistance against compulsory examination by the public and by politicians. In discussions about school curricula, many protagonists imagine mathematics to be an almost emblematic example of school education detached from life. It is still seen by many to be a highly abstract exercise of the mind that serves to classify children as 'talented' or not, and which does not prepare children for anything useful which may serve them in their later life – perhaps with the exception of simple calculations similar to everyday protoarithmetic. Recent discussions amongst psychologists, among others, show that the relation between mathematics at school and its influence on the mental development of the individual (child) is far from understood. Various implicit epistemologies of mathematics are shared by didacticians and teachers, and transmitted, through teaching, to pupils and, indirectly, to their parents. In the Czech context, as in many other contexts worldwide, parents, politicians, students, teachers and psychologists argue about the role of mathematics in the school curriculum from different epistemological standpoints, and these each have different consequences for the conception of mathematics in school.

What I want to deal with first are the basic epistemological approaches inherent in educational work in school. Those approaches reveal different answers to essential questions: what is mathematics? Or: what does it mean to be 'doing mathematics'?

In the recent past this type of implicit questioning gave rise to an often shared answer, namely: to be an efficient mathematics teacher/learner presupposes engaging in active methods, taking a constructivist view of learning, and understanding learning to be situated in particular contexts. Only then do mathematics and the knowledge it communicates make sense to the child. The idea of the child as an active sense-making individual within a social context, for instance engaged in solving problems or in mathematical games, has undoubtedly contributed to the history of teaching the discipline.

Nevertheless, I am going to attempt to show the limits of this approach. The idea that all learning is situated has been interpreted widely to imply that learning mathematics either needs to be based in everyday contexts, or is about recognizing its utility in a range of situations. The activity theory perspective of Leontiev, developed later by others (for example Clot's work-activity analysis, 1999) reveals the structure of cognitive activity in which mathematical concepts represent the tools to resolve specific tasks. But it goes further than mere utility; it also makes it possible to distinguish an instrumental 'managing of the situation' school mathematics task from the educational shift in which the apprehension of mathematical terms has contributed to the development of mental functions, and mental structures, and of the whole personality. This potential effect of learning mathematics in school is consistently underplayed in the implementation of the situated perspective, and also in attempts to apply Vygotskian theories to mathematics teaching and learning.

2. IMPLICIT EPISTEMOLOGY: WHAT DOES IT MEAN 'TO BE DOING MATHEMATICS'?

Charlot (1991a) analyzed three implicit epistemologies close to mathematics. It is well known that the most ancient epistemological conception of mathematics is the Platonic version of a certain *'celestial mathematics'* (Desanti, 1968). This conception is based on the idea that mathematical forms pre-exist the grasp of a mathematician, as if they exist 'in themselves'. This is a widespread conception not only in general but also among teachers; mathematics, they believe, is to be taught and learnt as coming to know universal truths and structures. Mathematical ideas are seen as pure and evident, and the mathematician (and the mathematics teacher) discovers them (their relations, structure, etc.). This world of mathematical ideas is basically independent of the mathematician's own activities; it is transcendent, and it is accessible by perception and contemplation. The French epistemologist René Thom (1974) says that according to this conception, mathematical structures are not only independent of humans, but people also have only an incomplete and fragmentary notion of them. In this view the task of school education consists of the teacher presenting the world of mathematical ideas with maximum clarity and assists the pupil in mastering the principles of abstract thought. The metaphors of light and perception used by Plato, where the pupil's mind stands for the 'eye of the soul', are still embedded in much mathematical pedagogical discourse. This implicit epistemological conception is the foundation of the so-called traditional education which focuses mainly on exposition followed by exercises, but can also be seen in exploratory tasks which are designed to lead to 'discovery' of mathematical 'truths' in the same way as a telescope guides discovery of planets. Memorisation and application of procedures are required to accumulate and enact such truths.

Another influential conception of mathematics may be described as *'terrestrial'*. It does not presuppose the existence of transcendent autonomous mathematical entities. Mathematical knowledge is seen only to reflect the structure of the natural and perhaps even the social world. The mathematician does not contemplate independent abstract entities; on the contrary, he abstracts the ideal, mathematical, structure of the world from the world itself. Again, mathematics exists outside of the individual, yet as a structure that he has to extract, not in the form of independent ideas. It is not transcendent but immanent. This implicit epistemological conception is the foundation of reformist education, i.e. pedagogy which endeavours to make the child discover mathematics above all (or only) by manipulation of particular mathematical 'objects'. Great emphasis is therefore put on the 'use' of mathematics in various practical situations. The child is thus shown that (a) mathematics is useful, i.e. can serve a purpose in practical life and that (b) mathematical concepts, laws and structures exist, have a rationality of their own and that it is important to learn to operate with this rationality as the authorities can do. In this case, doing mathematics means to rediscover that which is already given. Yet, this time, analytic manipulation, rather than perception, is both the method of discovery and also the method of using what has been understood.

The third conception of mathematics can be described as *'instrumental'* – mathematical knowledge represents tools which serve the solution of problem situations. Mathematics does not pre-exist either in the skies or hidden in the world around us. To do mathematics is not to discover but to create. The main conclusion is that mathematics is a historical creation by particular people under certain conditions, by people who themselves sought answers to particular problems. In this conception, mathematical activity consists in the generation of particular instrumental operations and, at the same time, in the establishment of a certain field of operations, of their interconnected network. (Note that here I use 'instrumental' in a broader sense than that used by Skemp (1976). His use was limited to the use of given tools, where mine implies a need to create, adapt, and interconnect a toolkit, this including relational understandings.)

This epistemological conception is the basis of education which relies methodologically on the belief that learning is the result of a successful demonstration that mathematical knowledge serves as a tool in the solution of initial problem situations. Such situations need not always be concrete and based on everyday experience. It is assumed that learning is related to the pupil's invention of a concept or of a rule which makes it possible to find a solution for such a situation. At the same time, the child does not have complete freedom to create *any* thing, for the situations in question need to have a potential for the creation of mathematical instruments, need to display inner normativity, or need to display constraints on the activity that can be performed in the situation. The child cannot therefore simply play or disrespect the limits of the situation.

Furthermore, as Charlot points out (1991b), the metaphor of light and vision related to perception ('to see a solution', 'to clarify the assignment') leads us to fairly unproductive schemes of interpretation. For instance, to explain why some 'see the light' and others do not it is normal to as assume gifts and talents, be it in biological terms: 'he's got a genius for it' or 'she is a mathematics prodigy'; or socio-cultural: 'he lacks the cultural capital of the abstract code'. The perception metaphor of light and vision gives us no mechanism to deal with those who 'do not see'. It is a positive feature of more active metaphors of learning and, above all, the instrumental conception of mathematics, that learning is related to mental work or activity, compared to the deterministic interpretations of 'talents' and 'capital'. The activity metaphor includes both the activity of mathematicians in history in particular situations which they had to resolve, and the activity of the child during the learning process.

In recent decades, the changing views of mathematics, and subsequently of the teaching/learning of the discipline, have led to the dominant Platonic epistemology being increasingly complemented by play-oriented active methods and, occasionally, by instrumental or constructivist conception using situations close to everyday life. The idea that to learn mathematics means 'to be doing it', i.e. to create, produce, and make mathematical concepts and procedures as tools for the resolution of tasks and problem situations, is now generally recognized. However, this acceptance is mainly in the discourse of didacticians and mathematicians. In schools, the application of this notion can be hesitant and often fails.

Rather than attribute such failure only to the inability of teachers to use these methods effectively, this should lead us to consider whether learning mathematical terms and techniques, which is still the aim of most education systems and assessment regimes, in problem situations that are modelled after everyday experience is the most efficient procedure. If the goal of education is to introduce learners to formalized mathematics, should it not rather be the task of teachers to underline the specificity of formalized mathematics as opposed to the situated methods which arise in everyday mathematics? Whatever else schools are for, one goal of school socialization in the cognitive domain in general is to initiate the child into an intellectual activity and to contribute through this to the development of mental functions, mental structures, and hence to the development of the child's personality in ways which would not be possible without school.

Furthermore, we are led to consider whether 'activity' or rather, cognitive activity, should not deserve a more differentiated analysis than that suggested by the three conceptions above. The history of relative failure of teaching suggests that learning mathematics is a complexly compounded activity which may encompass both the memorizing of definitions and routines, as in the traditional view, and also the difficult formulations of hypotheses in problematic situations, as in the reform view.

In my search for answers, I rely above all on the cultural-psychological tradition of Vygotsky, and the activity theory of Leontiev.

3. LEARNING CLOSE TO PRACTICAL CONTEXTS AND SITUATIONS

After years of domination by an individual-psychological approach to cognition and learning, the last few decades have seen renewed interest in the socio-cultural character of human cognition and of mental development in general. Much of this renewed interest has been due to, and reflected in, the availability of works in many languages influenced by Vygotsky.

This emphasis has risen remarkably in prominence since the 1980's. It draws on earlier inspirations: the unachieved work of Vygotsky from 1925-1934, followed by the work of Luria and Leontiev. Unfortunately, these were published relatively late and translated into foreign languages only from the 1980's onwards, remaining virtually unknown till then. In the 1970's and 1980's cultural anthropological research and theoretical work in the field of intercultural psychology began to focus on the influence of formal schooling on the mental development and ways of thinking of people within traditional cultures in Africa and other parts of the developing world. Cole, Gay, Glick, and Sharp, (1971); Scribner and Cole (1981); Lave (1977, 1988) and above all Cole (1996), an admirer and indirectly the pupil of Luria, are particularly identified with this shift in focus.

This led to a boom of literature about so-called situated or distributed learning. Significantly, this translates into French as 'learning in context' (apprentissage en contexte), a somehow inaccurate expression, but one which makes explicit reference to an important dimension of situated learning, that of context, which in turn reminds us of the other necessary term of the relation, 'text', making salient the 'text – context' relationship.

This turn towards situated learning, towards forms of cognition and learning in practical situations (of Lave's Liberian tailors; of seafarers in the Pacific), towards learning in practice (e.g. everyday arithmetic in the research conducted by Scribner (1986) and by Rogoff, (1990)) led to a full appreciation of cognition as a set of cultural practices. At the same time, it may have led to the overestimation of this form of learning at the expense of the importance and function of school forms of cognition and learning. In conjunction with the reviving educational reformism and a return to student-centredness, this, in my perception of post-communist countries during the 1990's, led to the overall negation of the developmental significance of school forms of cognition. Situated learning in contexts of practical life of the individual was placed on a pedestal, almost as a model for learning at school, in some educational discourse. Active and reformist (student-centred) conceptions of teaching/learning are strongly nurtured by this conception.

I shall now show that there is a substantial difference between situated learning, i.e. learning in the extra-curricular, everyday (e.g. family) context, and school learning, which was described by Vygotsky in *Thought and Language* (1976) as learning 'scientific' concepts.

Analyses of situated learning had the power to re-orient those educational conceptions which were still under a strong influence of the individual-cognitivistic tradition. What do these analyses stand for? The pivotal idea is that learning, apprehending knowledge, can only be construed in a 'situation', and is dependent on the pupil's participation in social and material contexts, the person and his/her world being mutually constitutive. This idea underlies, according to Moro (2002), the following theories: learning as *apprenticeship* associated with the works of Lave (1977, 1988) and Lave and Wenger (1991); learning as *guided participation* associated with the theoretical work by Rogoff (1990) and learning in the person-tool(s) system usually described as distributed learning, associated with the names of Hutchins studying pilots in a cockpit or subway dispatchers in work, (1995, 1990) and Resnick (1987).

All theories of situated learning redirect our attention towards the analysis of the situations in which learning takes place. Each in its own way puts the emphasis on one or other of the elements of Vygotsky's cultural-historical approach towards psychological functions, namely the prime importance of social activities; the inter-psychological nature of psychological functions; the key importance of mediation and the role of the adult-expert; and the formative effect of the artefact-tool. Thanks to these theories, Leontiev's concept of activity and the importance of the unit of analysis in examining psychological phenomena become prominent. What, then, is the problem? Why not merely arrange the theories of apprenticeship, guided participation, and learning in person-tool systems within the socially-mediated approach to learning and make use of them at school, seen as a social situation?

First of all, it is necessary to note that these theories:

(1) localize the dynamic of learning almost predominantly into the world of everyday experience and neglect the importance of activities provided and made necessary by the school, i.e. of activities directed specifically at reflection and abstraction. Thus, they hinder investigations into the differences and tension between an item of knowledge in its everyday form and one which is formalized, and therefore bypass the decisive moment of the cognitive and personal development of the individual.

(2) overestimate the formative influence of artefacts and situational configurations on mental functions, as if these were embodied in tools. This is because they fail to distinguish between the capacity to operate in context on the basis of the tool, and the mental work of an individual transforming particular psychological functions.

(3) fail to dispel the impression that in their psychology of situations the psyche in fact belongs to situations, thus only mechanically transposing mental *gestalts* which are originally localized in the minds of individuals into the situations.

These objections need to be overcome to fully appreciate the role of school in mathematical learning.

4. LEARNING IN THE SCHOOL CONTEXT

The theory of learning in everyday practical contexts differs significantly from the approach of the Vygotskian school in its conception of the unit of analysis and in its conception of mediation. Along with Leontiev, in using the term *unit of analysis* I refer to the isolation of units of enquiry which enable the objectification of psychological facts in their inter- and intra-psychological dimensions. 'Participation in apprenticeship' can help grasp activities in the socio-cultural framework and can substitute for the mechanical understanding of internalization; however, the nature of the *intra*-individual activity itself largely escapes it. On the other hand, mediation is considered by Lave and Rogoff above all as communication between individuals, and the prospective zone of proximal development as a communicative-relational network. The cognitive activity itself, that is the apprehension of knowledge *qua* apprehension of norms of activities with a given item of knowledge, is left aside. Similarly, Hutchins' treatment of 'tool' in the pilot's cockpit is admittedly instrumental and mediating; however, it is not what Vygotsky meant by a psychological tool, since it cannot demonstrate how permanent transformation of psychological functions and the development of the individual come about. For Vygotsky, the use of psychological tools transforms the psychological and intellectual functions of individuals. Finally, many users of situated cognition theory are insensitive to the fact that learning at school is also learning in a context with its own specificities, a context which represents a community of practices largely derived from a concept of scientific knowledge. A comparison with extra-curricular contexts makes it evident that the objective of school is epistemic. It aims at the transformation of modes of thinking, of experiencing, and of the self. This requires a clear conception of the relations between spontaneous learning (the kind of learning we do, and what it is we learn, in everyday contexts) and education, formal learning and development. What are, in Vygotsky's terms, the main differences between apprehending spontaneous concepts and those which are scientific (acquired mainly at school)?

4.1 Utilitarian vs. epistemic attitude to the world and to language

We will be helped by thinking about the classic comparison between the apprehension of spoken language and language learning at school with the support of writing.

Formalized learning can start where spontaneous learning in contexts of everyday life reaches a limit of what is possible. Spontaneous learning stands on *instrumental* usage, knowing how to say something; for example, knowing how the notion 'brother' works, or who is a particular brother, to make oneself understood. Bourdieu (1996) says that in practical action the word used fits the situation. Formalized learning paves the way for *reflection* and builds on it. This leads to knowing why something can/cannot be said in a particular way; what is essential about the structures of 'kinship' and why a 'sister' is the same as a brother according to the law of language, even if this is sheer nonsense in the context of everyday usage.

Although formalized learning of decontextualized 'scientific' knowledge makes use of spontaneous learning (and indeed is based on it), the important thing is that it transforms the substance of the knowledge thus acquired. Due to formal learning and its tendency to decontextualize, the child is brought to reflect upon and realize the specificities of language, and to the necessary generalization of linguistic phenomena. By means of this new attitude towards language, the child's attitude towards the world changes into one which is epistemic and not merely practical. This in turn opens new horizons in other domains of knowledge.

Olson and Torrance (1983) introduce another striking criterion. In their view, both the *context* and the *text* are available to people in their practical attitude to the world. But the situation of spontaneous learning forces them to give priority to information from the context, that is to rely on what is most probable in the given context. Olson and Torrance cite the following example:

They observe that according to classical Piagetian tests children up to 8 years of age understand instructions contextually (and proceed in their thoughts on the basis of such understanding). The critique of these tests features the classical example of a logical 'sub-class/class' relation (there are 9 flowers in the picture, 6 of them tulips and 3 roses). The question is: 'Are there more tulips or more flowers in the picture?' Children answer on the basis of comparing the sub-class 'tulips' with the sub-class 'roses' and conclude that there are more tulips than flowers. Olson points out that children answer not on the basis of text but depending on the context, i.e. on their everyday experience and act as is common in such contexts. We usually compare sets of the same kind or level. For example in everyday life, we might ask if there are more girls than boys in a class, but, as Brossard reminds us (2004), we rarely ask if there are more girls than pupils in a class. The child is thus guided by the context and not by the linguistic contents of the question and its logical structure, i.e. the 'text'. To follow the text, the child must undergo another type of learning than the more or less 'spontaneous' reaction recorded by Piaget.

This observation is especially important in mathematics, where questions designed to elicit techniques or applications of particular facts are often phrased in ways which relate very little to any well known context, even when a pseudo-everyday context is being cited. But the problem goes much deeper than that. At school, meanings and interpretations are not merely practised; writers and readers engage in reflection on meanings themselves. The processes of learning written knowledge are thus the decisive factor in the change of ways of thinking. Olson (1994) cites a Vygotskian distinction to that effect, namely that, thanks to writing, we have moved from thinking about things to thinking about the representations of things. Vygotsky himself says that spontaneous notions are generalizations about things, while scientific concepts are generalizations of these generalizations. This is what Vygotsky, in *Thought and Language,* describes as the key effect of school teaching/learning (1976). School education brings about (a) a rupture in and (b) the 'intellectualisation' of mental functions.

What does this rupture consist of? The aim of spontaneous everyday learning is to deal with a practical situation in life. The child that enters school has thus already mastered some knowledge, say in arithmetic. This is proto-arithmetic knowledge: he/she can divide marbles into two even parts, knows how many people there are in the family, can compare his/her own age to that of a sibling, can add and subtract from the number of objects and so on. At school, this spontaneous knowledge serves as a basis for the child to develop real operations of addition and subtraction with the help of a teacher; the child constructs (abstracts, rather than extracts) numerical properties of empirical objects. Whether we deal with marbles, apples or books is of no importance; in any situation, it is true that $2 + 1 = 3$ and $3 - 2 = 1$. The child performs a decontextualization based on generalization as an empirical abstraction of the concept of quantity. According to Vygotsky, this is the above-mentioned generalization of a lower order, a 'generalization about things'. Yet, arithmetic operations do not lie in (are not immanent to) the empirical situation, they are not additional properties of objects (besides colour or size, say). They are necessary non-empirical operations that the child must perform and these become the object of its learning at school.

However, an important breaking point occurs when, by virtue of these operations, the child discovers the properties of the decimal system (Brossard, 2004). At a certain stage of development, and depending on the school curriculum, the child begins to understand how the decimal system fulfils its purpose and how it works, and may even understand that it is also possible to count using other numerical systems (binary or other systems). From then on, the child understands the decimal system as a *particular instance* of other possible numerical systems, and that therefore it must have certain properties which, when understood, generate the whole system. This is a generalization of a higher order. This is the generalization of generalizations, a generalization based only on the relations between numerical entities.

4.2 Unreflected, or not consciously developed vs. planned and conscious procedure

The mastery of systems of higher generalizations makes it possible to make a relatively permanent developmental shift away from *particular tasks and situations* and, at the same time, to *realize* not only some particular forms of knowledge but also to realize an awareness of *one's own mental processes* and of oneself. This is exactly what Vygotsky calls the 'intellectualisation' of mental functions: it sets in when the mental function becomes dependent on the idea (concept) or is subordinate to it.

The example of intellectualisation of memory and of the relation between thought and memory is well known. A small child thinks by remembering. His/her representations of things and of ways of handling them are not conscious and organized systematically around a certain idea or concept. An older child or a teenager already remembers and recollects by (and thanks to) thought. The intellectualisation of memory consists in the organization of knowledge for the purpose of remembrance. The child thus increasingly works consciously and deliberately on his/her own memory processes. From a certain point on, the relationship between memory and intellect gets reversed. The introduction of conscious and planned (volitional) relations of the child towards his/her own mental processes is what cultural psychologists perceive as the criterion of a higher level of development.

It is valid universally that the emergence or discovery of the *relations of a higher generality* between concepts is the critical point (motor) of mental development. Situated perspectives cannot explain, nor do they have need of, intellectual relations at this level except by recognizing them as discursive patterns.

A remarkable geographical metaphor of Vygotsky's makes it possible to describe the concept as a geographical point at the longitude and latitude intersection, as Brossard (2004) points out. The 'longitude' of the concept determines its place on the meridian leading from the most concrete to the most general meaning. The 'latitude' of the concept then represents the point which it takes in relation to other concepts of equal 'longitude' (of equal generality) but relating to other points of reality. The combination of both key characteristics of the concept determines the extent of its generality. It is given not only by the concrete/abstract scale but also by the richness of connections to other concepts of the given conceptual network which form the domain in question.

Let us demonstrate this global metaphor of intellectualisation in relation to connections between arithmetic and algebra. The result of the operational development so far, e.g. the operations of addition and subtraction, becomes the 'source' of new processes, algebraic operations with variables and unknown quantities. The performance of a thought operation (to define, compare, factor out, divide etc.) presupposes the establishment of relations between various concepts within the corresponding conceptual system. A six year old child cannot 'define' an operator or a straight line, for instance, because the terms that he/she masters are not in relations of sufficient richness to other concepts. If, however, the child masters operations of the decimal system, an infinite number of means to express a concept, for instance of the number 'four', are available to him/her ($2 + 2$; $8 - 4$; $16 \div 4$ etc.). The concept of a higher order of generalization thus represents a point which makes possible several ways forward within the entire 'global' system.

School education plays a decisive part in this process of transformation of mental functioning. When learning 'close to everyday life' the child observes, discovers, considers, argues, and so on (Brossard says, that he/she 'coincides with the significations he/she practices' (2004)). It is due to school education that, along with all of this, the child also focuses his/her attention on mental processes which he/she performs when observing, discovering, considering etc. The child works on 'pure meanings' which are the main object of his/her reflection. Thus, the ability to define the number 'four' in several different ways involving various operators and their combinations necessarily places comparative reflection, the analysis of one's own attention and memory, knowledge about one's efficiency, and so on at the forefront. Such processes would never come about if the child were struggling with ignorance and the absence of automatic fluency with the elementary operator.

However, the child is very unlikely to reach this reflective activity 'spontaneously'. It requires a teacher, a plan, a logic of the curriculum and of the teaching process, a programme which is at first only external to the child. Mathematical concepts of a higher level of generality are especially distinguished by the necessity to introduce them from the outside; these are 'top-down' conceptualizations. Intellectualisation stands on an increasing subordination of individual operations to the higher organizational principle (with the two characteristics expounded in the above-mentioned geographical metaphor). From this point of view, mathematics represents activities in which, with a growing generality of a concept, the motive of the introduction of the concept is always 'external' in respect to the child and his/her 'spontaneous interest'. The 'new' conscious learning at school is guided by the requirements of the curriculum, or by the object of the cognitive activity. The pupil studies the 'programme' to learn a type of thinking whose observance is guaranteed, to some extent, by the institution of the school and the teacher. If we put this in Olson's terms, the 'textual' approach is exercised at school, sometimes with success, sometimes less so, in an approach which is supervised, systematic and planned. School mathematics which is supposed to fulfil its developmental psychological function must provoke that which is seen to be of greatest value: tension between various levels of conceptualization (the development level achieved by the pupil to date vs. the elaborate form of conceptualization constructed in a didactic school situation in co-operation with a teacher). Brossard (2004) talks about the internal motor of development acting alongside the external (socially motivational) motor.

While this is true to some extent in all school subjects, it is especially important in mathematics, since, as the Vygotskian school recognized, an advanced understanding of mathematics is as a system of signs with their own inner logic which cannot be encountered in everyday activity (Volosinov, 1973).

I have repeatedly been using the terms operation, task, activity and so on. Learning in a school context is however characterized by certain specificities which can be better understood with the activity theory model of Leontiev (1978).

4.3 Learning as a relation of operations, tasks and the object of cognitive activity

Leontiev points at the hierarchical and internally differentiated structure of every activity, including the cognitive activity. He understands activity as a fairly molar unit consisting in partial levels represented by tasks or actions. Every task is formed by operations at a subordinate level (1978).

For Leontiev, it is above all the contents of the given activity, i.e. its object, that is crucial. What is also important is whether the cognitive activity makes sense to the pupil (and what sense it makes). It is of less importance who is setting the task to the child or whether the form of the activity is playful or utility-focused.

The interesting things about this conception of activity are the relations between different levels of the activity and their functions, as demonstrated in the analysis of work-activity by Clot (1999). This points to the necessity of distinguishing between the relations of *efficiency* in practising operations and fulfilling tasks and the relations creating the *sense* of the activity as such. It is also necessary to make sure these relations are mutually interdependent. Examples of these activity levels can be laid out in the table below:

Activity	Object	Function
I. Molar activity: e.g. algebraic transformations	Motives: mastery; aesthetic experience; to be good at mathematics	Encouragement (initiative-provoking): to persist in efforts to overcome obstacles and difficulties arising at level II and III
II. Tasks: the calculation of functions of different types; the solution of a rider/theorem; the solution of a system of equations etc.	Goals: to find the correct solution; to identify the value of the unknown etc.	Orientation: correct input analysis of the task; good 'preparation' of the solution; the layout of steps, their sequence and time allocation etc.
III. Operations: Multiplication; reduction; position record; the discrimination of symbols; managing operations using memory	Means: material tools; symbolic instruments including cognitive processes (memory, attention, arithmetic operations)	Execution: material traces (notes, schemas, auxiliary calculations…); necessary technical support (infrastructure) of the operations

The relation between the quality of operations managed (III) and the quality of the solutions to tasks (II) expresses the efficiency of the cognitive activity/learning, usually in the form of microgenetic improvements. Automatization, the repetition of invariants of an activity, is an exemplar of such microgenetic developments which involves more abbreviated forms of an operation; it opens the way for a higher level of generality of the operational concept used and for the extension of the range of tasks which can be solved in the same domain.

The relation between the nature, frequency, complexity and above all interdependence (articulation) of tasks (II) and the essence of the activity expressed in its object (I) defines the sense of learning.

Learning a mathematical concept is therefore a complexly structured activity which may involve such activities as memorizing definitions, routine practising and consolidation of operations, as well as the difficult formulation of a hypothesis *vis-à-vis* a problem situation. The provision of pertinent tasks complemented only by verbal persuasion and model demonstration, without the elaboration of activities on levels II and III, cannot lead to success, since 'sense' cannot be enforced on the pupil from the outside; the pupil needs to possess tools to elaborate this sense for him or herself. It is the experience of a concept that comes from outside, never its sense. It is impossible to produce meaningful learning without efficient operations (including mental functions: attention, the memory of basic inference) and managed tasks. This efficiency alone, however, cannot ensure that pupils will find meaning in that which they may consider as an illogical chain of unrelated tasks, or even as a purposeless drill of isolated operations. However, fluently performed tasks *may* have a relatively positive effect, for example producing a functional solution of a task situation, but situations which can be resolved this way will fail to contribute to the development of intellectualisation (see above). Thus both efficiency and sense-making have to be thought about when designing and managing pedagogic tasks.

5. PERFORMANCE IN THE SITUATION VS. DEVELOPMENT

The difference between a *performance in the situation* (performance of a function), consisting of the repetition of invariants of an activity in a variety of situations, on the one hand, and *development* on the other is stressed by French psychologists Béguin and Clot (2004). Spontaneous learning first and foremost pursues efficient performance of a function in a situation whose boundaries are not transcended (such as to calculate correctly a subtraction;

or, more generally, 'giving correct answers to the questions'). The results of spontaneous learning are often preserved even within learning of scientific concepts at school, and can be resistant to the requirements of scientific knowledge. However, spontaneous concepts represent the basic starting point for the subsequent conceptual work. In Leontievian terms, we have to deal with a situated level of operations (manipulating 'tools') and tasks. Their incorporation into a routine is a sort of an organizational condition for the cognitive activity itself (this is especially true of the memory automatism regarding certain algorithms, e.g. arithmetic ones). However, such 'practical' learning (in regard to the school context) rarely goes beyond the level of the performance of a function in a situation, such as getting the answer. Hence, there is often no opportunity for a more scientific apprehension of a concept so as to open the way for development. Such learning, learning to perform in a situation, can fail to grasp the object of cognitive activity itself.

On the other hand, effectively mediated learning of the concept paves the way for development of the pupil's thinking, and hence of identity (as becoming someone who can think at a more abstract level than before). This requires that routine tools be used in a variety of tasks (actions), and that the tool use in various situational contexts enriches their functionality (e.g. basic mathematical operators should be practised in the context of calculus operating with both one-digit and double-digit numbers, in the context of tasks in arithmetic and tasks in geometry). Only such cognitive work, learning, enables a relevant generalization going beyond the limits of particular situation. Only thus could operations in decimal systems become, at least for some, a special particular instance of a more general set of conceptualizations. Learning which releases knowledge from a context without ignoring functionality in particular situations renders development possible: firstly the development of the child's thinking; then the development of other psychological functions. For example, we memorize better those things the inner logic of which we have understood. Finally the development of the personality of the pupil follows, through developing a feeling of mastery over self and knowledge, and hence becoming harder to manipulate by others, and less likely to fall victim to biased information.

Activity theory shows how education influences the process of intellectualisation and the transformation of mental functions. For this reason, we should be warned against the reduction of the learning activity to mere operations and tasks which are close to the child's current situation, and to superficial attractiveness and playfulness and for immediate sense. Of mathematics is this especially true, for it has an exceptional potential to contribute to the development of mental functions of the child and his/her personality; not merely to the broadening of his/her knowledge and capability in everyday situations, but also to develop mental functions. There is simply no 'immediate' (non-mediated) connection between mathematical concepts or questions and social problems in the lives of people. It is futile to search for and incorporate this connection artificially into the education of mathematics under the pretext of its becoming more attractive. This connection exists only as highly mediated. Instead, it is a key function of mathematics to contribute to the developmental emancipation of a young person by way of intellectualisation, as I explained above. Through mathematics, students can become able to deal with generalizations of generalizations, and other relations between generalizations, and become more skilled at engaging in other forms of mediated activity in which the questions to be addressed are not those of everyday social existence. What I want to emphasize is that mathematics and its didactics should not lose their developmental-psychological potential by accepting the reductionism of an active, constructivist and problem-situated attitude towards education in the discipline.

In conclusion, I recognize that the situated perspective has much to offer in increasing our understanding of the whole complex picture of classrooms, and the essential differences between learning mathematics in school and learning and using it out of school. But it also alerts us to the limitations of integrating out-of-school methods into the classroom. Such attempts ignore an important part of the essential nature of mathematics as an abstract discipline, and there are many who would say that this is not a problem – such knowledge is not necessary for all students. However, I have tried to point out that one of the powerful purposes of having mathematics as a school subject is that, taught well as a scientifically organized subject, it contributes directly to learners' development of higher psychological functions, and hence to the development of their personality, and their identity as mature people.

REFERENCES[2]

Béguin, P., & Clot, Y. (2004). Situated action in the development of activity. *Activités, 1*(2), 50-63.

Bourdieu, P. (1996). *Teorie jednání*. Praha: Karolinum (orig. La Raison pratique).

Brossard, M. (2004). *Vygotski. Lectures et perspectives de recherches en éducation*. Lille: Preses Universitaires du Septentrion.

Clot, Y. (1999). *La fonction psychologique du travail*. Paris: PUF.

---

[2] Where the text referred to is a translation into Czech the more widely known title of the translated text is cited in brackets.

Charlot, B. (1991a). Les contenus non mathématiques dans l´enseignement des mathématiques. In R. Bkouche, B. Charlot & N. Rouche (Eds.), Faire des mathématiques: Le plaisir du sens (pp. 129-138). Paris: Armand Colin.
Charlot, B. (1991b). L´épistémologie implicite des pratiques d´enseignement des mattématiques. In R. Bkouche, B. Charlot & N. Rouche (Eds.), Faire des mathématiques. Le plaisir du sens (pp. 171-194). Paris: Armand Colin.
Cole, M. (1996). *Cultural psychology. A once and future discipline*. Cambridge, MA: The Belknap Press of Harvard University Press.
Cole, M., Gay, J., Glick, J. A., & Sharp, D. W. (1971). *The cultural context of learning and thinking*. New York: Basic Books.
Desanti, J. T. (1968). *Les idéalités mathématiques*. Paris: Seuil.
Hutchins, E. (1990). The technology of team navigation. In J. Galegher, R. E. Kraut & C. Egido (Eds.), *Intellectual teamwork: Social and technological foundations of cooperative work* (pp. 191-210). Hillsdale, NJ: Erlbaum.
Hutchins, E. (1995). *Cognition in the wild*. Cambridge, MA: MIT Press.
Lave, J. (1977). Tailor-made experiments and evaluation the intellectual consequences of apprenticeship training. *Quarterly Newsletter of the Laboratory of Comparative Human Cognition, 1*, 1-3.
Lave, J. (1988). *Cognition in practice*. Cambridge, UK: Cambridge University Press.
Lave, J., & Wenger, E. (1991). *Situated learning: Legitimate peripheral participation*. Cambridge, UK: Cambridge University Press.
Leontiev, A. N. (1978). *Činnost, vědomí, osobnost*. Praha: Svoboda (Activity, Consciousness, Personality).
Moro, C. (2001). La cognition située sous le regard du paradigme historico-culturel vygotskien. *Revue Suisse des Sciences de l'Éducation, 23*(3), 493-510.
Olson, D. R. (1994). *The world on paper: The conceptual and cognitive implications of writing and reading*. New York: Cambridge University Press.
Olson, D. R., & Torrance, N. (1983). Literacy and cognitive development: A conceptual transformation in the early school year. In S. Meadow (Ed.), *Developing thinking* (pp. 58-81). London: Methuen.
Resnick, L. B. (1987). Learning in school and out. *Educational Researcher, 16*(9), 13-20.
Rogoff, B. (1990). *Apprenticeship in thinking: Cognitive development in social context*. New York: Oxford University Press.
Scribner, S. (1986). Thinking in action: Some characteristics of practical thought. In R. J. Sternberg & R. K. Wagner (Eds.), *Practical intelligence* (pp. 13-30). Cambridge, UK: Cambridge University Press.
Scribner, S., & Cole, M. (1981). *The psychology of literacy*. Cambridge, MA: Harvard University Press.
Skemp, R. R. (1976). Relational understanding and instrumental understanding. *Mathematics Teaching, 77*, 20-26.
Thom, R. (1974). Mathematiques Modernes et Mathematiques de Toujours, in *Pourquoi la mathématique?* Paris: Edition 10/18. (http://michel.delord.free.fr/thom74.pdf)
Volosinov, V. N. (1973). *Marxism and the philosophy of language*. New York: Seminar Press.
Vygotsky, L. S. (1976). *Myšlení a řeč*. Praha: SPN (Thought and Language).

Chapter 3

Participating In What? Using Situated Cognition Theory To Illuminate Differences In Classroom Practices

Maria Manuela David and Anne Watson
Universidade Federal de Minas Gerais, University of Oxford

Abstract: This chapter looks at intentional teaching in detail, drawing out significant distinctions in whole-class interaction sequences which may, at first glance, look similar. Such episodes are sometimes analysed only according to the amount of participation, or the patterns of participation, rather than the mathematical qualities of participation. We find the notions of *affordance*, *constraint* and *attunement* helpful in looking at classroom interaction in terms of how mathematical activity is structured in such interactive sequences. These ideas allow differences in mathematical learning to be understood within a situated perspective by asking 'what are the specific mathematical practices engendered in this lesson?' As well as offering a powerful frame for 'getting inside' interactive sequences, this approach gives insight into how learners' mathematical identity might develop in subtly different contexts.

Key words: mathematics teaching, mathematical activity, mathematical practices, whole class teaching, classroom interaction, affordances, constraints, attunements

1. INTRODUCTION

Theories of situated cognition focus our attention on the social structures and activities in the learning environment, and suggest that learning is a movement from novice, peripheral, participation towards mature participation in the practices of communities. Learning, in these models, is the process of participating in an increasingly expert manner, and knowledge is a property of people and their interactions and activities in situations. In this chapter we focus very closely on the nature of mathematical participation afforded in classrooms, as one aspect of complex learning and participation in such situations.

For Greeno (Greeno and MMAP 1998) all cognition is seen as situated, learning and awareness being essentially framed by practice to which new members become attuned. He describes, in language derived from Gibson (Gibson, 1977; Greeno, 1994), *affordances* as "qualities of systems that can support interactions and therefore present possible interactions for an individual to participate in" (Greeno and MMAP, 1998, p. 9). In other words, 'affordances' are the possibilities for interaction and action offered in a classroom. Within systems there are norms, effects and relations which limit the wider possibilities of the system, that is *constraints* which are seen as "if-then relations between types of situations … including regularities of social practices and of interactions". In classrooms, these might be the usual ways of behaving and interacting, and how one kind of action triggers a particular kind of response. Individuals acting with this system demonstrate *attunements* which are "regular patterns of an individual's participation … for example, well-coordinated patterns of participation in social practices". So in classrooms we see individuals responding and acting according to their predispositions, which have developed in their previous classroom, and other, experiences.

In this chapter we are going to use these concepts of affordance, constraint and attunement to observe the complexities of classrooms, and to understand the process of teaching mathematics as the non-simplistic generation of environments, including language and tasks, in which those activities generally seen as mathematical can be established as normal practice. Here we use 'practice' in the sense of 'sustained pursuit of shared enterprise' (Wenger, 1998 p. 45) including a shared repertoire of tools, rules and representations necessary to the enterprise. However, there are a number of associated problems.

Firstly, situated perspectives of the classroom can easily focus on the generation and nature of social and mathematical behaviour, but not so easily on the conventions of subject content which the classroom is supposed to present. However, the subject content of mathematics cannot easily be separated from its modes of enquiry and engagement. For example, learning about shapes involves identifying similarities and differences as a mode of enquiry, yet the resulting classification is also traditional, non-negotiable, subject content. Similarly, wondering about the values of a variable in an equation is the mode of enquiry which initiates the 'subject content' of solving equations. Such modes of enquiry can be seen as participatory activities, and hence learners can be seen as being drawn into conventional mathematical understandings through becoming increasingly engaged in mathematical activity. When such mathematical activities are seen as the practices (sustained activities of shared purpose) of a classroom community, a situated perspective seeks out non-conservative features, since new members can be seen to bring with them patterns of participation and insights which have developed in other, similar but not-quite-the-same contexts. Whether their participation is seen as peripheral or central depends on what kinds of classroom interaction the teacher and learners see to be legitimate (see Kanes and Lerman, this volume, for discussion about what peripherality might imply).

Secondly, situativity needs to explain how different individuals respond differently in the same situation. Bereiter (1997 p. 287) explores this in a thought experiment which resonates with us as educators. A student he calls Flora imbues her mathematical work with meaning while, in same situation, Dora merely rote learns how to do her mathematics. These different attunements could be explained by focusing on the development of identity over many situations, rather than on isolated incidents in isolated situations, and his experiment points to a need to articulate the non-determinism of situated perspectives. This sense of learners developing their identities through participating in a range of practices emerges strongly in this volume (e.g. Winbourne, this volume) as well as in the work of Boaler and others (2002).

Thirdly, it follows that we can ask: if individuals can act differently, why is a situated perspective useful at all? Rogoff (1995) provides a dynamic view of community-of-practice as a site of learning which makes sense of students' learning as participation which transforms practice, rather than through apprenticeship to existing practices. Lave (1996) then suggests the usefulness of focusing on changes in the participation of learners in the community of practice, and this does not presuppose that identical action is the goal of teaching at all but points to the mutual shaping of participants and the community.

The last problem we wish to introduce is related to those above and will be the focus of this chapter. Mathematics has often been chosen as the site for general discussions about situated cognition (e.g. Bereiter, 1997; Lave, 1996) and those discussions have generally been about drawing distinctions between the kinds of mathematical activity undertaken in different contexts (Lave, 1993; Nunes et al., 1993; Saxe, 1999). The aim of these discussions has been to raise questions about the nature of mathematical knowledge, learning and expertise from the point of view of socio-economic activity. Another strand of work uses situated perspectives to pose questions about the nature of school mathematics learning, and to draw contrasts between classrooms which are based on traditional transmissional beliefs and those which are based on discussion, problem-solving and exploration (e.g.

Boaler, 1997). Rarely has a situated perspective been used to focus on the *nature* of the mathematical participation, the engagement with mathematical practices other than that required to resolve everyday problems out of school or to 'transfer' between educational settings. Indeed, some articulations appear to force distinctions which are uncomfortable from a mathematical perspective. Engestrom and Cole for example, describe symbolisation as a mediating tool for action (Engestrom and Cole, 1997). For mathematicians, however, symbols are very often the subject matter with which they are expected to engage, and their use can also be seen as a practice – the mathematical practice of symbolisation (Hoyles, 2001, p. 278). For Flora, in Bereiter's experiment, symbolisation might be a mediating tool enabling her to engage with cultural meanings, but it might also be a transparent mediating tool for creating her own meanings (Bereiter, 1997). The practice of decontextualising is also a situated practice: this is what we do in mathematics lessons and other mathematical activities (Lave, 1993). Other authors focus on the social practice of mathematics as if there is nothing additional to the social, a perspective which makes some sense if teaching is seen as enculturation into such practices, but still leaves us with the question 'what, then, are the practices of doing mathematics?'

It seemed to us that, if a situated perspective cannot talk about mathematics as it currently is seen in schools and classrooms, as they generally are for most students, then it is not as useful a view as it could be. At the same time, the image of individual learners positioned in dynamic communities, gradually becoming more and more expert in participating, is a powerful one for explaining how teachers can talk about 'gaining control' and 'making them think' and 'getting them to behave how I want them to behave' and so on. Similarly, it also explains powerfully that some classrooms can break down from the teacher's point of view, and instead become places in which alternative mathematically-related practices can become established (Houssart, 2001).

In this chapter we do not focus on the students' *change* of participation, but on some of the affordances and constraints of participation. Furthermore, we take a non-conservative view of what is going on the class, seeing affordances and constraints as dynamic characteristics of a situation which can be modified according to evidence of unexpected attunements of some students. We examine new possibilities for the situated perspective to develop further mechanisms to illuminate classroom practice, reaching the level of the detail of mathematical practices by using a situated perspective with a finer grain than is usual in mathematics classrooms.

2. OUR POSITION ON SITUATED COGNITION

For us, the theoretical view of cognition as situated is a valuable tool in conducting fine-grained analysis of classroom events, so that we can come to a better understanding of differences among classrooms which are superficially similar; and of similarities among classrooms which are superficially different (see David, Lopes, and Watson, 2005).

Winbourne and Watson (1998) developed an articulation of *local communities of mathematical practice (LCMP)*. This invites complex descriptions of classroom incidents in which learning mathematics can clearly be seen as some form of participation. These can exist, they claim, in particular places, at particular times, with particular people and have the following features:

1. students see themselves as functioning mathematically and, for these students, it makes sense for them to see their 'being mathematical' as an essential part of who they are within the lesson
2. through the activities and roles assumed there is public recognition of developing competences within the lesson
3. learners see themselves as working purposefully together towards the achievement of a common understanding
4. there are shared ways of behaving, language, habits, values and tool-use
5. the lesson is essentially constituted by the active participation of the students and teachers
6. learners and teachers could, for a while, see themselves as engaged in the same activity.

The establishment of LCMPs is not necessarily a 'good' thing *per se*. We are not claiming that all classrooms in which these things are happening are 'effective' mathematics classrooms, irrespective of the conception you have of 'effectiveness'. LCMPs are just a way to describe certain incidents in which personal mathematical development, identity, is in tune with the mathematical classroom practices. Depending on the sorts of things you value and the conceptions you have about mathematical education, an LCMP does not guarantee 'good' mathematics is going on, or that the learning is mathematical in a 'deep' sense. One could have an LCMP which is solely about constructing ways to cheat in an exam, the view of 'being mathematical' limited to 'getting a high grade by any possible means'.

There has been a tendency for the notion of communities of practice, and even for local communities of practice, to be taken up by people with a progressive or reform idea of education, yet it is important to note that the kinds of environment which gave rise to the idea were often highly organised and rule-driven apprenticeship situations. As Wenger points out (p. 45) 'community of practice' is an illuminative description of a situation, not a structure to be imposed with expectation of some idea of improvement. The power of the LCMP tool is in the systematic examination of all 6 characteristics in classrooms. We see point 6 as being fairly crucial in distinguishing between classrooms in which students are always expected to acquiesce to the teacher's instructions and those in which their understanding of mathematics is sometimes explored jointly by the teacher and students, or where teachers and students are all questioning each others' statements mathematically. On the other hand, if the purpose is to get highest test score possible, 'engagement in the same activity' could be claimed in an acquiescent classroom, and acquiescence could be seen as a productive mode of participation.

3. CASE STUDIES

We will present three cases of UK teachers (all of whom are successful in achieving good test results and hold senior positions in their schools) engaging in whole-class interactive episodes which all conform to current UK accepted practice. The three lessons, rather than providing obviously contrasting practices such as 'reform' versus traditional, or transmissional versus problem-solving, offer excerpts which are superficially rather similar to the observer. All three lessons were observed during October, when the teachers were getting used to new classes. All three classrooms had pleasant, friendly, atmospheres. We assume that the teachers and students are still in a phase of establishing the practices of the classroom. We ask whether it is possible to see how differences in classroom *mathematical* practices are being established through the way teachers structure these episodes. We are not claiming that we can fully characterise these classrooms from these episodes, nor that we can compare them in any value-laden way; rather we are showing that there are tools within a situated perspective which can reveal ways to account for differences in mathematical participation. Further, we can imagine that if these episodes are typical of these classrooms, then over time different patterns of participation (attunements) would develop.

We deliberately chose interactive whole class sequences with similar classes in similar schools, where the teacher is at the front of the class presenting tasks and orchestrating activity. All the schools were comprehensive, with a wide socio-economic spread of students, predominantly white, with many bussed in from rural areas. The schools all perform around the national average in terms of exam results. The classes vary in age, however, due to opportunity sampling.

Norma is teaching a group who are 7 months away from high stakes examinations; Roisin is teaching 11-12 year olds new to secondary school; Susan is teaching a group who are two years away from high stakes examinations.

In focusing on interactive episodes we are looking at intentional teaching, not at apprenticeship models of learning, but it is through interaction that novices' attention is drawn to the distinctions, definitions, nuances, which are traditionally made in the culture of conventional mathematics (St Julien, 1997). In all of them students freely participate within certain social and mathematical constraints. Analyses of vignettes will be offered according to the six features given by Winbourne and Watson (1998), using the vignettes as illustrations of possible practices.

Knowing that we were merely observers, we developed the following analytical questions from Winbourne and Watson's definition of LCMP above:

1. How do students seem to be acting in relation to mathematics? What kind of participants do they seem to be within the lesson?
2. What developing mathematical competence is publicly recognised, and how?
3. Do learners appear to be working purposefully together on mathematics? With what purpose?
4. What are the shared ways of behaving in relation to mathematics: language, habits, tool-use, values?
5. Does active participation of students and teacher in mathematics constitute the lesson?
6. Do students and teacher appear to be engaged in the same mathematical activity? What is the activity?

The question throughout the analysis is: what are the mathematical practices in which learners are participating? As a starting point, we take a very broad conception of 'mathematical practices' meaning: sustained pursuit of shared enterprises seen by the teacher, administrators and/or students as related to the school's mathematics curriculum. The three cases below describe the current mathematical practices in three classrooms but a deeper understanding and a better characterization of these practices is going to be possible only through the analysis that follows.

3.1 Case 1: Norma

Norma was teaching a whole class of 14 year olds and had agreed that the two of us could observe the lesson. Students seemed comfortable and many were willing to call out contributions and suggestions throughout the lesson. The interactive sequence was composed of a monologue which was frequently interrupted by students, not only when she posed questions to them. She seemed happy for this to happen. The following excerpts have been selected to illustrate typical interactions which we have observed in this and other lessons of hers. She had written a list of numbers on the board:

1, 2, 3, 3, 4, 4, 5, 6, 9

and had asked students how they might identify the mode and median. The median was correctly identified as 4, and she then went on to tell them how to calculate the range. The observers had the impression that this was revision from previous lessons, as range was not defined and the meaning of these measures was not mentioned. Norma went on to say that the quartile was 'half the median' and asked what it would be for these numbers.

Student 1: Is it 2.5, half way between 2 and 3?

Norma agreed that it was but proceeded to show how to reach it by counting from each end of the numbers 1,2,3,3. She counted with two fingers from each end of the line of numbers as she said:

Norma: Just like you found the median, you count inward, there and there.
Student 2: You don't count the median?
Norma: Why, you've just counted it!

This question was not pursued further. Later in the lesson Norma had presented them with a new list of six numbers for finding medians and quartiles. In this case the median and quartiles all lay between the actual values in the data.

Student 3: Why did you choose that data?
Norma: Because that's the data I chose.
Student 3: But why?
Norma: Because I chose it.

Later, after more work on quartiles and the definition of the inter-quartile range, Norma asked them what the inter-quartile range might show them about the data.

Student 4: Middle
Norma: What do you mean? Middle is perfect.
Student 4: Is it just like the middle of the numbers, is it 5?
Norma: Well, it's 4 ½

Student 5: Roughly the median
Norma: Not always

In the example which was being discussed at the time, the median and the interquartile range were both 5, there was no further discussion of the value in this case.

Norma: I am after what the interquartile range actually shows us.
Student 6: It shows an average
Student 4: It is the middle of the data with no extremes
Norma: Yes, the middle of the data with no extremes, the range of the middle half of the data
Student 5: Can we plot it on a graph?
Norma: Yes, a cumulative frequency graph, now everyone look at my wonderful diagram

She showed a diagram she had prepared earlier, not a cumulative frequency graph. There was no further mention of such graphs in this lesson.

3.2 Case 2: Roisin

The following interaction took place with a whole class of students. In our experience it is fairly typical of Roisin's teaching. They responded by putting hands up to offer an answer. Roisin would wait until several hands were up before picking someone to answer. In general, about a third of the students would put their hands up, and it was mainly the same people again and again. Roisin says that she follows up everyone during the individual written work which follows to ensure they have all understood. This was a year 7 class so had only recently come to the school, hence we do not know how much any of them knew before about this topic.

Roisin: Has anyone any good ways of working out percentages? If I was to ask you what is 10% of 230, who could tell me?
Student 1: Move all the digits one place to the left
Roisin: I'll write that up and follow your instructions carefully (she goes to the board to write '10% of 230'; as she did so Student 1 called out 'to the right')
Roisin: You want me to move all the digits to the …
Student 1: To the right
Roisin: Done that

```
230
   230
```

Student 1: …and then you knock off the zero
Roisin: I knock off the zero; what's the mathematical equivalent of knocking off the zero?
Student 2: Put a decimal point in
Roisin: (she puts the point in the right place) Is that what you want me to do?
Student 2: Yes
Roisin: OK. (To everyone) Be careful what you say. If you say 'move the digits to the right or the left' you could end up with them over there (points to extreme right of the room) So 10% of 230 is 23 point zero. Can I work out what 20% is? (She writes: 20% of 230)
Student 3: Move the digits two places to the right.
Roisin: So you think (writes it up) 20% of 230 is 2.30? That's not quite right. (She erases it and speaks to the student, quietly) Listen carefully to the next bit to see where we are going with this. Who else can explain what 20% of 230 is?
Student 4: (Inaudible answer – Roisin said afterwards that she didn't understand what was said)
Roisin: Let's have another think
Student 5: (calling out) Can you use 10% and then double it?
Roisin: Yes. We found out that 10% was 23; how much bigger is 20% than 10%?
Student 6: (calling out) twice
Roisin: Yes it's twice as big isn't it? So 20% of 230 is twice as big as 10% was. So can you tell me what 20% is? (asks S3)
Student 3: 46 (R writes 46, begins to write .0 but changes her mind and erases it)
Roisin: What about 50%? What's the easiest way to get 50%?
Student 7: Halve it
Roisin: Yes, 50% is half. Well done, so 50% you just halve it. Have any of you got other percentages you can do? We have 10% and 50% …
Student 8: 75%
Roisin: How can you get that?
Student 8: You half it, and half it again, and add them …..

3.3 Case 3: Susan

These excerpts from a lesson on fractions are typical of Susan's practice and the norms of her classroom. We give several excerpts for this lesson because they each illustrate practices which we had not seen elsewhere. In the first excerpt she talked about an overhead projection of a sheet containing several blank diagrams of what she called 'whole sticks'. They were all the same length, but chopped into different numbers of blocks. For example, this is a '24-stick':

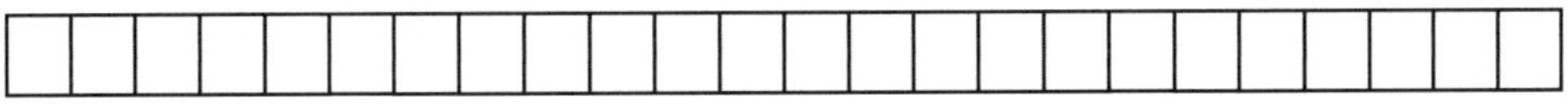

In the introduction to the lesson Susan demonstrated three fraction sums in similar ways. In this one she shaded seven blocks in red, linking the written symbol $\frac{7}{12}$ to the diagram by saying 'I have shaded 7 on the 12-stick'. She shaded three more blocks in blue on the same strip and wrote the fraction in blue. She then added and simplified, and also simplified before adding. This diagram shows a worked example with arrows to show the connections she made in the exposition:

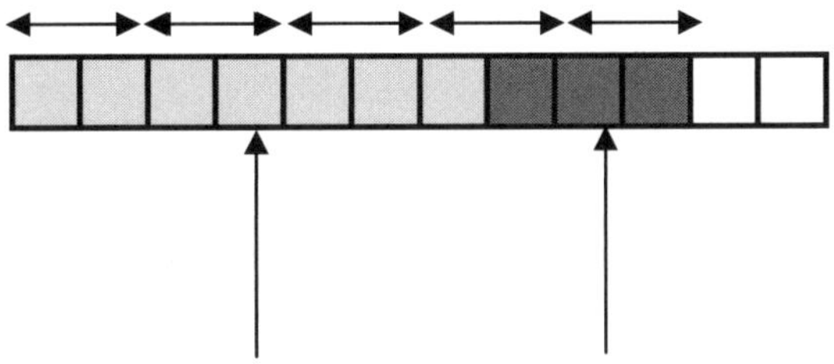

$$\frac{7}{12} \quad + \quad \frac{3}{12} = \frac{10}{12} = \frac{5}{6}$$

$$\frac{7}{12} \quad + \quad \frac{1}{4} = \frac{10}{12}$$

During the sequence above she talked prolifically about what she was doing and no one else said anything. She then asked them to make up some examples of their own.

Susan has devised a low-tech way in which students' work can be displayed on the overhead projector. Paper sheets printed with fraction strips and spaces for working are tucked inside clear plastic wallets; students write on the plastic wallets with dry-wipe pens. When Susan wants to display their work she removes the inside paper sheet and tucks a transparent version of the sheet in its place. This produces a transparent version of the student's shaded strips and workings which she can display on an overhead projector.

After the students had been working a little while on their own choices of fractions, using these transparent wallets, Susan stopped them:

Susan: … and look this way. What I would like are a couple of your examples.

Student 1: Me, me, me!!
Susan: I need all of you looking this way so you know what to do next. (long pause while she prepared someone's work for the overhead) What you need to do is work out how this person got from the shape to the fractions.

She pointed to the last example on the sheet, which was also the most complicated:

$$\frac{16}{24}+\frac{6}{24}=\frac{22}{64}$$

$$\frac{8}{12}+\frac{3}{12}=\frac{11}{12}$$

She pointed to the $\frac{8}{12}$

Susan: Can we cancel this a bit more than eight-twelfths?
Student 2: (called out) Four-sixths (Susan paid no attention to this)
Student 3: Two-thirds
Susan: Yes (she wrote this) and can we cancel three-twelfths?
Student 2: one quarter
Susan wrote $\frac{2}{3}+\frac{1}{4}=$ as the next line.
Susan: On your individual pieces of work you should have cancelled down everything which could cancel. What I want you to do now is see if you can work backwards from two-thirds add one quarter. Decide which stick you need, shade in two-thirds of it and one-quarter of it and see what the final fraction will be.

She went around the room helping students do this (we did not have access to her individual interactions). She collected a variety of examples from the students and wrote them on the board for others to work through. She told them to choose only the ones they felt they needed to do to help their understanding.

Later, after some time working on these, she wrote on the board:

$$\frac{1}{4}+\frac{7}{12}$$

and said: 'I am going to give you a hint about which stick you need. This is a 12-stick. What's a quarter of twelve?'

Student 4: 4 (she ignores this)
Student 5: 9 (she ignores this)
Student 6: 3
Susan: Three, so that's how many to shade in

She writes: $\frac{3}{12}+\frac{7}{12}$ and asks 'how many have I shaded in?' However, she gave no diagram given for this one; she was still using the language of 'shading in' but not actually doing it.

4. ANALYSIS OF CLASSROOM SEQUENCES AS LOCAL COMMUNITIES OF PRACTICE

We will show the strengths and limitations of LCMP used as an analytical tool to provide articulation of differences between profound and superficial interactive routines. The analysis which follows is a conjecture, since we only have excerpts here and have only seen a small number of lessons. Hence we do not know the full complexity of these classrooms, their affordances and constraints, their norms and expectations. Nor do we know about the students' perceptions of mathematics. Instead, we bring mathematical knowledge to bear and analyse the affordances and constraints in terms of potential mathematical participation, based on these excerpts. We do not claim that these are as perceived by the students or the teacher, nor are we claiming that any teacher is better than any other teacher. Rather, we look for the mathematical invariance and relationships which are afforded in the classroom, and whether students seem to be attuned to reacting to these, and how they do so, these attunements being necessarily shaped by past patterns of participation.

4.1 How do students seem to be acting in relation to mathematics? What kind of participants do they seem to be within the lesson?

4.1.1 Norma:

Students appear to be an audience for the teacher's exposition; but the six students in the excerpts are also acting as thinkers and contributors, calling out their ideas. We can see that the situation allows them to participate verbally. However, all their ideas are ignored so it is hard to understand their role in the lesson. They can contribute what they already know; answer closed questions if they can. Otherwise they watch – a few choose not to watch.

4.1.2 Roisin:

At least one student seemed to be making sense of what he was seeing and hearing in the lesson by seeing and using patterns, but this was not explicitly encouraged by the teacher in this sequence. He was told he was wrong. Later he was asked to answer a simple question about what 'twice 23' is, and answered correctly. Most students are watching, the majority offer contributions which suggest matching what is said and written to previous experiences with percentages, place value and numerical processes.

4.1.3 Susan:

Students do a variety of things at the same time: writing, discussing with each other, listening to the teacher and contributing ideas. There is a workshop atmosphere in the lesson, with very little direction about when to listen, when to discuss, when to write and so on. They are supposed to be aligning their meanings and ideas to those of each other and the teacher. Also they match symbols to diagrams and to words.

4.2 What developing mathematical competence is publicly recognised, and how?

4.2.1 Norma:

While the dynamics of the lesson are about interaction, in these excerpts the teacher rarely responds positively to students' ideas. The word 'middle' is praised. The developing competence seems to be about guessing what the teacher wants. Other forms of thoughtful contribution are not publicly recognised, but expression of them is allowed.

4.2.2 Roisin:

We do not know what the students already could do, so it is not clear what developing competences are valued. However, we do know that answering closed questions and stating correct methods and thus contributing correctly to the exposition are valued. The one instance which is clearly a development (since it depends totally on what is said and done in the sequence) is not seen as competence although it is based on generalising from a perceived possible pattern.

4.2.3 Susan:

Offering ideas and examples, making acceptable links between representations, and participating by articulating – these are all valued. But mathematically incorrect or incomplete offerings are ignored.

4.3 Do learners appear to be working purposefully together on mathematics? With what purpose?

4.3.1 Norma:

No collective task is identified, unless we say that the learners are collaborating in allowing the teacher to make progress with the exposition and watching it; however, some of them call out questions in a way which might impede the teacher's progress through the planned exposition.

4.3.2 Roisin:

Use of 'we' by the teacher from time to time implies collaboration to contribute to correct exposition on board of methods to calculate standard percentages.

4.3.3 Susan:

The purpose is to be able to do typical questions, and to achieve this by understanding where such questions might come from. Students' own examples are used as the raw material for whole class work, so individual work is purposeful in the group. The whole class works as a group guided by the teacher.

Since we were focusing on whole class interactive episodes, 'work together' here is taken to mean 'engaging with shared purpose in the dynamic of the whole class'. In all these three cases this dynamic was orchestrated and controlled by the teacher.

4.4 What are the shared values and ways of behaving in relation to mathematics: language, habits, tool-use?

4.4.1 Norma:

Students listen to the teacher and call out answers to open questions about meaning. It is expected that learners will accept a list of numbers as 'data' but some students challenge the meaning of the list. The data is written on the board and Norma makes marks on it to delineate median, quartiles, inter-quartile range and so on as a tool for understanding these terms.

4.4.2 Roisin:

Students listen to the teacher; they adopt the mathematics register; they use a 'hands up' protocol; some students answer public questions; they give reasons when they are correct. They work with symbolic representations and talk in terms of procedures. The teacher uses the board to select and filter students' responses to create correct written examples which will be used as tools for further work.

4.4.3 Susan:

Tools have been created which allow students' ideas to be made public. Patterns of speech connect symbolic and iconic modes. Multiple representations are offered and used. Everyone is supposed to be connecting the different modes for themselves. We also see signs of implicit negotiation in this episode, in that she seems to be trying to share meanings with the class, as evidenced in her suggestion: "What you need to do is work out how this person got from the shape to the fractions". Susan, too, needs to engage in this shared practice of interpretation.

4.5 Does active participation of students and teacher in mathematics constitute the lesson?

4.5.1 Norma:

Trivially yes, but only insofar as it concords with teacher's chosen direction. More overt participation by students is ignored and does not contribute to the lesson.

4.5.2 Roisin:

Trivially yes. Students' answers and reasoning, if correct, influence the exposition as the teacher repeats correct them; students' ideas and prior knowledge are sought to be included in the exposition.

4.5.3 Susan:

Substantially yes. Participation is by joining in the language and tool-use of linking representations and thus making sense; another form of participation is by contributing examples. However, there is also strong guidance, for example, some 'wrong' answers are ignored by the teacher with no comments made.

4.6 Do students and teacher appear to be engaged in the same mathematical activity? What is the activity?

4.6.1 Norma:

Insofar as we can conjecture from our experience of Norma's lessons, the shared activities are: getting best possible grades; constructing 'good' answers to future questions.

4.6.2 Roisin:

The activity seems to be about getting some exposition and examples about percentages onto the board, which will be used as tools for further work. At times students might be listening to each other, but for the flow of the interaction it is only necessary for the teacher to listen to individual students and all students to listen to her.

4.6.3 Susan:

Susan's stated aim is that they should all get the best grades possible, through understanding the underlying structures of mathematics. In this excerpt, they all, including Susan, seem to do this by: linking representations; using forms of language which make these links; devising examples for discussion; trying to understand each other's reasoning. We consider all of these as valuable mathematical practices.

In all these cases we were able to identify features of local communities of mathematical practice, and in that sense all were successful in the shared enterprise of enabling learners to participate. Therefore, we would say that, to some degree, all three classrooms could be seen as LCMPs. The analytical questions associated with LCMP helped to illuminate some similarities among these three different classrooms but they also allow us to lay out differences between these teachers, and indicate places to look for the different kinds of mathematical engagement afforded. That is, the analysis of classroom sequences as local communities of practice gave us a general perspective of the practices and of the degree of the students' participation in the classroom, but we shall now pursue a finer perception of the mathematical practices going on and of the attunements of the students with these practices.

5. ANALYSIS OF THE AFFORDANCES, CONSTRAINTS AND ATTUNEMENTS OF MATHEMATICAL ACTIVITY IN EACH SITUATION

If participation is an adequate metaphor for learning we need to look at the extent to which the participation was mathematical in order to say anything about mathematical learning. It is not enough to say that learners participated in discussion, nor even to analyse the discourse in terms of emergence of meaning. Within the analysis of practices, therefore, we had to make some judgements about what kinds of participation were afforded and constrained within each situation. We are going to look for the kinds of mathematical activity which are afforded in some classroom interactive sequences, and how these are constrained by the teacher. The sequences are all co-constituted by the active participation of students, though not all students. However, as well as a variety of normal classroom constraints of expected behaviour, signals of participation (mainly putting one's hand up), one person speaking at a time and so on the constraints we choose to focus on are those which relate to mathematical activity.

In the first case, Norma's choice of numbers allows students to feel that they were on familiar ground: small positive integers. However, meaning is constrained to numerical value as the numbers do not represent anything, nor were they collected in answer to some question, hence they afford only the development of calculation techniques and abstract meanings. One student seems to be trying to find out if they are random or not; devoid of context they might as well be, unless they are specially constructed to draw attention to features of the calculation. Students are asked about mode, median and range, so that all the possible ways they could look at this data are constrained by their understanding of these three words. Median and range are worked out, so that this data set then affords further experience of the meanings of words. A new word, quartile, and a definition are offered and students are asked what it would be. It seems that this classroom regularly affords learners the opportunity to attempt new ideas without always having models to follow. A student suggests an answer, and the teacher presents a physical image of how this answer might be found. This image, of counting from both ends, becomes one of the affordances, the possibilities, of the situation, and students are reminded that they already have experience of a similar image. A student's question alerts us to the fact that the situation also affords confusion between discrete and continuous data. The teacher's reply does not illuminate this confusion, but instead offers what sounds like a rule – you don't count the median as you have already done so. Mathematically this makes no sense, and at this point the myriad possibilities offered by the data set and the image become constrained so that some students may have the impression that they have to follow a model and rules which might seem arbitrary. One student asks 'why did you choose this data?' suggesting attunement to attaching meaning to the example. Later in the lesson it turns out that both the median and interquartile range are both 5, thus affording confusion between the two, a confusion which would also be possible linguistically from the frequent use of the word 'middle'. Even with this closing down of possibilities, one student shows attunement to making connections with other experiences by suggesting a graph, but the sequence is then constrained by the teacher to the diagram she had planned to make already. Overall, the situation affords the possibility for some learners to relate the words, definitions, and data in ways which provide models of the practice of interpreting those definitions for different data sets, yet it also affords the possibility for others to attune themselves to rules for counting and omitting the median as future practices. In addition, although the situation allows students to make suggestions, those suggestions do not always become part of the mathematical development. This is potentially confusing. Are learners supposed to participate by clarifying ambiguity in examples or not? Are they supposed to make links to other forms of representation or not? Are they supposed to learn to be more precise about the use of 'middle' or not? (Recall that Norma says 'middle is perfect'). These forms of mathematical practice seem to be deviant in this context rather than shared. The shared enterprise in this sequence does not appear to be the development of meaningful mathematical ideas, although this could be afforded in this mode of teaching, but instead is to produce the social appearance of shared mathematical development while mathematical practices are constrained by the teacher's non-mathematical defence and cursory response to learners' ideas.

Roisin's lesson has some similarities to Norma's, but it is clearer in hers that one aspect of participation is to generalise from the examples offered, as well as suggesting methods both with and without having models to follow. It is also clear, from her return to S3, that mathematical participation includes reviewing personal ideas in the light of subsequent information. Learners can also participate by offering examples and explaining how to work on them. Thus there is shared development of mathematics which is gradually inscribed on the board to produce a record of acceptable practice, and personal development which is not publicly inscribed. Although there is not the depth of questioning from learners that there is in Norma's class, there is more overt participation in the mathematical practices of exemplifying, generalising from examples, and constructing methods for new situations by adapting previous experiences. These are afforded locally in Norma's questioning in this sequence, but must also be generally afforded in the practice of the classroom for learners to be attuned to participating. In Roisin's case, this was not always afforded. For example, the boy who exhibits the mathematical practice of looking for, expressing and using patterns has this followed up with a simple closed question constraining him to try to engage in the teacher's intentions, with no further work on using patterns. This could indicate that Roisin is aware of the limitations of over-enthusiastic inductive reasoning; indeed other episodes show that she pays more attention to deductive responses and encourages their articulation.

Susan's lesson affords participation in the mathematical practices of exemplification, generalisation, shifts between representations, presentations of equivalence, and the possibility of abandoning drawn spatial images for mental and linguistic substitutes, which might lead to symbolic fluency. One could say that these possibilities for mathematical practice are afforded in all mathematics lessons, and that is why a few students become mathematicians even with very constrained teaching. However, in her lessons these are shared developments, overt features of the practice of the community rather than deviant or occasional behaviour. Participation in mathematical practices is also constrained by using her materials, tools and representations, and there is no place for learners to use images or techniques to which they may have been earlier attuned by other teachers. Furthermore, learners are drawn into the centre of practice by being on the periphery until they know how to participate; contributions which do not fit with the central development of mathematics are ignored rather than picked up to be explored as conflicts, as Roisin tried to do. Norma, too, ignored contributions which did not 'fit' and drew learners into maintaining her own exposition. The emphasis in Susan's lesson is that they all, including her, work hard on a central strand of to-be-shared meaning.

Cases 1 and 2 share getting an exposition onto the whiteboard, to which learners can refer for subsequent work; they also share a focus, with strong explicitness in Norma's case, on coverage of content for the next examination. Cases 2 and 3 share the use of students' contributions to shape the lesson, case 3 using them as the material for shared work. Case 3 also focuses explicitly on two-way shifts between representations, where case 1 uses different representations as one-way illustrative shifts.

All three use repetition, probably as a tool to establish habitual responses. In these excerpts no teacher uses direct revoicing (Hoyles, 2001) although Roisin appears to recast and restate the content of what is said to structure her next question. Further deliberate pedagogic tool-use is seen in the carefully constructed visual layout of symbols to give transparent access to mathematical structure. In Susan's case special tools for diagrammatic representation also give access to mathematics. Single lessons are not enough to conjecture whether these tools become the focus in themselves, rather than giving transparency to the learning task (Adler, 1998).

On the level of considering whether students are well-behaved participants or not, in terms of attending to what is going on and the trying to produce artefacts (answers and workings out) which the teacher recognises as 'good', case 1 does rather better than case 3, but maybe well-behaved participation is less important than what participation entails. In terms of considering the extent to which mathematical discourse is shared, all three cases do well although case 2 does rather better than 1, where the responses are very limited by the teacher. In terms of the participation being distinctively mathematical, and engaging learners into activities which are mathematical, and hence acting mathematically, case 3 does better than the other two. It is also the only case in which the explicit aim of the teacher is to draw learners from peripheral participation in mathematics to playing a central role in activities which are similar to the teacher's activities, in this case: making examples, switching representations, articulating relationships and understanding each other's reasoning.

6. INFLUENCE OF STUDENTS' CONTRIBUTIONS

We find that the nature of participation varies in the three cases. For example, in one classroom, the teacher focused on techniques and questions, and participation was to fill gaps in a lively monologue to provide a chorus for the teacher's exposition. In another, the teacher employed several well-known strategies to engage discussion and recognise all contributions, and progress from the periphery to full participation was indicated through giving answers and explaining 'own' methods. It is noticeable that these three teachers give different amounts of attention to students' contributions which are not what they had in mind. Although all of them generally drive the task strongly along their own lines of development, if we look closer at the participation of the students, we can see some subtle differences in the selected episodes regarding how far the teachers allow the students to actively participate in the constitution of the lesson. In spite of the subtlety of these differences, we believe they can make a lot of difference in terms of how the students see themselves as participants in these classrooms and, therefore, they are worth further investigation.

During our analysis of these episodes, we began to sense that Susan somehow gives a different quality of attention to students' contributions than the teachers in the other two episodes. On analysing further episodes we found that, at times, Susan changes the line of mathematical thought according to the students' influence, allowing them to participate in orienting the lesson. We now present an example of this with another class of students. They had been working with Susan for about two terms so we can assume that some shared classroom practices have been established during this time.

6.1 Susan's class on 'Equations'

In the beginning of the lesson Susan gave the students this diagram:

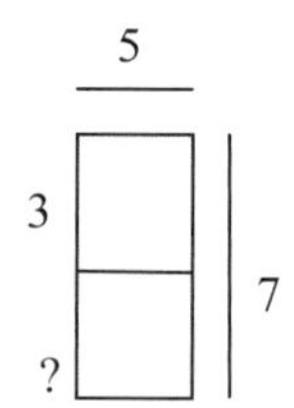

and asked them about the area of the top rectangle, then the total, then the bottom area, always relating the drawing and the calculation. She used other similar diagrams, always giving them the total height of the rectangle and also the height of one of the parts into which it was divided. They could find the missing number without thinking about areas and only relating the heights.

After a few given examples she asked the students to make up some and show them on the board. Gradually, led by some of the students' examples, the situation changed slightly to finding the missing 'bottom height', being given the total area, the 'top height' and the width of the rectangles. For example, in this diagram the total area is 12 units, and the 'top height' and width are both 3 units (Susan encouraged algebraic thinking by not using scale drawings or even neat diagrams; students' diagrams, like hers, were all rough sketches. She talks explicitly about this with her students.).

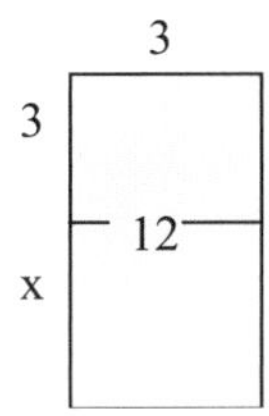

Now students had to make use of the formula of the total area of the rectangle to find the total height, and thence the missing x. The mathematical affordances and the constraints of the problem had been changed by the students' examples

Later on, one of the boys came up with this example:

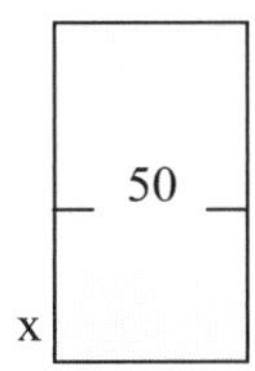

He gave the total area (50), but not the total height or width. Apparently, students spontaneously try to minimize the amount of given data. Susan explained to the boy that it was necessary to give the height of the top rectangle, to which he replied: '3'

Susan asked how much the total height would be, and how much x would be. The boy answered 7 (for x) and she said: "OK, let's jump on that" and asked him "why is it 7?"

Following his correct explanation involving a width of 5, Susan constructed another example, similar to this one, with the total area, but not with the total height, making the problem harder for the students but shifting the focus to 'areas' involving two dimensions and not just one. When she asked them for a 'hard' area some call out '450'. She suggested something smaller like '27' but they insisted on '450' so she went ahead with it. We are fairly sure that she was planning to move to this kind of situation anyway and she moved to it because of the student's example, maybe earlier than she had expected.

This extra episode shows Susan's openness to change affordances and constraints of her work by the influence of some students' unexpected interventions, when perceived as relevant. This episode also strengthens a non-conservative view of what is going on in the class, open to the possibility of seeing affordances and constraints as dynamic characteristics of a situation that can be modified according to unexpected attunements of some students.

To sum up, in the third classroom, we saw learners being gradually drawn into public mathematical practices which originated with the teacher, as if she was conscious of students' progression from legitimate peripheral participation towards central roles in the practices. Of course, we see this to some extent in all three classrooms, but only in the third case were social practices augmented with complex mathematical behaviour such as symbolising, representing, generalising, controlling variables, exemplifying and so on. While all teachers are establishing practices which afford mathematical activity of a certain kind, Susan is deliberately establishing practices which afford new-to-them mathematical participation for these students. Thus, students are not merely rehearsing familiar practices (such as providing gap-filling answers, listening to the teacher (or not)) but are also being drawn into new practices which we can describe as mathematical activity. Even short excerpts, such as we give here, show these as affordances and constraints of the situation, and indicate some attunement.

7. CONCLUSION

We found that the structure of local communities of practice enabled us to analyse mathematics classrooms unencumbered by the weaknesses we identified in a situated perspective when used at a coarser-grained level. While LCMP helped us to analyse practices by laying them out to be compared, it was the concepts of affordance, constraint and attunement which helped us magnify how mathematical practices differ in each case. Thus we were able to focus on specifically mathematical ways to act, such as looking for pattern, invariance, relationships, properties, structures, and imbuing symbols with meaning by engaging in specifically mathematical activity with these. We were able to focus on the unique ways in which mathematics students engage with symbols, and use subject-specific tools

(Hoyles, 2001). Moreover, we did this within an understanding of classrooms as complex social settings in which microcultures of mathematical practice emerge. By using short, opportunistic, excerpts we illustrated the power of the analytical tools to focus on mathematics, and hence suggest how superficially similar classroom sequences shape very different attunements to mathematical participation.

The chapter contributes to the general perspective of situated cognition theory by devising a more fine-grained approach capable of illuminating subtle differences in mathematics classroom practices and students' participation in these practices. It contributes to attempts to analyse classrooms beyond obvious social factors and allows mathematics, as a research perspective, to be incorporated into the insights afforded by a social perspective. Furthermore, it also contributes to a dynamic view of classrooms as possible locations of LCMPs where the afforded participation of the students may, as in the final episode, transform the mathematical practices instead of merely reproducing the existing practices.

ACKNOWLEDGEMENTS

We would like to acknowledge the particular contributions made by Alison Tomlin and Jackie Fairchild to this chapter, and also the contributions made by pseudonymous teachers and their students.

REFERENCES

Adler, J. (1998). Lights and limits: Recontextualising Lave and Wenger to theorise knowledge of teaching and of learning school mathematics. In A. Watson (Ed.), *Situated cognition and the learning of mathematics* (pp. 161-177). Oxford: Centre for Mathematics Education Research, University of Oxford Department of Educational Studies.

Bereiter, C. (1997). Situated cognition and how to overcome it. In D. Kirshner & J. A. Whitson (Eds.), *Situated cognition: Social, semiotic, and psychological perspectives* (pp. 281-300). Hillsdale, NJ: Lawrence Erlbaum Associates.

Boaler, J. (2002). The development of disciplinary relationships: Knowledge, practice and identity in mathematics classrooms. *For the Learning of Mathematics, 22*(1), 42-47.

David, M. M., Lopes, M. P., & Watson, A. (2005). Diferentes formas de participação dos alunos em diferentes práticas de sala de aula de matemática. In *Anais do V CIBEM – Congresso Ibero-Americano de Educação Matemática,* (pp. 1-13). Porto/Portugal (CD ROM).

Engestrom, Y., & Cole, M. (1997). Situated cognition in search of an agenda. In D. Kirshner & J. A. Whitson (Eds.), *Situated cognition: Social, semiotic, and psychological perspectives* (pp. 301-309). Mahwah, NJ: Lawrence Erlbaum Associates.

Greeno, J. (1994). 'gibson's affordances'. *Psychological Review, 101*(2), 336-342.

Greeno, J., & MAPP. (1998). The situativity of knowing, learning and research. *American Psychologist, 53*(1), 5-26.

Houssart, J. (2001). Rival classroom discourses and inquiry mathematics: 'the whisperers'. *For the Learning of Mathematics, 3*(21), 2-8.

Hoyles, C. (2001). From describing to designing mathematical activity: The next step in developing a social approach to research in mathematics education. *Educational Studies in Mathematics, 46*(1-3), 273-286.

Lave, J. (1993). Situating learning in communities of practice. In L. B. Resnick, J. M. Levine & S. D. Teasley (Eds.), *Perspectives on socially shared cognition* (pp. 17-36). Washington, DC: American Psychological Association.

Lave, J. (1996). Teaching, as learning, in practice. *Mind, Culture, and Activity, 3*(3), 149-164.

Nunes, T., Dias, A., & Carraher, D. (1993). *Street mathematics and school mathematics*. Cambridge: Cambridge University Press.

Rogoff, B. (1995). Observing sociocultural activity on three planes: Participatory appropriation, guided participation, and apprenticeship. In J. Wertsch, P. d. Rio & A. Alvarez (Eds.), *Sociocultural studies of mind* (pp. 139-164). Cambridge: Cambridge University Press.

Saxe, G. B. (1999). Cognition, development, and cultural practices. In E. Turiel (Ed.), *Culture and Development: New Directions in Child Psychology*. SF: Jossey-Bass.

St. Julien, J. (1997). Explaining learning: The research trajectory of situated cognition and the implications of connectionism. In D. Kirschner & J. Whitson (Eds.), *Situated cognition: Social, semiotic and psychological perspectives* (pp. 261-279). Hillsdale, NJ: Lawrence Erlbaum Associates.

Watson, A. (Ed.). (1998). *Situated cognition and the learning of mathematics*. Oxford: Centre for Mathematics Education Research, University of Oxford Department of Educational Studies.

Wenger, E. (1998). *Communities of practice: Learning, meaning, and identity*. Cambridge: Cambridge University Press.

Winbourne, P., & Watson, A. (1998). Participation in learning mathematics through shared local practices. In A. Olivier & K. Newstead (Eds.), *Proceedings of the 22nd International Group for the Psychology of Mathematics Education* (Vol. 4, pp. 177-184). Stellenbosch, SA: University of Stellenbosch.

Chapter 4

Social Identities As Learners And Teachers Of Mathematics

The situated nature of roles and relations in mathematics classrooms.

Mike Askew
King's College London

Abstract: Case-studies conducted as part of a longitudinal study of mathematics teaching and learning in primary (elementary) classrooms were originally intended to shed light on children's mathematical understandings. It became clear, however, that mathematical understanding and attainment were inseparable from the social identities as learners of mathematics that children were able to adopt. Such identities are not 'givens' but are situated and made possible through the affordances and constraints the classroom cultures. In the first half of the chapter I explore the emergence of social identities as learners of mathematics and in the second half examine how shifting our attention from identities to relations has implications for how classrooms might be organized to allow more children access to a social identity as a successful mathematician.

Key words: primary (elementary) mathematics, social identities, situated learning relations, classroom culture

1. INTRODUCTION

"Show me."

Twenty-eight seven-year-olds held up 28 whiteboards – marker-pen and board equivalents of chalk and slate – showing their answers to "nine times two". Some held their boards up high; some held them low, close to the back

of the child in front; some waved them about. Meg held her board high and steady and, given her central position on the mat, it caught the teacher's eye.
"Well done. Meg, how did you work it out?"
"I did like what George did yesterday. I doubled and took away."
"Well done."

From where I was sitting I had seen Meg count out nine fingers, make nine nods of her head while looking at each finger in turn and mouth something along with each nod. I'd observed her do this before and was confident that she has counted on in twos, rather than doubled 10 and taken 2 off, as George had explained the day before.

Meg looked over and caught my eye. She smiled widely.

2. PUPILS' SOCIAL IDENTITIES AS LEARNERS OF MATHEMATICS

Meg was one of our case-study pupils tracked as part of the five-year longitudinal Leverhulme Numeracy Research Programme (LNRP. See Millett, Brown and Askew, 2004, for an overview of the research). We (myself and colleagues Margaret Brown, Hazel Denvir and Valerie Rhodes) carried out case-studies of 60 children in their careers as learners in mathematics lessons; 30 from Reception to Year 4 (from being 5-year-olds to 9-year-olds) and another 30 children as they progressed from Year 4 to Year 6 (8-year-olds to 11-year-olds). The above incident is taken from field notes made when Meg was in Year 2.

Our original intention had been to investigate, through these case-studies, the development of the children's mathematical knowledge across the primary years and to shed light on progression in understanding. A 'core' project tracked a sample of around 3000 children by assessing them twice a year and observing their lessons once a year; our 60 case-study pupils were selected from this larger sample. The 'core' data would give us broad-brush-stroke pictures of progression; the case-studies would fill in finer details.

As the research unfolded, however, our emphasis changed and expanded. It became clear that the children's learning could not be understood separately from the roles and relations that were played out in classrooms. The children's learning of mathematics was inextricably bound up with the variety of 'social identities' that individuals (children and teachers) took up within the classrooms. For example, some children, Meg included, clearly placed a high priority on being perceived by their teachers as 'good pupils': they strove to provide 'right' answers; they made efforts to be called upon to answer questions; they presented work in ways that they knew would elicit teacher praise. Some children achieved a 'good-pupil' identity through engaging with the mathematics but others we observed were skilful in drawing on the work of their peers ('copying' in some cases!) to create the outcomes necessary to be seen as 'good'. Meg would occasionally copy answers, but was often observed to use counting procedures to carry out calculations. When publicly called upon to explain her methods she displayed considerable skill in appropriating explanations that she had heard her peers use and which she knew would elicit praise from the teacher. If nothing else, 'in my head', was an effective, praise inducing, response to 'how did you do it?'

This is not to suggest that Meg, or other children who acted similarly, was necessarily aware of any contradictions in her actions. Over the years, as we got to know Meg, it seemed that for her there was no tension between what she did to figure things out and her post-hoc explanations. These were simply two different practices; one was done 'privately', the other was what you did 'publicly'. (In fact, as Meg grew older, the mismatch between these two practices diminished and it became clear that the strategic explanations she gave did mirror the methods that she was using. While her previous mismatches might have been thought of as 'cheating' and so discouraged, it may be the case that such 'rehearsal' of explanations provided the foundation for actually adopting them.)

Not all of our case-study pupils displayed the desire to be 'good-pupils'. For example, some occupied the position of being a 'quiet-worker' and acted in ways that minimised attention being drawn to themselves. Like the 'good-pupils', there was no simple connection between being a 'quiet-worker' and mathematical understanding. Some 'quiet-workers' appeared simply to want to 'get-on' with the mathematics undisturbed; others did not want to draw attention to themselves, it seemed for fear of being seen to be having difficulties.

Other children established identities as 'top-in-maths', George was firmly established within this identity (and note how Meg drew upon George's social identify to validate her own). Others included the 'disruptive-worker', or the 'chatterer'. None of these identities, we suggest, were 'givens' but were situated and made possible through the affordances and constraints of the classroom cultures. Thus the construct of 'social identities as learners of primary mathematics' emerged as central to our understanding of how and why children were making differential progress in their learning.

3. LEARNERS' SOCIAL IDENTITIES

Notions of 'identities', 'learner identities' and 'social identities' are increasingly present in the literature but are far from being worked through as a coherent and consistent collection of theoretical constructs. For example, Rees, Fevre, Furlong, and Gorard (1997), drawing on the work of Weil, define 'learning identity' as:

> The ways in which [individuals] come to understand the conditions under which they experience learning as facilitating or inhibiting, constructive or destructive. Learner identity suggests the emergence or affirmation of values and beliefs about learning, schooling and knowledge. The construct incorporates personal, social, sociological, experiential and intellectual dimensions of learning as integrated over time (Weil, 1986, p. 24).

They go on to point out that:

> learning identity is not simply a matter of success or failure at school; it is also the product of more complex processes. For example … forms of curriculum, pedagogy and assessment. … The learning identities that educational institutions aspire to engender in their students vary both between different types of institution and also historically (p. 493).

In line with Carr (2001) a 'social identity' is a:

> culturally and personally located social schema that has the potential to be transacted, redefined and resisted and, like discourse, called upon when the moment is – to the learner – opportune (p. 527).

Our choice of 'social identities as learners of primary mathematics' indicates a view that the identities that children adopt and develop are neither fixed nor something that they bring with them to the mathematics class. They are a product of the interplay between the cultures of the classrooms, the relations set up within these cultures and individuals' personal resources.

Attending to 'social identities' rather than 'inherent' traits or 'personalities' of individuals shifts the focus to actions. Social identities have to be established and maintained through the ongoing actions of all participants. Individuals' actions and transactions between classroom participants become basic analytic categories.

> When action is given analytic priority, human beings are viewed as coming into contact with, and creating, their surroundings as well as themselves through the actions in which they engage. Thus action, rather than human beings or the environment considered in isolation, provides the entry point into the analysis. This contrasts on the one hand with approaches that treat the individual primarily as a passive recipient of information from the environment, and on the other with approaches that focus on the individual and treat the environment as secondary, serving

merely as a device to trigger certain developmental processes (Wertsch, 1991, p. 8).

The situated nature of social identities – different classroom cultures constrain or enable different identities – means that they cannot exist at in isolation: for someone be 'best-at-maths' someone-else must be 'worst'; the 'quiet worker' can only exist in juxtaposition with the 'noisy' or 'disruptive' pupil. Thus questions are raised about the 'natural' order of classroom and the inequities that may (unintentionally) be maintained through normalised classroom practices.

The normalising of the range of classroom identities can be seen through accounts that treat the 'ecology' of classrooms as posing a number of difficulties or threats that need to be dealt with (usually through eradication). Pupils' misconceptions have to be corrected, differing levels of attainment require particular grouping strategies (usually according to some measure of 'ability'), behaviours need to be 'managed'. To pursue a horticultural metaphor, if classrooms are gardens, then they are in constant need of weeding, bedding out, and training of plants to grow in particular ways.

Social identities are the result of transactions between the learner and the sociocultural context. It is not simply the case that children are 'positioned' as being particular sorts of learners of mathematics nor is it the case that individual agency can completely overcome the dynamics of lessons or the structures of the mathematics classroom. But despite the moves to see identities as not fixed, such shifts in perspective have not loosened the grip of the focus on the individual. Gergen summarises the pervasiveness of the technology of the 'individual' in models of 'good education':

> Good education, we say, will prepare the individual to participate productively in society, and to contribute as a responsible citizen to the democratic process. Such beliefs are also tied closely to our teaching practices. We hold the individual student responsible for his or her own work, we chart the progress of the individual, we evaluate and assign marks to individual performance; individual student scores are arrayed hierarchically for purposes of rewarding superior and correcting the deficient. (Gergen, 2001)

I want to go further than arguing that identities are 'shaped' by external circumstances and look at how a shift away from the individual (the 'owner' of identities) to a focus on relations between individuals may be a more productive way forward.

4. FROM IDENTITIES TO RELATIONS

Despite the introduction of 'situatedness' into the discourse of mathematics education, much is still focused on the individual. For example, Greeno's argument for affordances as "qualities of systems that can support interactions and therefore present possible interactions for an individual to participate in" (Greeno and MMAP, 1998, p. 9) still comes down to the individual being 'supported' by a system. Individuals are 'in' a situation, rather than being 'in' the relational activity of continuously co-creating the situation.

A relational view suggests that we need to look at how different relations provide differential opportunities for individuals to respond: individuals are located within a network of relations – with the teacher, peers, mathematics – which make it non-meaningful to speak of them being 'in the same situation'. A focus on relations brings the interactions, the dance, to the foreground.

My starting point in beginning to move away from the focus on the individual is through appropriating Vygotsky's view of 'method' as being both tool and result. In his original work Vygotsky was referring to the search for 'method' in the study of psychology:

> The search for method becomes one of the most important problems of the entire enterprise of understanding the uniquely human forms of psychological activity. In this case, the method is simultaneously prerequisite and product, the tool and result of the study. (Vygotsky, 1978, p. 65).

What does it mean to treat pedagogy as 'simultaneously prerequisite and product'? That teaching approaches are both the tool-and-result of classroom activity? Extending this view of method as 'tool-and-result' to encompass pedagogy means that cognition and identity are not so much 'situated' within particular pedagogical approaches but dialectically bound up with the chosen teaching methods. Teaching method is not simply something applied to an already existing 'content' and then 'done' in lessons where students 'learn'. All (pedagogy, content and learning) are situatedly bound together and mutually co-constructive. It makes no more sense to talk of the 'teaching method' as distinct from the 'content' or the 'learning' as it does to talk of taking the 'orange' out of 'orange juice'.

In traditional narratives of teaching, the activities of schooling are treated as being able to be decomposed into separate 'objects' – this includes the 'individuals'. Texts and courses are provided on teaching 'methods', different texts and courses address the curriculum 'content'. Teacher knowledge can be 'separated' in 'content knowledge' and 'pedagogical content knowledge' and 'pedagogical knowledge' as though the activity of teaching is based on such separate 'knowledges'. As Holzman argues

> Knowledge is not separate from the activity of *practicing method*; it is not 'out there' waiting to be discovered through the use of an already made tool. ... Practicing method creates the object of knowledge simultaneously with creating the tool by which that knowledge might be known. Tool-and-result come into existence together: their relationship is one of dialectical unity, rather than instrumental duality. (Holzman, 1997, p. 52, original emphasis).

A particularly powerful and persistent story of 'instrumental duality' is that of students having different 'abilities' in mathematics, and that such 'differences' have to be catered for or dealt with, either through grouping practices that reduce the range of difference or through pedagogic approaches that 'meet' individual needs. This can lead to a pedagogy of 'taking care of' (adjusting the level of difficulty so that it is well 'matched' to individual 'needs) as opposed to a 'caring relation' in Noddings' (2005) term.

While not denying that students have different levels of experience and expertise in meeting the demands of the classroom, different narrative of 'ability' might lead to different outcomes. 'Ability' is tool-and-result, is something co-constructed within the ongoing discourse and practices of classrooms, rather than a fixed 'given' that needs to be solved. As Varenne and McDermott (1999) argue, schooling is 'filled with instructions for coordinating the mutual construction of success and failure. ... low achiever ... learning disabled ... (are) positions in education that get filled by children.' (p. 152)

While agreeing that such positions are occupied by children, I find the language of 'positioning' too deterministic. I can position a knight on a chessboard, but children have considerably more free will than chess pieces. So while not denying the current occupation of the positions of 'gifted', low attainer', 'dyscalculic' or whatever is the latest fashionable label, I want to argue for a more dynamic, emergent model of classroom identities: teachers and students co-constructing the available 'positions' (although with different power bases) and that by changing the narrative there is the potential for different configurations to be developed, configurations arising from attending to relations rather than individuals.

Newman and Holzman argue that acknowledging both the indeterminacy and the improvisational nature of activity opens up space for alternatives.

> we engage in the self-conscious, ongoing historical activity of creating culture, which in turn determines who and how we are; our nature is,

fundamentally, at once both socially constructed and emergent (that is, not fixed). (Newman and Holzman, 1996, p. 139)

The task then is to help children develop and so engage in "*new ways of being*" in schools "where children come not in order to know – but in order to grow." (Holzman, 1997, p. 126, original emphasis).

Relations within classrooms – relations between pupils and teacher, pupils and pupils, and between both teachers and pupils and the mathematics – are central to pupils' ongoing negotiation of new ways of being; the 'social', as emerging in and through relations, is at the heart of classroom interactions.

I now examine relations through some reflections on a recent design research project (Mathematics, Teachers and Children (MaTCh)) that set out with the explicit intent of creating classroom cultures with a 'flatter' range of possible identities, where the normative range of identities was challenged.

5. CHANGING CLASSROOM RELATIONS

One unanticipated, and somewhat disappointing, aspect of the LNRP findings was the limited range of types of lessons that we saw, and consequently the range and types of relations that came to be established in classrooms. Our original expectation had been that given the number of classrooms and teachers that we would visit (some 600 of each over the course of the five years) there would be a range of pedagogic practices that we would observe. In the first year of the study this was indeed the case. The majority of the lessons we saw that year could be described as traditional in that they took the form of the teacher setting up and explaining a task to the pupils who then went off and worked, usually alone, on completing it. We did, however, observe a substantial minority of lessons that differed from this pattern, most notably by teachers engaging the pupils in an extended problem or investigation that required little teacher direction at the beginning and which pupils might explore over more than one lesson.

1997, the first year of the LNRP, coincided with the setting up of England's National Numeracy Project (the NNP, piloted in a small number of education authorities or districts). Toward the end of that year it had become clear that the NNP was going to be rolled out nationally (becoming the NNS, the National Numeracy Strategy). Although the NNS was not officially due to start until 1999, transfer of knowledge from schools involved in the NNP to other schools was spectacularly rapid, particularly with regard to the style of mathematics lesson that the NNS was going to expect: three distinct parts comprising oral/mental starter, main teaching and plenary. Although the non-mandatory NNS was not due to be in place until the third year of LNRP, by the second year of our research nearly all of the lessons that we observed conformed to this 'three-part' model.

This rapid and extensive take up of the model was impressive but also surprising. In the first year of LNRP teachers were being told to teach literacy through a four-part lesson model and this had met with considerable resistance. It may be that teachers more willingly adopted the model for mathematics through having had to accept the model for literacy lessons (through draconian inspections that castigated teachers for deviating from the model) and so lost will to resist. But interviews with teachers later in our research suggested a different reason. Many commented on how the National Numeracy Strategy had given them 'permission' to teach mathematics in the way that they had always thought it should be taught. While the model for literacy lessons conflicted with what teachers thought was good practice, this was not the case for mathematics. Teachers, it seemed, were happy to teach mathematics largely through what they interpreted to mean a transmission approach.

As a researcher this was all very interesting, but personally disappointing. Having seen some practices at the beginning of the research that chimed with the literature on inquiry-based learning, by the end of the study this had all but disappeared. Was it true that researchers and theorists were too idealistic? Were different visions of mathematics teaching and learning simply too difficult to put in place, given the everyday constraints of schools and classrooms?

I decided the way to address these questions was to work alongside teachers in school actually trying to develop alternative practices: practices that would challenge the range of 'normal' social identities that had been apparent in the LNRP classes, in particular that mathematics is something that you are naturally good at, or not. My original intent was to find a local 'friendly' school that was 'middle-of-the-road': not so successful that changes to practices might be seen as threatening to that success but also not in circumstances that suggested that the other aspects of the school might need to worked on before changes could be made to the mathematics teaching (for example, high staff turn-over). In the event, however, the education officer from the authority that I approached did suggest the latter type of school. Judged by national test results and inspection reports, the school was considered to be in the bottom 7% of schools in England. A new headteacher (principal) had been appointed and as she was a specialist in literacy then some outside support in mathematics might be welcome.

Results in mathematics were low and the general sense amongst the staff was that this was an intractable problem. In meetings to set up working in the school, teachers commented on the lack of ability of 'these children' and how difficult it was to promote 'interactive' teaching. In particular, any emphasis on oral mathematics was problematic; since many of the children only spoke English as an additional language and the children from white working-class backgrounds had difficultly expressing themselves, then there was little point in asking the children to discuss mathematics. There was an 'empty-vessels' perception of the children as learners – they did not bring anything to mathematics lessons worth building upon and, thus, teaching needed to be 'input' model focused on the teacher explaining. Relations were very focused on teachers and pupils being separate, the former delivering to the latter.

The following account looks at our attempts to change classroom relations (a colleague Penny Latham joined me in working with the school), and in particular the pressures from relations with and between pupils for us as teachers to conform to normalised teacher social identities.

6. CLASSROOM RELATIONS

Taking a social-cultural-historical stance on how we create cultures forces us to see that there is nothing 'natural' about how people participate with each other – and that includes within classrooms. Nevertheless, by the time a group of children been together in the same class for a few years, many relations will have become sedimented. For example, a dominant pupil social identity in many of the classrooms we worked in could be described as 'passively helpless' (or in some cases 'aggressively helpless'). Asked to solve a problem, discuss some mathematics, or work on a strategic solution, then pupils would either sit quietly (the passively helpless) or demand attention (the aggressively helpless). Non-challenging, procedural tasks were what the children would work on (albeit sometimes reluctantly).

When a class has an established range of social identities, the primary newcomer is the teacher. Teachers, just as pupils, have to act to establish a social identity and this will be informed by pupil expectations and the pupil-teacher relations. If the relations that pupils experience over time are somewhat similar, these can easily be mistaken for the range of 'normal' or allowable. The challenge to Penny, myself, and teachers in the school was to establish teacher-pupil relations (and pupil-pupil and pupil-mathematics) in the light of resistance from the established norms.

In line with the recommendations of Holzman (1997) and Gergen (1991) we challenged sedimented relations by directing our attention to the group rather than to individual pupils. We shifted from attending to individual 'cognitions' to group relations, from individual behaviours to group performance. By attending to the development of the group, we believed that individual learning would follow (rather than the commonly held reverse view of attending to the 'needs' of the individual first).

Given the history of the children, this shift in attention was not easy or immediate. For example, we were committed to use a lot of paired work to encourage everyone to articulate their mathematical thinking and set up relations of pupils listening to each other. Such a change to the culture was initially met with strong resistance from the children.

The responses from a class of 8- and 9-year-olds were typical. I put up a calculation and launched into a 'think-share-pair' routine: work out the answer and share your method with the person sitting next to you. The children begrudgingly engaged in the 'think' part, but when it came to them 'sharing' thoughts and methods with a partner, the atmosphere turned sour. Children argued over who was going to explain first – neither wanted to be the listener. Methods and answers were 'secrets' to be kept to oneself and not 'given away' to anyone else. In some cases, one partner would start to explain, but the other would sullenly turn away. Very few children were interested in what their peers thought and arguments flared up around the room.

Recent research by Jenny Young-Loveridge[3] reveals such attitudes as not uncommon. In interviews with her, children expressed contradictory views about the importance, or not, of peers explaining methods. Jenny asked the children whether they thought it was important for them to listen to other children explain methods. The majority of the interviewees indicated that they did NOT think this was a good idea. Their reasons for this included arguing that it would be 'cheating' because in maths it was important to work things out for yourself. Later in the same interview Jenny asked if they thought it was important for them to explain their methods to other children. Most of the children who had indicated that it was 'cheating' for their peers to explain were equally strong in their conviction that it was OK for them to do the explaining as that would help others. Children were happy to explain but didn't want to listen!

Penny and I began exploring ways to get over this resistance and to help children engage in, and appreciate, the benefits of co-operative activity. Rather than simply 'tell' the children to work together, we looked at how the ways the curriculum content was set up might challenge existing relations. Three approaches that proved successful in this were:

- Parallel calculation chains
- Solver and recorder
- Clue problems

[3] Talk given at Massey University, New Zealand, July 2005.

6.1 Parallel calculation chains

In looking at the dynamics of the classroom relations, it seemed that part of the difficulty might have arisen from asking children to talk about the same calculation. Although from my, the teacher's, point of view the benefits of explaining to each other were clear, these benefits were less obvious to the children. After all, if you've got an answer to the calculation, what was the point of hearing how someone else got there? We had assumed that the pupils would appreciate the benefits of relations based on helping each other, but this was too far removed from the prevalent culture.

Our way into changing relations was to have the children tackling similar yet different calculations. In pairs, children had to decide who would be 'A' and who 'B'. Two sets of calculations were put up on the board, the A set and the B set. The children had to find the answers to the calculations in their set. They could do this any way they liked, but they had to record something that would convince their partner that they had found the correct answer. They then had to explain their method and solution to a partner.

For example, while teaching a Year 5 class, we noticed that many of the children were still counting on in ones when adding a single-digit to a two-digit number. We decided to use parallel calculation chains (PCCs) to provoke the children into considering more efficient strategies.

A typical PCC that we set up was:

A	B
477 + 6	568 + 5
646 + 9	466 + 7
886 + 7	753 + 9

Figure 4-1. Typical parallel calculation chain

This time the children were less resistant. Since different calculations were being discussed, nothing was being 'given away' and there was a more genuine reason to listen to a partner. Not only were the pairs able to share different methods, but also from the written records we were able to select children to explain effective methods to the class. For example, Sam's use of the empty number line showed bridging through 10 and, by asking her to share this with the class we were able to being to set up relations where children's voices were valued.

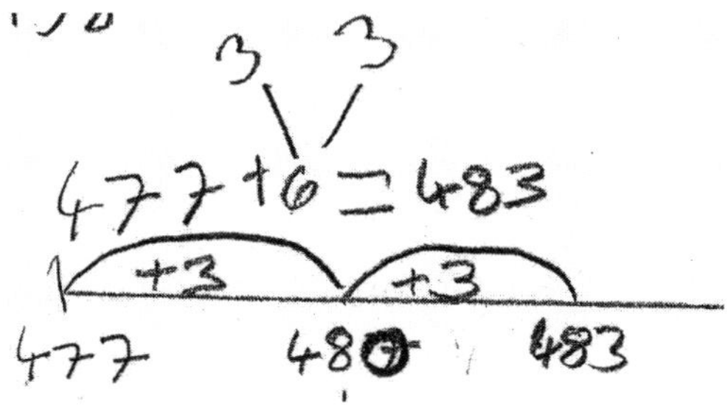

Figure 4-2. Sam's work

Such public sharing was, however, initially far from being a forum for discussing methods. Children in the class with established social identities of 'best-at-maths' would act disruptively when children who they considered did not occupy this position were invited to explain their methods. Relations between the 'best' and the rest were not supportive. It took several weeks of explaining, negotiating and, meeting some of these children 'half-way' by giving them 'voice' before an ethos was established and accepted whereby anyone might be called upon to explain.

6.2 Solver and recorder

Through parallel calculation chains and reflecting together on how they were coping with these, the children began to be willing to work together and a different range of relations began to emerge in the classrooms, based on cooperation rather than competition. We were then able to set up tasks that required pairs to listen to each other as they were actually figuring things out, rather than explain post-hoc. With only one piece of paper and one pen between two, the children had to take it in turns to be 'solver' – the one figuring out a solution method – and 'recorder' – the one recording on behalf of the solver (or the 'robot' as some children came to dub it).

An unexpected spin-off from this approach was that once a child had acted as 'recorder' they were often moved to offer an alternative strategy. For example, a Year 3 class was exploring methods of addition and subtraction through the teacher setting up a fantasy context of frogs having a jumping competition.

Jubel and Kirstie were figuring out the following: "Cath, the frog, jumped 50 cm. She jumped again and in total had then jumped 160 cm. How big was her second jump?"

Jubel, solver, asked Kirstie to draw an empty number line that started at 50. He asked her to record a jump of 40 to 90 and then a jump of 70 to 160, a total jump of 110. Kirstie noted that a jump of 50 from the first 50 would land immediately on 100 and then all that was needed was a jump of 60 – still a total jump of 110 cm.

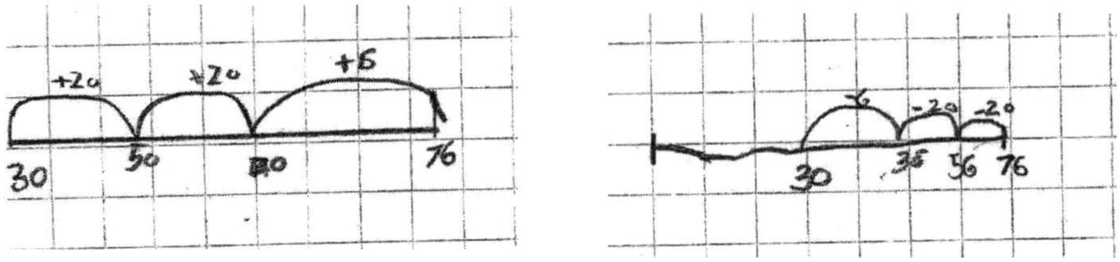

Figure 4-3. Jubel and Kirstie's work

Demi and Jay were finding how far a frog jumped from 30 cm to 76 cm. Demi got Jay to record jumps of 20, 20 and 6. Jay decided to check if 46 was correct by jumping back 20, 20 and 6 from 76.

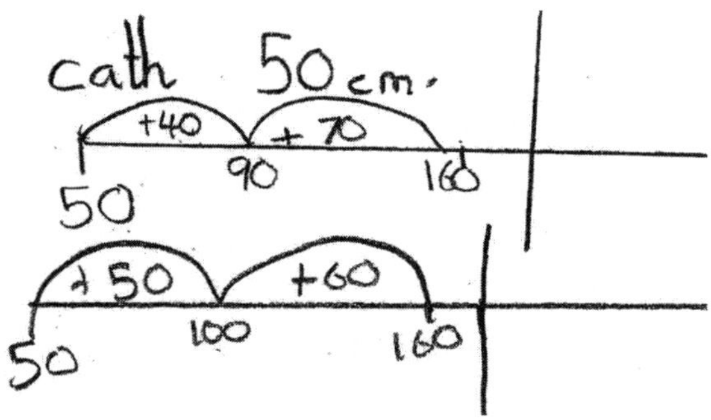

Figure 4-4. Demi and Jay's records

While happy that their answer of 46 must be correct, they called the teacher over. "We have 30 and 76 on each line, so why," Demi asked "do we land on different numbers in between?" The teacher invited Demi and Jay to the board to explain their methods and the class discussion helped the pair appreciate the reasons for landing on different numbers when counting back to those landed on when counting forward. Such discussions began to move the culture away from one of simply 'sharing' methods to one where children were beginning to think about, engage with and question each others' work, without feeling that this meant they were being criticised.

6.3 Clue problems

Our third approach required the children to co-operate in groups of two, three or four. Groups were given a joint problem to solve but individual group members were given only one ‘clue’ towards the solution. They were free to tackle the problem as they chose, but the basic ground-rule was that they were not allowed to show their ‘clue card’ to anyone else. They could read it out (and some wrote it down) but the information on card itself had to be kept out of sight of others. (Teachers often ask what the point of this rule is – after all, if a child writes down what is on card, then why not just show the card to everyone? Showing your cards is simply not as successful. Showing your card gives permission to others to ask to take it and letting go of your card seems to invoke ‘letting go’ of being part of the group; holding onto a hidden card keeps individuals involved).

For example, the following example was set up as a ‘clue’ problem for pairs of 8- and 9-year-olds to work on: “Robbie Williams is performing in London. How many tickets are still on sale?” (Clue 1: 5003 tickets were on sale; Clue 2: 4997 have been sold).

This problem was set up not only to further the relations of co-operation but also to see if pupils could appreciate the distinction between solving this following the ‘action’ of the story (take-away) or the more efficient mathematical model of counting on.

As we expected the majority of the children did some form of taking away. Some chose to use a vertical algorithm, and not always successfully (see example 1). Some used an empty number line (example 2). Only one pair of girls chose to count up from 4997 to 5003. When this was put forward as the third method shared with the class, a pair of boys, previously ‘the best mathematicians’ argued that this could not be correct as ‘it isn’t take-away, it’s finding the difference’. By this point in the year, we were able to ask the children to discuss, in their pairs, which they thought was the most effective method; everyone, including these boys, agreed on the counting-on method.

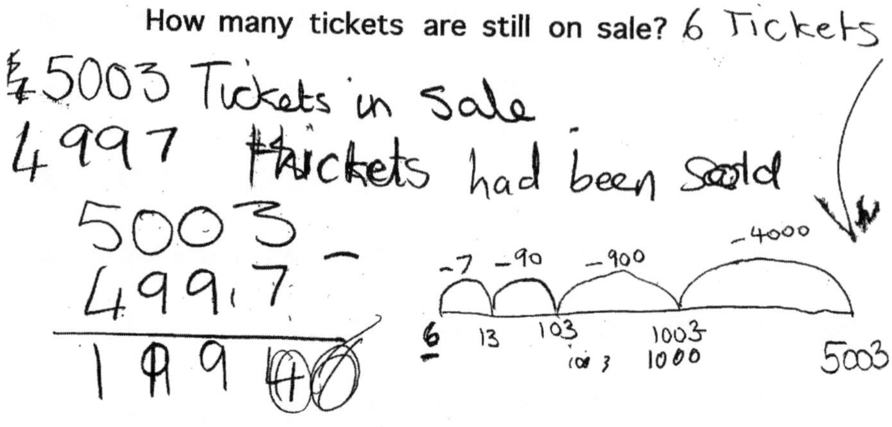

Figure 4-5. Clue problem 1

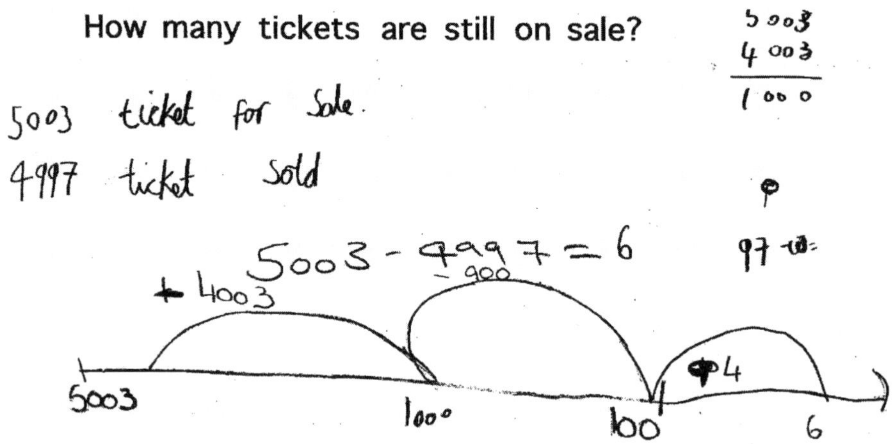

Figure 4-6. Clue problem 2

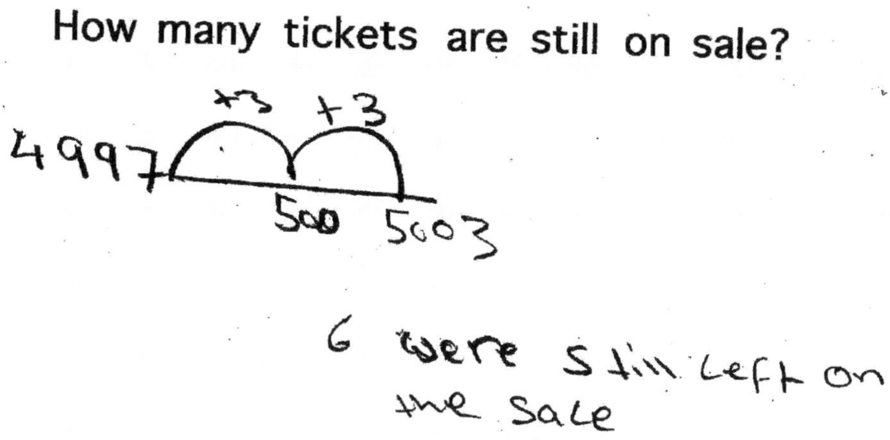

Figure 4-7. Clue problem 3

7. DISCUSSION

Changes to attainment in mathematics within the school have been dramatic and substantial. In 2003, the number of 11-year-olds attaining the expected level in National Test results was 54% (the Government's expectation is 75%). In 2006, this had risen to 86%. Last year, the school carried out an evaluation which included eliciting pupils' views and mathematics was unanimously chosen by the children as their favourite subject. Whilst not claiming that these changes are solely the result of our intervention, they do demonstrate that substantial changes in relations and attainment that can be brought about. One thing that has not changed in our years of working in the school is the curriculum; the materials of the National Numeracy Strategy still form the core of the teachers' planning. What have changed are the prevailing relations within the school; they are teachers of children who can do mathematics, rather than cannot.

At the end of the second year of working in the school, I discussed with Jan, the mathematics manager, what she thought had been the most important parts of the work. She identified children's views of themselves as learners as the most important:

Jan: I think this has been the most significant aspect of the work to date. Allowing children the freedom to say what they think and try to prove this has enabled children to be more willing to 'have a go.' Children regularly ask to answer questions and know to back up responses with 'because'. Children are more adept at working as a pair. They are more willing to share ideas and have a willingness to help each other to find a solution to a problem.

Although couched in terms that still focus on the individual, there is acknowledgement here of the power of shifts in relations. Linked to this was a move away from 'transmission' based pedagogy to teaching that recognised that children are not 'empty vessels' and that they do bring knowledge to the mathematics classroom.

Jan: Enabling children to 'have a go' rather than give a specific method in the first instance has allowed children to be willing to take risks in their work and offer ideas they would normally keep to themselves.

Note the use of 'enabling' (as opposed to, say, encouraging) – a focus on actions as opposed to attitudes. As we have tried to indicate, we worked on developing productive relations. This has involved a big shift in the teachers' perceptions of what is possible based on their previous relations.

In the early days of the project, our own experiences with the children could so easily have confirmed that, indeed, they could not work in ways that deviated from a very safe, teacher-centred, transmission pedagogy of being shown a procedure and then having to practise it. Attempts to get pupils to share their methods with a partner that resulted in widespread behaviour problems could have led us to abandon our attempts to change the relations, adopting instead the 'distant' relations that the children had come to expect. Instead we sought actions that would support the children as they struggled to cope with new relational expectations and the uncertainties that came with these.

8. CONCLUSION

Most research in mathematics education focuses on the processes and outcomes of pupils' cognitive changes. Even research that purports to locate learning within the 'social turn' (Lerman, 2000) attends, by and large, to the role of social interactions in promoting individual cognitive change. The 'social' in terms of relations tends to be taken as a given; if good social relations exist, these are in the background of accounts of mathematical teaching and learning. If there are problems in establishing effective social relations, then there is a wealth of texts on 'behaviour management' to turn to for advice; texts on mathematics education would not be the first choice.

I suggest a need to bring to the foreground the impact of classroom relations and have argued that this may have a substantial impact on pupils' learning. This is not to diminish the importance of cognition. But it is to argue that relations are just as 'situated' as learning and, moreover, that as relations play out different social identities emerge for pupils, and teachers. These identities, in turn, have an impact upon learning outcomes.

But the benefits in attending to relations may go yet further. As Gergen puts it

> Distance, alienation, competition, hierarchy...all may recede. In their place we might hope for relational dances that celebrate communion, invite exploration without fear, and enable a conjoint construction of better worlds. (I fear the words now become inflated, naively optimistic, sophomorically idealistic...but then again, if we are to live in meanings of our own making, why not chose zest?) (Gergen, n.d., original spelling)

REFERENCES

Carr, M. (2001). A sociocultural approach to learning orientation in an early childhood setting. *Qualitative Studies in Education, 14*(4), 525-542.

Gergen, K. (1991). *The saturated self: Dilemmas of identity in contemporary life*. USA: Basic Books.

Gergen, K. (2001). From mind to relationship: The emerging challenge. *Education Canada, Special Edition on the Shape of the Future, 41*, 8-12. Draft version downloaded from: http://www.swarthmore.edu/SocSci/kgergen1/web/page.phtml?id=manu22&st=manuscripts&hf=1.

Gergen, K. (n.d). Writing as relationship. Downloaded from http://www.swarthmore.edu/SocSci/kgergen1/web/page.phtml?id=manu17&st=manuscripts&hf=1.

Greeno, J., & Middle-school Mathematics through Applications Program. (1998). The situativity of knowing, learning and research. *American Psychologist, 53*(1), 5-26.

Holzman, L. (1997). *Schools for growth: Radical alternatives to current educational models*. Mahwah, NJ: Lawrence Erlbaum Associates.

Lerman, S. (2000). The social turn in mathematics education research. In J. Boaler (Ed.), *Multiple perspectives on mathermatics teaching and learning* (pp. 19-44). Westport, CT: Ablex Publishing.

Millett, A., Brown, M., & Askew, M. (Eds.). (2004). *Primary mathematics and the developing professional* (Vol. 1). Dordrecht/Boston/London: Kluwer Academic Publishers.

Newman, F., & Holzman, L. (1996). *Unscientific psychology. A cultural-performatory approach to understanding human life*. New York: iUniverse Inc.

Noddings, N. (2005). *The challenge to care in schools. An alternative approach to education.* (Second ed.). New York: Teachers College Press.

Rees, G., Fevre, R., Furlong, J., & Gorard, S. (1997). History, place and the learning society: Towards a sociology of lifetime learning. *Journal of Education Policy, 12*(6), 485-497.

Varenne, H., & McDermott, R. (1999). *Successful failure: The school America builds*. Boulder, CO: Westview Press.

Vygotsky, L. S. (1978). Mind in society: The development of higher psychological processes. Cambridge, MA: Harvard University Press.

Weil, S. W. (1986). Non-traditional learners within traditional higher education institutions: Discovery and disappointment. *Studies in Higher Education, 12*(3), 219-235.

Wertsch, J. V. (1991). *Voices of the mind: A sociocultural approach to mediated action.* Hemel Hempstead, UK: Harvester Wheatsheaf.

Chapter 5

Looking For Learning In Practice: How Can This Inform Teaching

Peter Winbourne
London South Bank University

Abstract: In this chapter I explore the implications for teaching of applying theories of situated cognition to the teaching and learning of mathematics in school. I use accounts of some learners' experiences of mathematics in schools to suggest that teaching, as planning for learning, might more usefully be conceptualised in terms of planning for the development of powerful, identity-changing practices than in terms of the achievement of a range of pre-specified mathematical objectives.

Key words: alignment, becoming, identity, community of practice, predisposition

1. INTRODUCTION

This chapter represents a continuation of my struggle to make sense of theories of situated cognition and communities of practice (Lave and Wenger, 1991; Lave, 1996; Wenger, 1998; Winbourne and Watson, 1998) as applied to schooling in general and, in particular, to the teaching and learning of mathematics. The attraction of this perspective for me has been its promise of liberation from a central assumption that runs through our discussion of teaching and learning: namely, that what is learned in formal educational settings is necessarily connected to what a teacher has planned to teach. In this chapter I want to explore further the consequences of applying to schooling a perspective from which all learning is accounted for in terms of the learner's participation in and developing sense of identity within some community of practice.

The main thrust of the chapter is that consistent application of this perspective is illuminating and of practical importance. I will suggest that through this 'lens' (Lerman, 1998a) practices do show up in the school context which suggest ways to re-conceptualise the function of the teacher and, indeed, of the school. I will use it both to describe what happens in schools and classrooms (in terms of teaching and learning) and to suggest how teachers and schools might best plan for their students' learning.

In this chapter I shall stay with the understanding of community of practice about which Anne Watson and I have written in some detail. We characterised a community of practice as follows:

1. participants, through their participation in the practice, create and find their identity within that practice (and so continue the process of creating and finding their more public identity);
2. there has to be some social structure that allows participants to be positioned on an apprentice/master scale;
3. the community has a purpose;
4. there are shared ways of behaving, language, habits, values, and tool-use;
5. the practice is constituted by the participants;
6. all participants see themselves as essentially engaged in the same activity.

(Winbourne and Watson, 1998, p. 94)

It should be noted that, in themselves, communities of practice are neither good nor bad; they are descriptive theoretical tools. I will not expand on the idea of community of practice here. However, I want to stress that it follows from this understanding of community of practice that there are many situations where people interact which neither constitute nor lead to the development of a community of practice (Wenger, 1998)

This understanding is consistent with Wenger's exposition of community of practice (1998). The idea of identity used here is essentially that of Lave and Wenger (Lave and Wenger, 1991; Lave, 1993, 1996) and it is consistent with recent interpretations by other members of the mathematics education community (Boaler, 1997). Holland et al. point out that Lave and Wenger's notion of identities as important outcomes of participation in community of practice is analogous with their notion that identities are formed in the process of participating in activities organised by figured worlds (Holland, Skinner, Lachicotte, and Cain, 2001).

From this perspective all learning is situated in practice and represents a progression from legitimate peripheral to more central, expert participation in that practice; learning can be seen as a form of apprenticeship (Lave and Wenger, 1991). So, where we see signs that learning has taken place, it makes sense to look for evidence of the practices in which it is situated. It would be naïve to assume that schools in themselves are communities of practice. It is not clear whether schools fit in at all with an apprenticeship model of learning within communities of practice, and if they do it is not clear how (Lerman 1998b). Of course, it is possible that some few schools might in themselves be communities of practice and I am sure that there are some communities of practice within some schools (including communities of mathematical practice of a kind to be found elsewhere in this book.) There is no single community of practice of school, or a single community of practice of mathematics and here my interpretation does seem to differ from that of others (Boaler, Wiliam and Zevenbergen, 2000.) It is, however, helpful to think of schools and classrooms in terms of multiple intersections of practices and trajectories where these practices extend in space and time well beyond the boundaries of the institution.

According to Boaler (1997) 'back to basics' and 'exposition and practice' approaches are fairly typical in mathematics lessons. I suggest that we can recognise such practices by the uncritical implementation of a mathematics curriculum requiring the achievement of a set of pre-specified objectives. Using 'back to basics' as a shorthand, I want further to suggest that within a 'back to basics' mathematics classroom, to be successful, a student needs to be, or to become the kind of person who does well in mathematics examinations. Successful participation in this kind of practice requires little reference to mathematical activity in itself, rather it requires reference to measures of success in mathematics. I conjecture that 'back to basics', 'exposition and practice' approaches depend very much for any apparent success on the predisposition of students to engage in such practices.

Indeed, Lave has said that

> school teaching [as instruction] has as a condition of possibility other aspects of learners' learning projects. (1996, p. 157)

From this perspective we can choose to differentiate between learners in terms of their identities. That is, we may 'see' learners whose sense of identity includes that of legitimate expectation of success within a practice of school mathematics. Similarly, we may see others whose identities in no way depend upon any school-based community of mathematical practice (except, maybe, negatively.) In this way we can reasonably talk about a 'measure of alignment'[4] relating to learners' learning projects. Using this

---

[4] This idea of alignment resembles that to which the Cognition and Technology Group at Vanderbilt University (CGTV) have referred (1996). They suggest (ibid.) that in order for children's competencies to reveal themselves a number of elements have to be properly aligned. For CGTV the computer can be seen as an element of a physical and social context which affords or enables 'early competencies' in young children's number. This

measure the former 'well-aligned' and the latter 'non-aligned' learners would be measured as some way apart. So, students bring multiple identities with them into the mathematics classroom, and we may judge the extent to which these identities are aligned with the practices that are valued within that classroom. The teacher in whose class sit lots of the 'well-aligned' learners has a different task, I suggest, from her colleague in a class of [5]non-aligned learners. Teaching in this latter case may seem necessarily to be a very intrusive business. Currently in the UK teachers feel that they must meet prescribed targets and children must meet the objectives set for them without much reference to their own interests. In this context teachers appear to have to change their students' very sense of 'who they are' and this raises very difficult questions about the nature of teaching. However, I have argued elsewhere that 'identity' should be seen as the aggregation of the smaller 'becomings' (or identities) identified with a learner's participation in a multiplicity of communities of practice, local and not so local, some of which are locatable within school classrooms and most not (Winbourne and Watson, 1998). With this view of identity it may be enough for teachers to aim to ensure that local communities of practice (ibid.) are established in their classrooms; within these learners can develop identities part of which involves seeing participation in those practices as legitimate. What more, after all, could they do?

Lave points to another consequence of applying theories of situated cognition to school teaching:

> ..if teachers teach in order to affect learning, the only way to discover whether they are having effects and if so what those are, is to explore whether, and if so how, there are changes in the participation of learners learning in their various communities of practice. (1996, p. 158)
>
> and
>
> teachers need to know about the powerful identity-changing communities of practice of their students, which define the conditions of their work. It is a puzzle, however, as to where to find them, and how to recognize them. (1996, p. 159)

---

provides a link between Watson and Winbourne's notion of LCP (1998) and the situated abstraction of Noss and Hoyles (1996). Just as they claim the computer provides domains which support students' abstraction, so Watson and Winbourne claimed LCP's support students' growing image of themselves as someone who is legitimately engaged in mathematical practice, as someone, in other words, who is becoming a mathematician.

[5] I prefer to think of non-aligned learners in classrooms as people who are actually learning something different from what the teacher intends.

It seems doubtful, to say the least, that teachers working with largely non-aligned students within school mathematics practices of the 'back to basics' type could expect to see much evidence of their effect on learning in terms of powerful identity-changing communities of practice. However, it is clear that there are many students whose experience of school mathematics practices is similarly limited to the 'back to basics' type and who are, nevertheless, successful[6]; learning does seem to take place in classrooms that are a long way from being communities of practice. The 'success' of these students and the possibility of learning in such classrooms needs to be explained, too. I want to suggest that this can be explained in terms of identity, alignment and communities of practice. I have set myself the task of providing such an explanation in this chapter, but, put simply, my argument goes something like this: if a child appears to be learning successfully in a school mathematics classroom, the classroom may, indeed, be the site of a mathematical community of practice; sadly it is more likely that it will be the site of a community of practice where the constitutive activity is that of learning to do well in examinations. Of course, some children do appear to learn mathematics in the latter kind of classroom; some children appear to learn in classrooms where there may be no community of practice centring on any kind of school learning. From my perspective evidence of such learning suggests participation in a community of practice that, whilst it may include the physical space of the classroom (simply through the child's presence), actually extends well beyond its spatial and temporal boundaries.

So, if teachers really want to promote learning – to teach – then an understanding of powerful identity-changing communities of practice should be helpful. It is to Lave's implied challenge of identifying powerful identity-changing communities of practice that I turn now.

2. IDENTITY-CHANGING COMMUNITIES OF PRACTICE

In this chapter I will take up Lave's challenge in two ways. Firstly, I will suggest that when you look closely from this perspective at the learning teachers plan to happen inside what I have called a 'back to basics' classroom, you tend not to see much evidence of powerful identity-changing communities of mathematical practice. Moreover, if you persist in probing for signs of powerful identity-changing communities of practice, the teacher's mathematical objectives recede into the background and effectively

[6] Success here is measured in terms of the A-C economy (Gillborn and Youdell, 2000).

disappear. As you work with learners to explain their experiences and look for evidence of learning – and in this chapter that means participation in communities of practice – what emerges are quite different stories about learners 'identities in practice'. These may explain learning, but the process does not seem to require much reference to teaching.

Secondly, I will suggest that you *can* find evidence of powerful identity-changing communities of practice in schools within which learners develop a sense of themselves as legitimate practitioners (in something like mathematical practice), but it helps if you are ready to look beyond classrooms. This is not to deny the possibility of achieving something like this within the classroom. Indeed, Anne Watson and I have argued strongly (Winbourne and Watson, 1998) that teachers can and should plan on encouraging what we called local communities of mathematical practice in their classrooms. In the same paper we also suggested that, without the presence of such local communities of practice, you needed to search beyond the classroom for evidence of powerful identity-changing communities of practice if you were to account for the perception that a student's learning has actually followed teaching.

I will illustrate what I mean by reference to two pieces of research. I will outline these briefly here and discuss each in turn in more detail thereafter. In the first of these pieces of research I look closely at a moment in a lesson that is typical of many teachers might recognise as common in a 'back to basics' or 'objectives-led' mathematics classroom. Here the starting point of my account is the teaching objective. I have called the moment in the lesson where the teacher draws the students' attention to this objective a *teaching moment*. I use this moment as the pivot for a series of accounts that seek to probe for some of the practices intersecting in the classroom at that time and within these some sense of the developing identities of students.

In the second piece of research, my starting point is the identification of a particularly powerful community of practice located within a school setting but outside of any identifiable *teaching moment*. This community of practice is mathematical in nature and happens to be in some sense supported by the learners' teachers, but it is located in a context that extends considerably beyond the mathematics classroom. The mathematics teachers have provided a technologically rich context that turns out to have been supportive of learning within a powerful community of practice, but that support is neither articulated nor planned. This practice appears to be one that contributes further to the alignment of students to the kinds of mathematical practices that teachers do seek to encourage. Here the starting point is some mathematics that has been learned. I have called this a *learning experience*.

In both pieces of research I have chosen to produce and read accounts of teaching and learning in terms of identity and community of practice. These studies were designed to provide maximum opportunity for probing into learners' identities in practice and for evidence of participation in practices whose intersection with school and classroom might be seen to be constitutive of the alignment to which I have ascribed such importance. My aim has been to produce rich descriptions of the learners' experiences. Wherever possible, these are the products of a hermeneutic process of interpretation and re-interpretation of data carried out together with the learners themselves (Van Manen, 1990).

3. A TEACHING MOMENT

I have chosen to probe deeply into this *teaching moment* precisely because I see it as an example of much that goes on in a 'back to basics' mathematics classroom. It is the very 'ordinariness' of the teaching moment that I take to be important. It is clear that teachers can and do organise their classrooms in ways that enable distinctive mathematical practices to flourish (Boaler, 1997; Winbourne and Watson, 1998). This appears to explain some of the success of a minority of students who get to participate in such practices. What interests me here is how students whose experience of mathematics is typified by this rather bleak mathematics *teaching moment* nevertheless manage to be successful in their school mathematics studies. Within what practices is their learning situated? How is their success to be explained?

There is a range of stories that can be told that centre upon any one moment. In the case of the *teaching moment* each of the students had many different stories they could tell that might include it. I invited them to explore their experiences of the *teaching moment* with me and together we probed the 'reality' of that moment for students (I shall discuss the methodology below). Using different aspects of these discussions I have assembled three different stories, each with a different focus. The first story is focussed on the *teaching moment* and presented in terms the mathematics content of the lesson. Here students talk about their experiences of the mathematics and there is little attempt to probe more deeply into other aspects of their experience of the moment. The second and third stories focus on one girl's, Kamalah's, experience of that moment. The second account is included to give a sense of the way in which I invited the students to talk about their experiences. In this case, I took one small 'event' captured on the tape of the lesson and used this to give the students a 'way in' to talking about their developing sense of self. The third account is broader; it situates the 'moment' in wider practices representative of those held to be formative of the learner's identity. This third account makes use of observations of the learner over a number of months and in a number of contexts within the school. These are linked to discussions with the learner over the same period. In some of these discussions she comments, with some of her friends, on the events containing the 'moment'. The third account is thus the result of this hermeneutic process of joint examination, interpretation and re-interpretation of that moment as recorded within notes and video.

3.1 Methodology: the teaching moment

The *teaching moment* is taken from an ordinary mathematics lesson in what is recognised by inspectors, parents, students and community as a good multi-racial school[7] for girls in south London. I had spent a considerable amount of time tracking six Year 8 girls as they worked together (in various combinations) and went about their daily lives in the school. I had been with them to all of their lessons. I would meet them in registration periods, sometimes in assemblies or the playground. I had made notes about what I had seen, looking for any material that might later speak to me, or the girls themselves, or to others, about their identities. This meant, of course, that there needed to be opportunities for people to show signs of who they were, of their developing identities as learners, of who they were becoming. This material was included with other notes, transcripts of video and audio tape that together became the text for interpretation. The students and I interpreted the text together. The aim of this interpretative activity was to describe these children's experience of what I have called the *teaching moment.*

The tracking began in March of 1999 and finished in July of the same year. During this period I spent nine complete days in the school. On the first seven of these (19, 23 March; 26, 27 April; 4, 10, and 14 May) I spent all of the day tracking the students. They were my guides and I went with them to all of their lessons, making notes of my observations.

On May 14 I tracked the girls for most of their lessons as usual, but I also used a video recorder to 'capture' what was thus a randomly chosen mathematics lesson. The camera was focussed on Afya, Dhanya, Kamalah and Priya; Esther and Josie were absent from school on that day. The *teaching moment* comes from this lesson. It was not selected for being

[7] In fact, the school is a Beacon School. Beacon status is awarded to schools in England by what was then the Department for Education and Skills. Most recently such awards are on the basis of of the school's identification by Her Majesty's Chief Inspector for Schools as an outstanding and consistently high performing school over a period of 3 to 4 years in relation to its circumstances.

particularly interesting, but simply as a moment where I thought the teacher's objectives were particularly visible.

I returned to the school on 21 May when Dhanya, Kamalah and Priya watched the video of the previous week's mathematics lesson with me (It was lunch time and Afya was busy elsewhere in the school.) As they watched, the three girls commented on what they saw. I had prepared some prompts for our conversation based on my interpretation of the video of 14 May, the most prominent of which was the reminder to myself that I must allow the girls freedom to say what they saw and encourage them to do so. I made a further video recording of our conversation as the four of us watched the tape.

I watched and analysed both videos by myself and then again with a colleague on 29 June. As we talked about the videos and what we saw, it became clear that each of us saw different things. It would be surprising had we not, but my colleague's response to what she saw, and the differences in the level of detail with which we described what we saw, suggested that I should ask the girls to watch the first video once again. So, on 9 July, Dhanya, Kamalah and Priya were good enough to watch it once more with me. Again they talked about what they saw as they watched, although on this occasion, at their request, I made not a video, but an audio recording of the conversation.

I should add that I had an additional, unplanned conversation with Kamalah when I met her in the corridor whilst visiting the school for a different purpose on 14 June.

3.2 Three stories including the teaching moment

The *teaching moment* is that part of the lesson where the teacher invited discussion of the following question:

Solve these inequations. Show all your working. You don't need to draw the balances

$2w+4<w+6$

Question B1, part b (SMP 1992)

3.2.1 Story number one

The first story focuses on mathematical 'events' within the lesson. It represents an attempt to evaluate the students' learning in terms that are fairly 'conventional' in the sense that they refer to an explicit objective of the lesson (understanding inequalities.)

As my colleague and I watched the tape of the lesson (of 14 May) together we noted how the teacher, Ms W, had likened the rule for equations to the rule for equivalence of fractions.

Ms W: Right, we're going to do these together. You've had trouble with equations before, so we're going to go through it very carefully. What has been our rule about equations, our golden rule? Yes, Priya?

Priya: Whatever you do to one side, you do to the other.

(Ms W goes through an example, solving the equation x+3=7)

Ms W: Whatever you do, you must do the same to the other side. When else do we do that? Remember fractions? How can I simplify fractions? I don't want to change the value of it. I want to just simplify it (writing 6/8 on the board)

Student: Divide by 2

Ms W: Right. Only the top?

Student: And the bottom

Ms W: Same rule applies here. If I want to keep the value of my fraction the same, whatever I do to the top, numerator, I must also do the same to the denominator··. ...So that rule always applies. Both sides of the equation are treated the same way. So, in inequations, the same thing applies. They draw you a nice little picture.

We paused the tape. My colleague and I wondered what the students had made of this. Had they been looking for meaning? We could see that the students were writing in their books as the teacher spoke. Were they, we wondered, simply marking?

I invited the students to talk about this when they watched the video again on 9 July. The following is a transcript of some of their responses to my questions.

Peter: Let me pause that just for a minute. So, I have paused the video at 12.31. Do you think there is sense then in which solving inequations is like working with fractions? Cos of the example that was given..

Kamalah: I hadn't thought so..

Dhanya: In a way yeah

Peter: Did it help you to understand it?

Kamalah: I think it kind of confused me a bit more..

Priya: Yeah, same here.

Peter: Did it?
Kamalah: You had to think of two things instead of just
Priya: the one
Kamalah: the single, one thing that you were doing.
Peter: Right. When you say you were thinking of two things, you didn't think they were connected?
Kamalah: In some of them I thought they were, but in the longer ones I thought it was bit more complicated.
Peter: What do you mean the longer..? which ones?
Kamalah: Like questions like B1b, it was much more longer than B1a.
Peter: Right. Yes. What's the difference between B1a and B1b?
Kamalah: Cos in B1b there's more working out, because on both sides there was a *w*

I will end the first story of the *teaching moment* here. I suggest that the scope of this story is like many we, as researchers, tell about classrooms. In its telling, I have made no attempt to probe the students' experiences more deeply or to look beyond the classroom to explain their responses. Note, how in Kamalah's last two utterances the mathematics appears as a distant, unfocused activity, as activity identified with the coding of its position in the book, as an activity unlikely to contribute significantly to the shaping of the learners' identities.

3.2.2 Story number two

This story brings together the results of some deeper probing into the students' experiences of the *teaching moment*. The data come from the videotaped discussion between the students and me as we watched the video of the lesson. We were sitting in a small room used both as a sixth-form classroom, as an occasional storeroom, and as a meeting room. As Kamalah, Dhanya and Priya sat around the monitor they commented freely on what they saw and ate their lunch at the same time. Having started the camera recording, I joined them. I wanted the students to feel able to comment on their experience in ways that might provide some insight into their sense of identity, their sense of themselves as they sat in the classroom and, if possible, some idea of the practices within which they had been participating. I had decided that I would start by inviting them to talk about the experiences of a fellow student in the hope that discussion might progress, via analogy, to their own. I had noticed how one of the girls in the class had kept her hand up for a long time, so started by drawing attention to her.

Time (in lesson)	Speaker	Words spoken	Event on tape at that time
12.26	Peter	Who's that with her hand up there?	Towards the back of the class, Katie can be seen to have her hand up.
	Kamalah	Katie	
	Peter	You watch her hand staying up..	
	Dhanya	It's going on..	
	Kamalah	That's me, you see.	
	Peter	What?	
	Kamalah	That's me as in, like, that would be me keeping my hand up until Miss would say something	
	'..	....	...
12.30	Kamalah	(Talking about Ms W) She'll say, 'What's 4 x 8?' She'll say, she'll say 'What's 4 x 8?' then she'll answer it herself.	Ms W: What has been our rule about equations, our golden rule?
	Dhanya	I remember that·	Priya gives the answer·
	Kamalah	(*sort of chanting*) What you do to one side you do to the other·	

In the first story, the students refer to themselves, but there is no real reference to any sense of their identity; no real suggestion that discussion of who they are is relevant. In this case, Kamalah literally identifies herself through her empathetic response to Katie's situation. I think that this conversation can also be taken to mark Kamalah's growing understanding of the hermeneutic process in which I was asking the students to engage. Kamalah showed further evidence of this in later discussion (see for example, Friday 14 May – Registration, below.)

3.2.3 Story number three

In the telling of the third story I have allowed the narrative scope to extend still further beyond the *teaching moment*; indeed, the story 'starts' some months before. My aim has been to situate this moment within Kamalah's wider experience of her schooling and to present data that give some idea of the wider practices in which Kamalah participated, both inside and outside of school. The story also reflects the continuing growth in Kamalah's awareness of the way I was asking her to look at her experiences. Here she shows her willingness to explore with me the notion that her learning might usefully be seen in terms of her sense of who she was and who she was becoming.

Friday 19 March

Lesson 3 English – Teacher: Ms D

Ms D leads a discussion of children's and parents' rights.

Kamalah, who, for this discussion, is in role as child, says that parents have a right to intervene and suggest what actions children should take in their lives.

Tuesday 23 March

Lesson 2 Religious Studies – Teacher: Ms B

Religion and Art.

Ms B tells the students that in this lesson they will be asking questions, 'talking without saying anything'.

Ms B talks about Dave, the deaf boyfriend of a friend. She describes him. He has a sign for himself. She makes the sign, placing the forefinger of her right hand to her brow, she sweeps it away, up and to the right in a salute that is not military.

It means, she says, 'I may be deaf but I am not stupid!'

She says that later the girls will devise signs for themselves.

Ms B asks the girls to make the sign for their own names.

Kamalah points to her forehead. She says, "This is me because I use my brain."

Tuesday 23 March

Lesson 4 Maths Teacher: Ms W

"Someone asked to repeat 'Brackets 1', so we're all going to do that one again".

Kamalah's hand goes up whilst the teacher talks.

The teacher says 'just do B3 for me, please'.

Monday 26 April

Lesson 1 Technology Teacher: Ms M

As the girls work on their project, they are able to chat informally. Kamalah tells me that tomorrow they have their Maths Challenge (set by Leeds University). She explains the marking system.

Kamalah is nervous. Ms W has made the challenge compulsory. Had she not Kamalah's mum would have wanted her to do it anyway.

Overhearing, Francesca asks, "What's this? I haven't heard of this?"

Kamalah says, "It's a maths challenge. Only for set 1."

Francesca says, “Oh, that explains it. I’m in set 3. I don’t like our teacher. He can’t teach anything”.

Tuesday 4 May

On the way to the IT lesson, Kamalah tells me how her Bank Holiday weekend has been spoilt. A man who lives on the streets had been causing a nuisance in her father’s shop. He has been swearing, feigning illness. This has been distressing for her and her little brother who is 4. Kamalah helps out in the shop.

Friday 14 May

Registration

I ask Kamalah, Dhanya, Priya, and Afya if they will watch the video to be made today during their lunchtime next Friday (21 May).

I explain that I will ask them simply to watch the video and tell me what they see.

Dhanya asks what this has to do with stories. Kamalah explains that I am trying to look at big pictures; to explain them in terms of the real stories underneath. They agree to watch the video.

Friday 21 May

The students watch the video for the first time. Some of their comments have been included in the first two stories; here are some that were not included there.

Kamalah: We marked [our homework] during the lesson but not everyone had finished it. It was only about three or four who had finished it.

Peter: So, when did you finish yours?

Kamalah: For homework. We had it on Thursday night, so I finished it on Thursday night.

Dhanya: Miss W usually says try and do a bit more, so she [Kamalah] done a bit more from the lesson before, cos she thought she might not get it finished in the lesson

Kamalah: No. That was in scale drawings you remember?

Dhanya: Oh, yeah, yeah

Kamalah: Cos, we ended up drawing, so I thought I wouldn’t be able to finish it, so I did a bit extra

Kamalah admitted to me later on that she had done the finishing-off and bit extra quite late at night after helping out in the shop.

Monday 14 June

I am visiting the school with a colleague in connection with some other research. I meet Kamalah in the corridor. She tells me that she got a bronze in her maths challenge. She was one of only two students to do this.

Friday 9 July

The girls and I watch the video for the second time

Peter: I've been just taking little tiny moments in lessons and asking you to talk about them? Any sort of comments or details you want to add?

Kamalah: ..you could have, like, taken us in other lessons as well as maths because I think that we're different .. in different lessons and you would have got a more wider view of what we're really like rather than just in a maths lesson.

Peter: Right…Right…Um.. Go on Kamalah.

Kamalah: But also, it depends on the classroom that you're in; like who you're with. I suppose the people in the lessons, most, cos in maths there's a few people that I don't get on with

Peter: Really?

Kamalah: and I don't feel comfortable around them, and they're in my tutor group as well..so it doesn't help..[..]

Peter: And that makes a difference when you're in maths?

Kamalah: Yeah [..]

Peter: You've said this just now, and we're coming back and back and back; we keep looking at these same things on the tape and you didn't say it first time around. Is there a reason for that?

Kamalah: I just didn't think that it had anything to do with it, but now I think that it does.

Peter: You do?

Kamalah: Yeah. Because obviously it had an impact on my behaviour in that lesson on that day, .. because I wasn't participating.

It is clear that Kamalah had become well aware of the nature of my research; after all, an important part of my methodology was that the students should join me in interpreting classroom events in a search for evidence of identity. For this reason, there may be questions about the validity of Kamalah's interpretation of what she saw: was she really having these feelings at this time? Perhaps her reading was influenced by subsequent events? I can only say that I take Kamalah's contribution to be an essential part of this story. From my perspective, what emerges strongly from Kamalah's interpretation and re-interpretation of this moment in this story is her awareness of herself as a participant in particular social practices, mostly originating outside and extending way beyond the classroom, that happen to intersect with this 'moment'. We see signs of practices outside school within which people are expected to work long and hard in family businesses and still succeed academically. We see signs of other practices within school, for example in some English and RE lessons, where a student's sense of 'who she is' is essentially and explicitly part of the activity. We glimpse signs, too, of practices within which Kamalah is positioned as a high achiever and others are not, for example the conversation with Francesca in the technology lesson. These practices are not in themselves mathematical. Indeed, with each further cycle of interpretation, the location of the mathematics learning appears to shift, seeming to recede from the 'moment' and re-emerge, diffuse and spread over time and place, in the varied contexts of the wider practices extending beyond the classroom in which Kamalah participates. However, Kamalah's successful participation in these practices, and her friends' doubtless similarly successful participation in others, might go some way towards explaining how the group of children experiencing the teaching moment came to be so apparently well-aligned. The relationship between the girls' participation in these wider practices and their classroom performance at times like the *teaching moment* is far from simple. I cannot and I do not mean to say that this is a causal or even a direct relationship. The stories of wider participation emerged from a research process to which Kamalah had become attuned and, in the research context, they were stories to which she attached importance.

This concludes the stories of the teaching moment. I shall return to discussion of these in the final part of this chapter.

4. A LEARNING EXPERIENCE

The students whose experiences are reported here studied in a prestigious, private school for girls in Auckland, New Zealand. They were all senior (16/17 years old) students taking mathematics as one of their higher level courses. They attended mathematics classes in which the teacher, Mrs. G supported and encouraged the use of graphics calculators making use of the technology in her own teaching as well as encouraging her students to become independent users. I do not mean to suggest that this school is directly comparable to the London school of the *teaching moment*. I did, however, find evidence here of strongly mathematical practices, originating outside and extending beyond the mathematics classroom. Amongst these mathematical practices there was, I think, an example of a powerful identity-changing community of practice; as Lave has pointed out (1996), such powerful identity-changing communities of practice are hard to find and to recognise.

This work was part of a wider project on which I had been working with other colleagues (Winbourne, Barton, Clark, and Shorter, 2001). As part of this, a colleague and I had discussed with Mrs. G our aim of probing her students' experiences for signs of the mathematical or calculator practices in which they participated. We had made clear our interest in evidence of mathematical practices which extended beyond the classroom. We had also made clear our hope that we would be able to interpret the students' responses to our questions in terms of their developing identities within those practices. Mrs. G had responded promptly by telling us what I shall call the 'biology story'.

4.1 Methodology – the biology story

A little while ago, a colleague of Mrs. G's in the biology department had asked his students to do a t-test on some data they had gathered. He had prepared an Excel template for this and showed the students how to do it. He had been surprised when one of the students told him that they could use their graphing calculators for this. He had told them that they were mistaken. Their calculators might be able to draw graphs but could not do the kind of statistics they needed.

The student had persisted. It certainly could do what they needed, she said. She showed the biology teacher and he had been convinced. Mrs. G added that these students were studying statistics, but they had not yet been taught to do t-tests in their statistics lessons.

The 'biology story' appealed to us because it seemed to us to suggest that there were, indeed, mathematical practices extending beyond the classroom. We felt that we could use the story within our interviews with students to orientate them towards aspects of their learning in which we were interested. The 'biology story' became our starting point for our planning of our interviews with Mrs. G's students. In the course of these interviews we planned to elicit from the students their interpretation of the story.

Melissa

Melissa was the student Mrs. G had been thinking of when she told us the biology story. Melissa had bought the graphics calculator because it was helpful and everyone else had one, so she got one.

Melissa used the machine all the time. She gave physics as an example, although here it was mainly just to add things. She used it in statistics and had had to use it in biology to do Student t-tests.

Before last year, Melissa had made some use of the school graphics calculators that the teacher had brought in. She had been confident from the outset that she would be able to use the graphics calculator. She was good at using it now.

Asked about use outside lessons, Melissa said that, at home her sister had a graphics calculator, but no one else was interested. Her mother was good at using computers and some of her mum's friends were also good. Melissa had never met anyone outside of her school who had or used a graphics calculator.

Melissa said that at lunchtime she might well take out the graphics calculator when helping someone else who might already have one. Recently, for example, she had shown someone how to use it to do the student t-tests they needed for biology.

I invited Melissa to tell me more about her use of the calculator in biology.

Melissa: I don't have Excel at home on my computer and I don't know how to work Excel that well: it's quite difficult and so and...uhm...[then] I heard from somewhere else that you can do it [on a graphics calculator]. And I was, like, cos it's got lots of functions and things on there and I was, like, looking through it and then I saw the student t-tests.

Peter: But you heard from somebody else?

Melissa: Yeah.

Peter: Where did you...? I am probing here because this is exactly the sort of thing I am interested in. So, you say 'I heard it from someone else', I mean, where were you? Can you remember where you were?·

Melissa: We're just in Stats: the girl that sits behind me. Like, the teacher hasn't come yet and so we're just talking. Cos, like, we were stressed out about our biology and cos everyone has to do t-tests. Cos we have to... (I don't know what it does...)

(Melissa and Peter laugh)

Melissa: and, um, yeah, I said 'I don't know how to use student t-tests on Excel' and she said, 'Oh, you can do it in your calculator' and stuff. And I said, 'Oh, really?' and then I went to calculus and I asked Mrs. G how to use it.

Melissa had gone on to convince her biology teacher that this was a legitimate use of the calculator.

We interviewed four other students in the same mathematics class, asking each if they thought that the biology story seemed likely in their experience of their school; was it to be expected? None of the four took biology. Annette and Zara agreed the story was, indeed, to be expected. Tara was doubtful and Laura non-committal.

Annette: had taught her mum about the graphics calculator. She found mathematics hard, especially as she was used to being at the top of the class. She was excited by the power of the graphics calculator, which has helped her a lot.

Zara: thought the plausibility of the story would depend very much on who the biology teacher was. She used her graphics calculator all the time in Physics. She didn't talk with others any more about the graphics calculator, though she had a lot at first.

Tara: said she was not good with graphics calculators, though she knew more about them than her curious father. If she didn't have a calculator with her she got tense when she had to do calculations.

The biology story suggests, I think, that there have been some powerful *learning experiences* here. It also hints at a community of practice participation in which has brought with it the kinds of identity transforming experiences and opportunities to which Lave (1996) refers. This practice is mediated and made visible by powerful personal technology, though the technology itself is not essential to this discussion. The practice appears to originate in the classroom/school, but it is seen to extend beyond this location. It is identified as a practice using criteria derived from reading of the work of Lave and Wenger and about which I have written elsewhere (Winbourne and Watson, 1998).

5. DISCUSSION: WHAT DOES THIS HAVE TO SAY ABOUT TEACHING AND LEARNING?

I want to discuss the implications of the *teaching moment* and the *learning experience* at two levels:

- The learning of school students
- The conceptualisation of teaching

5.1 The learning of school students

As educators and researchers plan for and research into learning they make decisions about *where* to look. Whilst there is growing acknowledgement of the importance of the 'social' in accounting for learning (Cobb, 1998) much research into the learning of school students confines its gaze uncritically within the walls of the classroom. From the perspective of situated cognition, this kind of planning and research fails to see that, for many of its subjects, the fact that they have already stepped as learners into the school arena is a major factor in their learning. Kamalah, for example, brings with her to the mathematics classroom a strong sense of herself as a disciplined and effective learner; this aspect of her identity is seen to be situated in practices originating outside and extending way beyond the classroom. These practices do, in some sense include the classroom – you don't stop being the person who, for example, stays up late to do homework just because you happen to be in the classroom. In this way, many of those learners who will be successful in mathematics bring with them, like Kamalah, identities that are well-aligned with the practices valued in the mathematics classroom. For many other learners, of course, the many overlapping practices in which they are legitimate participants bear little or no relation to those valued in the mathematics classroom and their identities are less well-aligned to the practice of succeeding in school mathematics.

I have provided descriptions here of a number of practices to be found in and around mathematics classrooms. In some of those associated with the teaching moment, there has been a lot of learning going on, but, I suggest, it has mainly been concerned with learning how to become a good or better student of mathematics or coping with everyday social pressures. Paradoxically, perhaps, on closer examination of these practices, the actual mathematics being taught has effectively disappeared, the content remaining only as some coded symbol of the life-forming significance of being a good, well-aligned student.

In my discussion of the *learning experience*, I have drawn particular attention to what might be called a graphics calculator practice; this is, of course, one of a multiplicity of overlapping practices participation in which forms the identity of a person as learner. Participation in the graphics calculator practice not only provided the students with experience that was mathematical in nature; it also gave the students opportunities to experience what it feels like to be at the very centre of a practice – to be and to be seen to be expert within that mathematical practice. I want to suggest that, given the opportunity, many (most?) students would be able to respond positively to situations which, like the one in the biology lesson, allow them to function as 'experts' in some practice in which they participate. In this case that turns out to be a graphics calculator practice. Melissa participated in this practice alongside her calculus teacher and some of her fellow students, some more and some less expert than she. Her identity encompassed the confident use of the graphing calculator; she had become someone disposed both to seeing the usefulness of the graphing calculator in a range of contexts and to seeing value in becoming more expert in its use through working with other participants in that practice. So, when her peers flagged up the potential of the graphing calculator for doing t-tests in her biology lesson, she was ready to grasp the opportunity this represented and keen to learn from her calculus teacher. In this story of the learning experience, the fact that Melissa's impulse to make use of the graphics calculator was prompted by a desire to avoid using Excel is evidence, I suggest, of the complex, changing and multilayered relationships between the practices in which her activity was situated and the identities formed within them.

I have every reason to think that the participants in this practice were, in any case, very well-aligned learners, but I want to go further and suggest that the calculator practice was, indeed, a powerful, identity-changing community of practice. I believe that participation in this practice also encouraged the development of predispositions in its participants to see themselves as similarly capable of becoming legitimate and successful participants in whatever mathematical practices might be on offer to them. The development of such predispositions is, I think, the best that can be hoped for when we talk of transfer. It is not so much that learners take decontextualised knowledge with them from one context to another; rather, as participants in mathematical practices, they carry with them identities that predispose them towards looking for and making use of mathematical knowledge in a range of contexts.

5.2 The conceptualisation of teaching

I cannot deny that there may be an element of caricature in my representation of the school and classroom practices in the stories I have told about the *teaching moment*. Within these stories, the picture of the classroom that emerges is one in which many communities of practice doubtless intersect. The communities of practice of most interest to me as a mathematics educator and researcher are those, participation in which might lead to success in mathematics. When we probe deeply into the students' experience of the *teaching moment*, the identities that the students seem to claim for themselves and which seem to deliver this success in mathematics, require little direct reference to mathematics or the teacher's teaching. The *teaching moment* stories suggest, I think, that, though plenty of communities of practice were present in the classroom, in the more mathematical of these the mathematics is mostly visible only indirectly and then in coded form ('Please do example B1b'). Using my criteria for community of practice, whatever these practices are, they appear to rely heavily on the identities that learners bring with them. Classroom practices like those containing the *teaching moment* are certainly strongly driven by the genuine desire to move learners towards the achievement of learning objectives; but the paradox is that the achievement of those objectives appears to be a product of participation in other practices. So, whilst teachers in such classrooms may wish to claim the credit for their students' success in mathematics, we cannot, in fact, associate that success with any planned action on the part of the school or the teacher.

Returning to the *learning experience*, we can see the powerful effect that the biology teacher, through acting in this context, was able to have on learners' participation in the graphics calculator practice and thence mathematical practice. This is strikingly different from what we might normally expect to see in terms of the effects of planned instruction: you don't plan this kind of thing and yet it seems to have been key in some powerful learning. That learning was a function of the existence of a community of practice that happened to be centred on the graphics calculator. It should be noted, too, that the calculator served to render this particular community of practice visible, acting, if you like, as 'window' (Noss and Hoyles, 1996) onto the learning situated within it. From the perspective of situated cognition, all learning reflects participation in practices; most of these are very probably less visible than the graphic calculator community of practice and fairly certainly uncharted.

So, what can you plan? From my perspective, teaching needs to concern itself with learners' identities in practice to be effective. I have drawn particular attention here to a small number of practices; these are, of course, only a small selection from a multiplicity of overlapping practices participation in which forms the identity of a person as learner. Moreover, none of the practices identified with powerful learning appear to have been the result of direct planning by teachers. I have to say that I am not entirely sure how you can plan for the development of powerful mathematical communities of practice in schools. Certainly, teachers would need to set up opportunities for practices to develop within which their students have a good chance of becoming legitimate participants with a very high chance of functioning as 'experts'. In the case of the students whose story I told in the *learning experience*, the graphics calculator practice provides the spur to what I have called alignment with practices valued in school; for other learners, the spur could well be sailing a boat or making a good cup of tea.

Perhaps the task of the teacher is best conceived as stroking into being 'the right kinds' of practice in their classrooms. At the planning stage, these practices may well have very little to do with the identification of clearly specified objectives, mathematical or otherwise. However, if they are lucky these practices may reach into and beyond the classroom and become the kind of powerful, identity-changing practices that Lave identifies.

ACKNOWLEDGEMENTS

This chapter is a revised version of an article that was published in *Ways of Knowing*, Vol. 2, No. 2 in 2002. I am very grateful to the editors of that journal and to the copyright holders for giving permission for this chapter to be published in this volume.

REFERENCES

Boaler, J. (1997). *Experiencing school mathematics: Teaching styles, sex and setting.* Buckingham: Open University Press.

Boaler, J., Wiliam, D., & Zevenbergen, R. (2000). The construction of identity in secondary mathematics education. In J. F. Matos & M. Santos (Eds.), *Proceedings of mathematics education and society conference* (pp. 192-202). Montechoro, Portugal: Centro de Investigação em Educação da Faculdade de Ciências Universidade de Lisboa.

Cobb, P. (1998). Analyzing the mathematical learning of the clasroom community: The case of statistical data analysis. In A. Olivier & K. Newstead (Eds.), *Proceedings of the 22nd International Group for the Psychology of Mathematics Education* (Vol. 1, pp. 33-48). Stellenbosch, South Africa: University of Stellenbosch.

Gillborn, D., & Youdell, D. (2000). *Rationing education: Policy, reform and equity.* Buckingham: Open University Press.

Holland, D., Skinner, D., Lachicotte Jr., W., & Cain, C. (2001). *Identity and agency in cultural worlds*. London: Harvard University Press.

Lave, J. (1988). *Cognition in practice*. New York: Cambridge University Press.

Lave, J. (1993). The practice of learning. In S. Chaiklin & J. Lave (Eds.), *Understanding practice: Perspectives on activity and context* (pp. 3-32). New York: Cambridge University Press.

Lave, J. (1996). Teaching, as learning, in practice. *Mind, Culture and Activity, 3*(3), 149-164.

Lave, J., & Wenger, E. (1991). *Situated learning: Legitimate peripheral participation.* New York: Cambridge University Press.

Lerman, S. (1998a). A moment in the zoom of a lens: Towards a discursive psychology of mathematics teaching and learning. In A. Olivier & K. Newstead (Eds.), *Proceedings of the 22nd International Group for the Psychology of Mathematics Education* (Vol. 1, pp. 66-81). Stellenbosch, South Africa: University of Stellenbosch.

Lerman, S. (1998b). Learning as social practice: An appreciative critique. In A. Watson (Ed.), *Situated cognition in the learning of mathematics* (pp. 33-42). Oxford, UK: University of Oxford, Centre for Mathematics Education Research.

Noss, R., & Hoyles, C. (1996). *Windows on mathematical meanings: Learning cultures and computers*. Dordrecht: Kluwer Academic Publishers.

SMP 11-16 (1992). *Inequations school mathematics project.* Cambridge: Cambridge University Press.

Van Manen, M. (1990). *Researching lived experience: Human science for an action sensitive pedagogy*. New York: State University of New York Press.

The Cognition and Technology Group at Vanderbilt (1996). Looking at technology in context: A framework for understanding technology and education research. In D. C. Berliner & R. C. Calfree (Eds.), *Handbook of educational psychology* (pp. 807-840). New York: Simon and Schuster.

Wenger, E. (1998). *Communities of practice: Learning, meaning and identity*. Cambridge: Cambridge University Press.

Winbourne, P., Barton, B., Clark, M., & Shorter, G. (2001). A community of practice associated with graphics calculators. In M. Van den Heuvel-Panhizen (Ed.), *Proceedings of the 25th International Group for the Psychology of Mathematics Education* (Vol. 1, p. 379). Utrecht, Netherlands: University of Utrecht.

Winbourne, P., & Watson, A. (1998). Learning mathematics in local communities of practice. In A. Watson (Ed.), *Situated cognition in the learning of mathematics*. Oxford: University of Oxford, Centre for Mathematics Education Research.

Chapter 6

Are Mathematical Abstractions Situated?

Mehmet Fatih Ozmantar and John Monaghan
University of Gaziantep, University of Leeds

Abstract: In this chapter we address the question: are mathematical abstractions situated? We first consider empiricist accounts of abstraction which see abstraction as a development process from the concrete to the abstract achieved through the recognition of commonalties isolated in a large number of instances. We discuss difficulties involved in empiricist accounts and propose an alternative approach which we call a dialectical account of abstraction. In this approach, an undeveloped initial idea develops through the use of mediational means and social interaction. This development is not from the concrete to the abstract but, rather, a dialectical *to and fro* between the concrete and the abstract. Unlike empiricist views, our approach regards context, in the formation of mathematical abstractions, as paramount. Although the construct 'context' is difficult to delineate precisely, we focus on the importance of students' personal mathematical histories, the tools and knowledge artefacts they work with, the people they work with and the tasks they work on. We exemplify the importance of these contextual factors through a study where two teenage girls worked collaboratively, with an interviewer assisting them, in completing tasks designed to generate abstractions in the field of graphs of linear absolute value functions.

Key words: abstraction, absolute value, context, dialectics, social interaction

1. INTRODUCTION

Here is an account of a learning activity:

> John is set a task sheet with dot patterns of triangular numbers. After counting the dots and seeing the geometrical patterns he abstracts a formula for the triangular numbers. Sometime later he applies this

abstraction when he realises that it provides a solution to the handshake problem[8].

The interesting thing about this account is that it is a fairytale; the abstraction appears 'by magic' from the pattern spotting and is 'transported' to solve a problem. This chapter attempts to tell a real story about abstraction, full of messy (but fun!) human twists and turns. We start the story by answering the title question. The answer is 'yes', mathematical abstractions are situated. Explaining this answer is the job of the remainder of this chapter.

We begin by giving an account of empiricist views of abstraction. We do not agree with these views and we say why we do not. We believe that abstractions arise in and are applied to contexts or situations but these terms are problematic and we devote a short section to considering problems with the terms 'context' and 'situation' before outlining what abstraction in context means. We outline several approaches to abstraction in context. There are differences in these approaches but there are considerable similarities. Of these approaches we outline one in further detail as this approach is, to us, 'operational' in the sense that we feel it allows us to be quite precise about developmental aspects of abstraction. The chapter then moves on to the central empirical study of two girls and an interviewer working on a task designed to lead to an abstraction to construct a method to obtain the graph of $f(|x|)$ from the graph of $f(x)$. The final substantial part of this chapter discusses aspects of abstraction in the light of the girls' work and with particular regard to mediation, people and tasks. These are very human themes: people do things together, with artefacts and for reasons. These themes are essentially interrelated, though we consider them separately. We conclude with a reconsideration of our 'yes' answer to our title question, are mathematical abstractions situated?

1.1 Empiricist views on abstraction

The term 'abstraction' is often linked with empiricist philosophy and stripping ideas from the material circumstance of their origin, their situation, in particular. This can be traced in writings from Aristotle to Locke to Russell. Locke, for example, wrote:

> … ideas become general by separating from them the circumstances of time and place, and any other idea that may determine them to this or that

[8] "There are n (say 10) people in a room. Everyone must shake hands with every other person in the room once. How many handshakes are there?"

> particular existence. By this way of abstraction they are made capable of representing more individuals than one. (1689/1964, p. 264).

We object to "this way of abstraction" and view the "circumstances of time and place" as integral to the process, "way of", and the product of abstraction. This 'way of abstraction' associates abstraction with generality achieved through the consideration of particulars and the formation of classifications. In much of the 20th century, particularly in the west, this view motivated (and provided a means for) investigations and discussions of the psychological process of abstraction until recent decades, when dialectical materialist and situated accounts of the process of abstraction were resurrected and/or constructed. The strong influence of empiricist views on such important figures as Dienes, Skemp and Piaget is apparent. Dienes (1963), for example, associates abstraction with the formation of a class by extracting "what is common to a number of different situations" (ibid., p. 57). In his view everyday objects are classified by visible appearance or known function but mathematical ideas are classified by forming an isomorphism ascertaining 'the same type of pattern' amongst different sets of materials. In a similar vein Skemp (1986, p. 21) viewed abstraction as a matter of classification:

> abstracting is an activity by which we become aware of similarities … among our experiences. Classifying means collecting together our experiences on the basis of these similarities. An abstraction is some kind of lasting change, the result of abstracting, which enables us to recognise new experiences as having the similarities of an already formed class.

The most influential work within this tradition comes from Piaget (2001) who views abstraction as an individual developmental process. This development involves three different forms of abstraction: empirical, pseudo-empirical and reflective abstraction. Empirical abstraction is concerned with information obtained directly from the properties of the world of physical objects such as the weight, colour and shape of a pebble. He argues that empirical abstraction is necessary for the categorisation of concrete objects. Pseudo-empirical abstractions are more advanced and concern actions on physical objects, e.g. counting the pebbles. The highest level of abstraction is reflective abstraction which concerns the extraction of a basic structure through a consideration of the interrelationships amongst actions, e.g. counting a set of pebbles and realising the unchanging number of the elements of the set irrespective of which pebble one starts counting and hence discovering commutativity. As Noss and Hoyles (1996) note, Piaget views the abstraction process as one in which learners become increasingly detached from the world of concrete objects and local contingencies and gradually ascend to the level of abstract thought, the 'ascent from the concrete to the abstract'.

These accounts of abstraction have two essential features, both of which are problematic. The first is that abstractions involve generalisations arising from the recognition of commonalties amongst a large number of particular instances. The problem with this view is the epistemological primacy of particular instances: that particular instances are epistemologically more basic than abstractions and that abstractions are produced from particular instances. The fundamental question here is: how can one recognise a particular as an instance of an abstraction without having at least a basic understanding of the abstractions itself? Ohlsson and Lehtinen (1997) argue that, "people experience particulars as similar precisely to the extent that, and because, those particulars are recognised as instances of the same abstraction" (p. 41). Thus abstractions, we hold, beget recognition of commonalties rather than vice versa.

The second feature of abstraction is that the product of abstraction is considered as a decontextualised, or 'pancontextualised', entity. An educational implication is that "knowledge acquired in 'context-free' circumstances is supposed to be available for general application in all contexts" (Lave, 1988, p. 9). Our own view is that the 'acid test' of an abstraction is that one can apply it in the context it arose from, e.g. if you have really abstracted the idea of an abstract group in mathematics, then it is not enough that to simply appreciate the axioms, you *should* be able to use it to demonstrate something group theoretic. This example encapsulates our view of what Davydov (1990) calls the 'ascent to the concrete'.

Later in this chapter we argue that the formation of an abstraction is dependent on the context in which teaching and learning activities take place, and hence that mathematical abstractions are situated. Before doing this, however, we briefly consider situation and context.

1.2 Situation and context

We both feel more comfortable using the term 'context' than the term 'situation' though we recognise this comfort comes from familiar texts which have been formative in our intellectual development. In this section we briefly consider problems with the terms.

Engeström and Cole (1997) describe the vagaries of the term 'situation':

> Behind the notion of situatedness lies the notion of situation. It is a deceptively simple notion: We all know what it means. But try to define it explicitly. Is a situation a moment in time? Is it a location, a place? Is it a life situation, a social situation, a configuration of relationships? (p. 301)

Gee (1997) considers that these vagaries arise from different contexts of use:

> The claim that thinking and meaning are situated … is now a popular one and stems from work in a variety of disparate areas, where the meaning of the word *situated* itself takes on somewhat differently situated meanings. (p. 235)

'Context' does not fare much better. Cole (1996, p. 132) describes context as "perhaps the most prevalent term used to index the circumstances of behaviour". He elaborates two suasive but problematic notions of context: that which surrounds and that which weaves together. The notion of context as that which surrounds situates, say, a learner doing a mathematical task in a peopled (classmates, teachers, others) set of institutions or communities of practice (classroom, school, community). Cole notes that "there is no temporal ordering. 'That which surrounds' occurs before, after, and simultaneously with the 'act/event'." (p. 134) and there is a dialectical, not a causal, relationship between activity in surrounding layers. But the learner-with-task can be seen as a thread winding between other peopled and institutional threads. Viewed in this way context weaves together rather than surrounds and:

> The boundaries between 'task and its context' are not clear-cut and static but ambiguous and dynamic … that which is taken as object and that which is taken as that-which-surrounds-the-object are constituted by the very act of naming them. (p. 135)

The surrounding and the weaving metaphors are illuminating but, like our focus on the meaning of these two terms, attempt to express generalities and, in doing so, fail to capture the wholeness of situated 'goal directed, mediated and peopled' activity. Our views here echo those of Lave (1988) on arenas "The supermarket as arena ... is outside of, yet encompasses the individual" (p. 151). To complete this brief consideration of context/situation we consider artefacts and mediation.

An artefact is an object or form and is both material and conceptual/ideal. Mediation is a feature of all cultural activities/actions; a person's actions towards an object operate through/with an artefact. "Mediation through artefacts applies equally to objects and people" (Cole, 1996, p. 118). Artefacts rarely, if ever, mediate in isolation; our experience of the world is through co-ordinated systems of artefacts. Our activity in writing this chapter is an example of this: we write in a language, to academic conventions, and with a word processor to a form/structure dictated by cultural history and the editors.

The upshot of this short section is that a consideration of context without a consideration of artefact-mediated human activity is vacuous. With these considerations of context and situation noted we move on to consider contextual, or situational, views of abstraction.

1.3 Contextual views of abstraction

Of the many accounts of mathematical abstraction (see Boero, Dreyfus, Gravemeijer, Gray, Hershkowitz, Schwarz, Sierpinska, and Tall, 2002) we consider: van Oers (2001), Noss and Hoyles (1996), Davydov (1990) and Hershkowitz, Schwarz and Dreyfus (2001).

Van Oers views abstraction as strictly dependent on the context: abstracting is a process of contextualising an experience, assuming a point of view (a relation, metaphor or image) for the construction of relationships amongst the situational objects. Assuming a point of view, however, does not happen in an arbitrary manner but is contingent upon the meaning attached to a situation or to the objects involved. This is, he claims, a discursive process which is only truly meaningful in an activity through which new meanings are negotiated interactively. In his account contexts are not pre-given entities but are created in communicative interpretative processes, which involve goal-directed, tool mediated human actions, organisation of objects, social and cultural influences under complex historical conditions.

Noss and Hoyles (1996) are interested in a theory of mathematical meaning which transcends a purely situated view and how artefacts structure and are structured by students' mathematical activities. They employ two important constructs: webbing and situated abstraction and see:

> learning as the construction of a *web* of connections – between classes of problems, mathematical objects and relationships, 'real' entities and personal-specific experiences. (p. 105)

Situated abstractions are their means of transcending a purely situated "mathematical limbo" (p. 119) and note that "all abstractions are situated" (p. 122). Their focus on meaning involves layers of meaning constructed through language and resource-mediated activity which can "point [the student] beyond the boundaries of that situation." (p. 122).

Situated abstractions are shaped by mathematical cultures (communities of practice) as well as by technology. Situated abstraction and webbing are complementary constructs which can assist in understanding how students make meaning, make connections and establish relationships and structure in educational settings; this understanding can help us design educational settings for meaning making.

Both van Oers, and Noss and Hoyles are opposed to a view of abstraction as an ascent from the concrete to the abstract. Their respective accounts differ but they share the view that abstraction is a dialectical development between the concrete and the abstract. But what are the concrete and the abstract? To empiricists the concrete is associated with physical everyday knowledge but the abstract concerns 'logical' and 'mental' structures, e.g. 'mathematics' (see Piaget, 1970; van Oers, 2001). This view is challenged by both van Oers, and Noss and Hoyles.

> The split between the concrete and the abstract actually creates a misleading divorce between the perceptual-material and the mental-conceptual world. Abstraction can never produce meaningful insights in the concrete world, unless there is some inner relationship between the concrete and the abstract (van Oers, 2001, p. 287).

> a standard description of the difference between thinking in lived-in experience and mathematical thinking is that the former is concrete, the latter is abstract. ... the history of Western thought has privileged the latter at the expense of the former. Abstract is general, decontextualised, intellectually demanding; concrete is particular, context-bound, intellectually trivial. (Noss and Hoyles, 1996, p. 45)

Abstraction is, to both van Oers and Noss and Hoyles, a process of making sense of a concrete situation by discovering new meanings to establish interconnections amongst the different elements of the whole. These ideas are compatible with Davydov's (1990) 'method of ascent'. In Davydov's account the source and the basis of all human knowledge is practical activity. During activity individuals draw upon features and potentialities of objects, artefacts and cultural-historical concepts. In activity human cognition acts at two different levels: empirical and theoretical thought. Empirical thought is necessary for the creation of everyday conceptions by means of establishing particular connections and relationships through sensory observations. The generation of scientific concepts, however, requires theoretical thought which is necessary to develop comprehensive interrelationships on the basis of mental transformations of the features and the potentialities of the objects.

Abstraction, to Davydov, starts from a simple undeveloped first form and ends with a consistent and highly structured final form achieved through theoretical thought. This is not an ascent from the concrete to the abstract but a dialectical relationship, a *to and fro*, between the concrete and abstract. The concrete and abstract are correlated with one another; the former refers to a structured, developed whole as a result of established interconnections and unification of different aspects of the objects. The abstract on the other hand has the characteristic of being devoid of differences, internally undeveloped and not yet particularised. Thus viewed, an initial abstraction can be considered as a simple relationship of concreteness and this abstraction ascends to the concrete by virtue of establishing real interconnections on the basis of the potentialities of the objects.

The main difficulty with Davydov's method of ascent, apart from the abstruse manner in which he often expresses himself, for us as mathematics education researchers is that, despite its explanatory power, it is difficult to see how to operationalise it to analyse the process of abstraction in an empirical investigation. His ideas, however, are incorporated into an operational model of abstraction in context by Hershkowitz, Schwarz and Dreyfus (2001). This model is concerned with abstraction of mathematical knowledge structures in social, cultural and historical contexts. Abstractions are generated in activity and require students to 'vertically reorganise' previously constructed mathematical knowledge into a new mathematical structure. The term 'activity' is used in the sense of activity theory (Leontiev, 1981) to emphasise that actions occur in a social and historical context. 'Vertical reorganisation' involves the individual integrating known mathematical elements/relationships and developing them into new complex structures. Complexity here relates to the 'depth' of a structure in the sense of having knowledge-components as parts (Ohlsson and Lehtinen, 1997) and to the demand on learners to assemble components together and forge new connections amongst them.

A new structure develops through three epistemic actions: recognising, building-with and constructing. 'Recognising' refers to the realisation of a mathematical structure; 'building-with' refers to combining mathematical elements to meet a goal, e.g. justifying a statement; and 'constructing' consists of assembling knowledge artefacts to produce a new mathematical structure. Constructing actions are related to the reorganisation of the previously acquired structures, which brings about the emergence of a novel structure; and recognising and building-with actions are important in achieving this. These epistemic actions are dynamically nested in that constructing involves building-with and recognising; and building-with requires recognition of already constructed structures.

Hershkowitz et al. view the genesis of an abstraction as passing through three stages: (a) the need for a new structure; (b) the construction of a new abstract entity through nested recognising and building-with epistemic actions with extant structures; and (c) the consolidation of the abstract entity or structure which involves recognising it and building-with it in further activities. Abstraction is a process leading to a product, an abstracted structure. Elsewhere (Monaghan and Ozmantar, 2006), we slightly amend this model and propose that the process of abstraction involves construction and consolidation (stages (a) and (b)) and the end product is a consolidated construction that can be used to form further abstractions. In this chapter we work within our slight amendment of Hershkowitz et al.'s model of abstraction and our primary interest is in the construction stage.

Although we could quite happily work with any of the contextual accounts of abstraction presented above we choose Hershkowitz et al.'s account because our empirical work largely corroborated this model (Ozmantar, 2005) and also the precision to which epistemic actions can be identified assists empirical investigations.

2. THE STUDY, THE TASKS AND PROTOCOL DATA

In this section we briefly describe an empirical study we undertook to investigate the formation of mathematical abstractions in relation to different aspects of human interactions, collaboration, and scaffolding, within the framework of Hershkowitz et al.'s model. We worked with 20 Turkish Year 10 students (16-18 years old) selected from 134 students who took a diagnostic test: we selected students with sufficient knowledge structures to embark on the tasks but who had not previously encountered the target constructions of the tasks. The students worked on tasks related to the graphs of linear absolute value functions. 14 worked in pairs and six worked individually. Four of the pairs and three of the individuals worked with the interviewer (first author) assisting them and the rest worked with minimal assistance, the interviewer only intervened when students expressed frustration. This interviewer assistance turns out to have been very important.

The students worked on four sequential tasks on the graphs of $y=|f(x)|$, $y=f(|x|)$ and $y=|f(|x|)|$ (referred to as $|f(x)|$, $f(|x|)$ and $|f(|x|)|$ hereafter). The first and second tasks aimed to construct a method(s) to draw/sketch the graphs of, respectively, $|f(x)|$ and $f(|x|)$, given the graph of $f(x)$. The third task was included to consolidate the constructions of the first and second tasks. The fourth task aimed to construct a method(s) to obtain the graph of $|f(|x|)|$ from the graph of $f(x)$. The four tasks were given to the students in paper-and-pencil format and all tasks were completed within one week over four consecutive days. Due to the space limitations and to carry out detailed considerations of the issues, we present protocol data excerpts from two girls (H and S), who worked together with interviewer assistance, on the second task – see Table 1 below.

Table 6-1. The second task

1. A function f is defined on the set of real numbers as f(x)=x - 4. Draw the graph of y=|x| - 4 and comment on any patterns or symmetries.
2. Do you see any relationship between the graph of f(x)=x - 4 and the graph of y=|x| - 4? Explain your answer.
3. The graph of f(x)=x+3 is given below. Can you obtain the graph of f(|x|)=|x|+3 from the graph of f(x)? Explain your answer. (NB: the graph is not shown here.)
4. There are four different graphs of f(x) given below. Find the graphs of f(|x|) by making use of the graphs of f(x).

A.

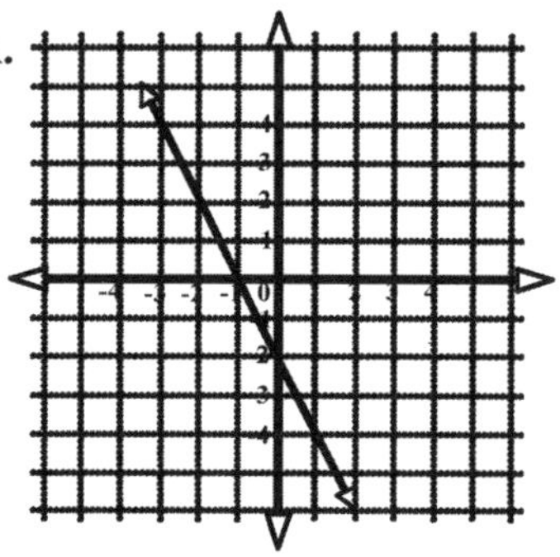

B.

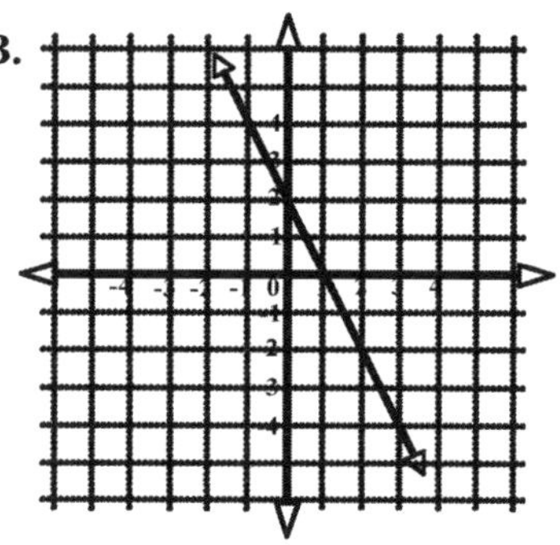

C.

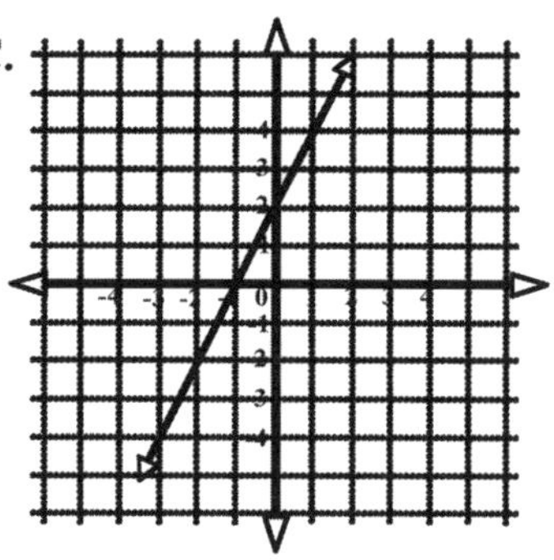

D.

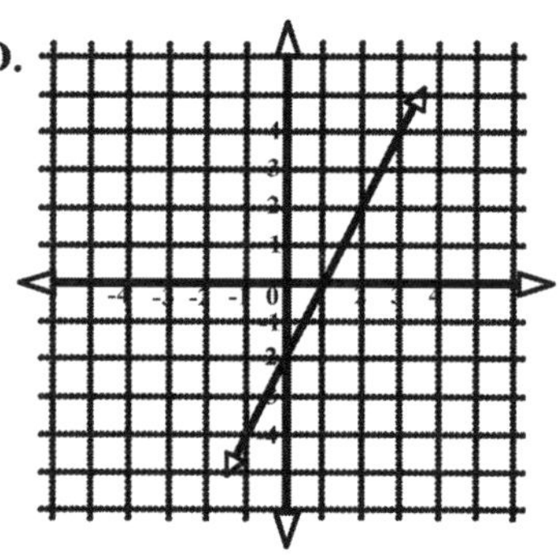

5. How would you explain to one of your friends how to draw the graph of f(|x|) by using the graph of f(x)? Demonstrate that your explanation is correct by using the above-given graphs.

Question 1 (Q1) was included to draw students' attention to symmetric relationships in the graph of $f(|x|)$. Q2 aimed to prompt students to establish (initial) interrelationships between the object graph and the graph of $f(x)$. Students were then expected, in Q3, to further develop the initial relationships observed/discovered in the first two questions. Q4 was intended to encourage students to develop a method to obtain the object graph without using the equation, i.e. using only the graphic representation of $f(x)$. Q5 aimed to elicit students' developing insights on the graphs of $f(|x|)$.

The protocol data below is composed of H and S's written work and verbatim transcription during audiotaped sessions. Speakers' (interviewer/ students) uninterrupted utterances were assigned a natural number. In the following 'I' refers to the interviewer and H and S are the initials of the two girls. Phrases in [square brackets] are explanatory text, not data. We divide the presented protocols into four episodes for the purpose of communication and provide short comments after each episode. Please note that we do not have space to comment on microgenetic interaction in terms of epistemic actions. However, episodes 1 to 3 contain only recognising and building-with actions but episode 4 contains these actions as well as constructing actions. We consider this difference in the next section.

2.1 Protocol data

The students obtained the graph of $f(|x|)$ accurately in Q1 by substitution and then compared, in Q2, the graph of $f(|x|)$ with that of $f(x)$. They focused solely on the conspicuous features of $f(|x|)$ and recognised that $f(|x|)$ has a line of symmetry. In Q3, they once again sketched the intended graph of $f(|x|)$ correctly by substitution and realised that this graph also had a symmetry line. We pick up the students' conversation as they start Q4.

2.1.1 Episode 1

43S: [S reads Q4]. OK, we don't have any equation this time.

44I: (…) You are just presented with the graphs without the equations. Are you planning to find the equation for each of the graphs?

45H: I think we should find the equation for the first graph and [then] with this equation we can develop a general pattern to draw the others…

H and S then obtained the equation for the first displayed graph and sketched the graph of $f(|x|)$ by substitution (see Figure 1).

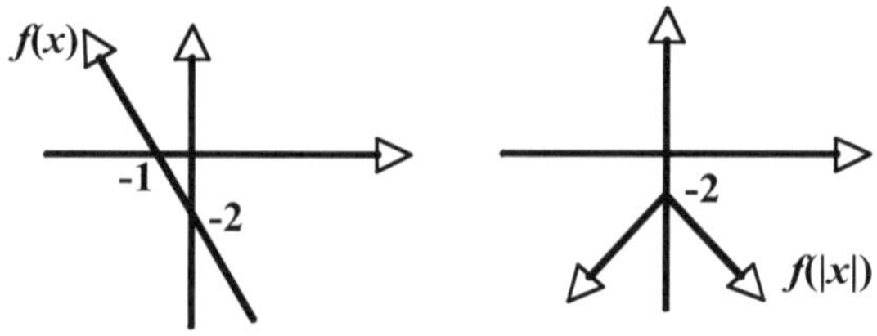

Figure 6-1.

Following this, H and S came up with an initial method, in S's terms, "the part of *f*(*x*) until the *y*-axis remains the same and then the remaining part is taken symmetrically" (see Figure 1). We call this the 'reflecting method'. It is somewhat ambiguous: where is symmetry taken, what is "the part of *f*(*x*) until the *y*-axis" and what is "the remaining part"? H and S then successfully applied this method to the second displayed graph and obtained an accurate graph of *f*(|*x*|). Following this, the interviewer intervened to draw H and S's attention to the ambiguity of their method.

2.1.2 Episode 2

76I: …which part's symmetry did you take?
77H: In the line of y=2
78S: According to the line… the line is passing when x=0
79H: According to this line [y=2]
80I: (…) But this is not what I was wondering. I want to know which part of *f*(*x*) is taken symmetrically?
81H: Oh, it is… we have problems to express it…
82I: OK, you can say for example, the part on the right or left side of the *y*-axis.
83S: We take the symmetry [of the part of *f*(*x*)] on the left of the *y*-axis according to the line of y=2

The interviewer, 76I, explicitly asked H and S which part of *f*(*x*) was reflected but they misunderstood the question and talked about the symmetry line (77H to 79H). The interviewer rephrased the question in 80I but H and S's answers remained specific to the graph at hand. H and S could not see the issue from the interviewer's perspective and hence did not clarify which part of *f*(*x*) was reflected.

2.1.3 Episode 3

The students moved on to the third displayed graph and drew the graph of $f(|x|)$ using the reflecting method but they obtained an erroneous graph (see Figure 2). S realised that something was wrong and suggested reflecting the part of $f(x)$ for which $x<0$.

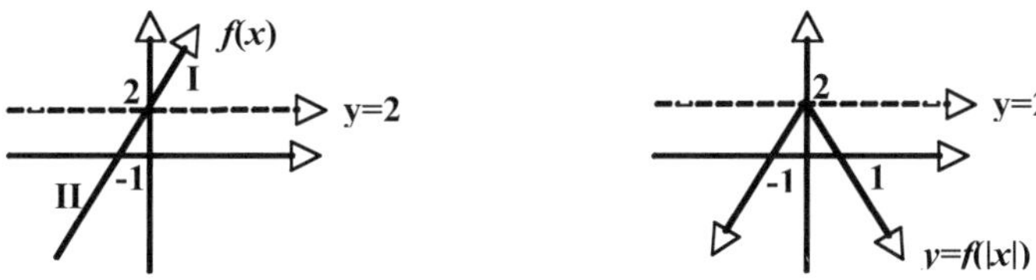

Figure 6-2.

109S: … shouldn't we take the symmetry of this part [of $f(x)$ at $x<0$]? Look at this graph [see Figure 1] the part of $f(x)$ until the y-axis is unchanged and the remaining part is reflected. I mean if I name the rays as I and II [she writes] shouldn't II be reflected in $y=2$.

110H: So are you saying that this graph should be like that? [She refers to the symmetry of II in the line of $y=2$].

111S: Yes it should be like that and these two rays should be symmetric in the y-axis… All other graphs [of $f(|x|)$] were symmetric in the y-axis… shouldn't it be so here too?

112H: Hang on… wait, I get confused (…) let me think! [they pause] … You are saying that two rays should be symmetric in the y-axis, right?

113S: Yes, it should be so

114H: OK, look at the graph… these [the two rays in the graph of $f(|x|)$] are symmetric in the y-axis anyway…

115S: Yeah, I know but I think something is wrong with this graph [of $f(|x|)$]…

Although H and S collaborated well in that they attended to one another's ideas and contributed to this episode almost equally, they were unable to realise the deficiency in their reflecting method or justify the accuracy (or inaccuracy) of the graph. This suggests to us that the intended structure of $f(|x|)$ was, at this stage in their development, beyond the students' unassisted joint efforts. Having realised this, the interviewer intervened in the next episode.

2.1.4 Episode 4

116I: (…) I would like you to (…) examine the graphs that you have drawn so far… and discuss which part or parts always change and which doesn't!
117S: Look, in this graph… the graph of $f(x)$ until the y-axis didn't change and after y, it has changed…
118I: Do you mean that the part of $f(x)$ at the positive x, which is always on the right of the y-axis, doesn't change?
119S: Yes (…) also on the right side of the y-axis, all of the values of x are positive…
120H: Did it remain unchanged in all of the graphs?
121S: Yes, let's have a look ... [They look into other graphs].
122I: S, you were saying that all values of x are positive on the right side of the y-axis!
123S: Oh, yes… they are positive and so we don't change them…
124H: Positive values don't change?
125S: I mean… we are drawing the graph of $f(|x|)$, right? (…)
126H: Yes, so?
127S: Umm… the absolute value sign is always outside of x, I mean it is $|x|$
128H: Positive values don't change in the absolute value sign…
129S: Exactly, it changes the negative values
130I: (…) [but] negative values of x, which are on the left side of the y-axis…
131S: They have to change… I mean in the absolute values sign, negative values change…
132H: Yeah, I agree… I mean absolute value sign is outside of x… so… so negative values of x must be different
133I: From what?
134H: Well, I mean there must be difference between the graph of $f(x)$ and $f(|x|)$ at the negative values of x
135I: Because?
136S: Because positive values of x remain unchanged in the absolute value sign, but negative values of x must be multiplied by minus to go out of the absolute values sign… thus (…) whatever changes occur in the graph of $f(x)$, it must be at the negative values of x
137H:(…) the part of $f(x)$ on the right hand side should remain the same, no matter what. And the part on the left hand side should change...

In this episode, H and S amended their ambiguous reflecting method which appeared to have been greatly assisted by the interviewer's interventions, e.g. 116, 118 and 122. This method was new to the students and constructed through the utilisation of existing knowledge artefacts that H and S had at their disposal such as features of absolute value, linear functions, symmetries and Cartesian grids in terms of x- and y-intervals. Both students appeared to use similar epistemic actions and undertook constructing actions which resulted in a complex structural knowledge artefact, the reflecting method.

3. ABSTRACTION: MEDIATION, PEOPLE AND TASKS

Mathematical abstractions develop in personal-cultural-historical space. They develop from artefacts, including knowledge artefacts, available to a person in a culture at a time – the knowledge artefacts available to the young Newton were distinct from those available to the young Archimedes. This view is expressed by Davydov (1990) amongst others:

> The individual must act and produce things according to the concepts which exist as norms in the society *beforehand* – and he does not create them, but accepts or assimilates them. ... they are already assigned to him as crystallised and idealised, historically developed human experience (pp. 252-253).

Our principal interests in this discursive section of our chapter is to return to the ideas of abstraction in the light of the protocols and address three very human themes: mediation, people and tasks.

3.1 Mediation

Mediation involves artefacts and people. We consider both separately below, with respect to H and S's development during the second task, with the understanding that they are intertwined and that this separation is made for reporting purposes only.

Artefacts include basic physical tools and representations of, and modes of action of using, these tools. An artefact never mediates alone. We live in material and mathematical worlds rich in coordinated systems of artefacts. Students' developing constructions and hence abstractions depend on the mediation of knowledge artefacts. With regard to the mediation of H and S's epistemic actions one can see, in the protocol excerpts, that H and S's recognising, building-with and constructing actions are all mediated by the knowledge artefacts at the students' disposal. Examples of these artefacts involve sketched graphs, features of Cartesian grids, e.g. negative/positive

x- and y-axis intervals, symmetry, features of absolute values and properties of Euclidian Geometry.

These artefacts cannot be separated from H and S's constructions and we would not expect H and S to make the observations and connections which they did make without these artefacts – epistemic actions are artefact-mediated actions. But artefacts cannot be separated from the person/people using the artefact, there is "an irreducible tension between agent and cultural tool" (Wertsch, 1998, p. 30). These considerations point to the situatedness of people-using-artefacts. Wertsch (1998) expresses this nicely:

> virtually all human action, be it on the individual or social interactional plane, is socioculturally situated; even when an individual sits in solitude and contemplates something, she is socioculturally situated by virtue of the mediational means she employs (ibid., p. 109).

We now turn to human mediation. We take it as given that H and S's developing construction was mediated by the interviewer's interventions. Note that we are not saying that this mediation was good (or bad), simply that interviewer mediation occurred: the interviewer acted as a knowledge artefact which the students made essential use of to produce the construction of the reflecting method. But how was this construction mediated by the interviewer and were there particular times at which this mediation was particularly important?

It could be said that the interviewer facilitated the students' mathematical actions but, although we consider this to be true, it seems somewhat facile and does not take an analysis far. There appeared to be times, however, when the interviewers' interventions brought about important transformations in the students' ways of seeing, talking and acting. To exemplify these, we consider episodes 3 and 4 as they represent times when interviewer interventions were not particularly important and important respectively. In episode 3 the students were acting at the recognising and/or building-with levels and the interviewer withheld his assistance. In episode 4 the students were acting at the constructing level, they achieved the construction of the reflecting method and the interviewer's six interventions appeared to play an important role in the activity.

In episode 3, H and S were talking about whether the absolute value graph obtained from the third displayed graph was correct or not, about which segment of $f(x)$ (at $x<0$ or $x>0$) should be reflected and about the reasons for these. After a lengthy debate they could not reach agreement. Their difficulty stemmed from the deficiency of their reflecting method with regard to which segment of $f(x)$ was reflected in the horizontal line intercepting the y-axis. Having realised this, at the beginning of episode 4, the interviewer intervened and asked H and S to look at earlier graphs and decide which segment of $f(x)$ had changed and which had not (116I). The intention of the intervention was to get the students to recognise changes in the graphs and to correct their ambiguous method. S's response "the graph of $f(x)$ until the y-axis didn't change" (117S) was ambiguous[9] and did not meet the interviewer's intention. The interviewer again intervened in 118 and 'rephrased' S's response by clearly indicating the important properties of this segment of $f(x)$ ("part of $f(x)$ at the *positive* x, which is on the *right*…"), recognition of which appeared essential for the students to overcome the ambiguity of the initial reflecting method. Immediately following this, S, in 119, reproduced the interviewer's utterance almost completely but this was not a simple copy of what was said to her, this 'came from her' (because she was now able to use it, see 123, 125 and 127) and it was mediated through the interviewer's intervention, an example of the 'mediation of recognition'.

Following the initial three interventions (116I, 118I and 122I) an important transformation in S's *way of seeing* is apparent (see 123, 125 and 127). She was now looking at a graph of $f(|x|)$ and seeing in it a particular line segment ("on the right side of the y-axis"; 119S) with specific qualities (positive xs, 119S); seeing in it relation to the absolute value sign (127S); seeing a 'link' to the expression of $f(|x|)$ (125S); and seeing a reason in it for being unchanged (123S). The extent of this transformation becomes clearer when one looks into S's way of seeing these graphs in episode 3 where she merely recognised and built-with ostensible features such as some 'parts' and 'symmetries' (e.g. 109S). H later quickly attuned to S's perspective (128H). Throughout episode 4, particularly from 127 onwards, there was a dramatic change in the way that these students talked about the graphs of $f(|x|)$. To better appreciate this change, first consider episode 3 in which H and S's talk was chiefly based on their beliefs and expectations (it is interesting to note the frequent occurrence of the auxiliary verb 'should' in their talk); and their reasoning, when stated, was related to similarities and differences of the graphs (e.g., 109 and 111). In episode 4, however, their talk involved explanations with mathematical reasoning rather than their beliefs and was almost always expressed with certainty, e.g. the use of "must" in 132H and 136S. During this episode the interviewer sustained H and S's way of talking and seeing by initiating reasoning steps or requesting reasoned responses (131I, 133I and 135I).

These differences in H and S's ways of seeing and talking are reflected in their epistemic actions. In episode 3 H and S were acting at the recognising

---

[9] Ambiguity is celebrated in many disciplines, e.g. literature, but is anathema to mathematicians. The interviewer, as a mathematician, we assume, mediates in different ways to mediators in other disciplines.

and/or the building-with level as they were usually concerned with justifying their statements without drawing on 'deep' mathematical features of the involved structures, e.g. 109S and 111S. In episode 4, however, constructing actions occur and they see and talk about deep structural relationships using knowledge artefacts, e.g. 132H, 134H, 136S.

3.2 People

We focus on students' personal histories (prior understandings of mathematical content knowledge and of social practices) and of the education system, as a culture, of which the students had experience.

As mentioned, the students in this research were selected by a diagnostic test to ensure that they had the necessary knowledge (of symmetry, of linear functions and of absolute value) to carry out the tasks. During their work in constructing the reflecting method H and S recognised and built-with features of absolute value and linear functions, and of symmetries and reflections. All of these features were available to the students prior to working on the tasks and were part of their personal histories. H and S's construction, then, is situated within the students' personal histories, i.e. their utterances are linked to and dependent on their personal experiences in mathematics – an obvious but an important point.

Further to this the microgenetic (moment by moment) development of H and S's construction was composed of 'strands that weaved in and out' over the course of the task; and the task itself was situated in their ontogenetic development. We can see, in 111S, 120H and 121S for example, H and S returning to examine previously sketched graphs, observing their similarities and differences. Their developing insights were built on these observations as well as on prior knowledge, e.g. "right side of y-axis, values of x are positive" (119S), "so we don't change them" (123S) and "positive values don't change in the absolute value sign" (128H). It was in this chain of utterances that H and S established that the graph of $f(x)$ in the positive x-axis interval remains the same when transformed into the graph of $f(|x|)$ (see 137H).

We now turn to students' prior understandings of social practices. Social practices involve cultures, customs, value judgements, societal norms and belief systems. For the present purpose we limit our consideration of social practices to the students' understanding of collaboration. To do this it is necessary to provide a brief background on H and S. These students described themselves as 'very good friends'. The school they attended had a dormitory and they lived in the same room and sat next to each other in class. They usually worked together in and out of the classroom. On getting involved in this research they expressed a wish to work together rather than working alone or with someone else. They were paired up and worked quite harmoniously during all of the four tasks. Their eagerness to work together, we believe, had a considerable influence on their evolving interactions and was reflected in the way that they, without restraint, shared their insights (e.g. 123S to 129S), confusions (112H), concerns (115S) and difficulties (124H and 126H) with one another.

These observations echo Forman's (1989) work on peer interaction in the construction of mathematical knowledge. She employed two girls who were best friends and observed that the students were not as motivated when they were working individually as they were when working collaboratively. Whether or not this would have been the case if H and S had worked alone we cannot say, but we strongly believe that had we paired either with someone they did not feel comfortable working with, then a very different pattern of interaction would have resulted: the girls' evolving work would likely have been adversely affected by the nature and quality of their personal relationships (Goldstein, 1999).

Close friendship alone, however, is not sufficient to ensure successful collaboration and communication, the quality of the collaboration is important: H and S attended to one another's ideas and proposals, shared their concerns and difficulties and respected one another's needs. It was through such collaboration that they became aware of the insufficiency of their initial reflecting method, of the inaccuracy of the sketched graphs (episode 3) which helped them move closer to the attainment of the reflecting method. We worked with other students in this study whose manner of communication during their work appeared to hinder their progress. We illustrate this with the case of T and K on the same task as H and S. T and K also described themselves as 'good friends' but their understanding of 'working together' appeared different to that of H and S. The excerpt below occurred when they were working on Q4. T argued that the symmetry in the graphs of $f(|x|)$ was related to the slope of the original graph of $f(x)$:

T: I think the symmetry is related to the slope (…) slope is everything…
K: In the first and third graph (…) negative value … err… positive…
T: No, no it's not about positive or negative… it's about the slope…
K: But we ignore [he mumbles inaudibly]…
T: Look, I now understood well… Yes! One should know about the slope… otherwise it's impossible...
K: Shall we substitute?
T*: Look, I am telling you… if the slope is positive then we take the symmetry of positive part [i.e. f(x) at x>0] but if the slope is negative

then we take the symmetry of negative part [i.e. f(x) at x<0] in the y-axis …

K: But positive values … [K mumbles and T interrupts him]

T: Look, I am telling you the rule… the graph depends on the slope, yes it is definitely right…

This pattern of communication was repeated a number of times over the four tasks. T appears as dominant, impulsive and hasty in his generalisations and acted as a 'higher authority' ("I am telling you the rule"). K appeared to accept the authority of T and his voice was barely heard. Such factors do play a part in knowledge development for if T had attended to K's suggestion of sketching the graph using substitution then he might have seen the inaccuracy of his arguments, expressed in T* – the way in which students communicate with one another and the mode of their interaction can hinder as well as enable their progress.

We now turn to students' understanding of education, the situated education system that was their knowledge of education. Ozmantar and Monaghan (2005) coin the term 'pedagogic resonance' to highlight the importance of the cultural background of the students and interviewers. Pedagogic resonance refers to teacher-students' mutual understandings of the social context of pedagogy (cultural reproduction-production). Throughout our educational life we adapt to certain customs and are instilled with expectations regarding teaching and learning. If, for example, a teacher adopts an 'open' approach to teaching, e.g. tries to avoid leading the student, and an adolescent learner has been taught in a 'didactic' manner from early childhood, then the teaching/learning activity may be frustrating and/or fruitless for the leaner. This is an example of what we call low pedagogic resonance. A brief excerpt which occurred in our study may clarify our point here. The excerpt below is taken from the dialogue between an individual student (L) and the interviewer working on the graphs of $f(|x|)$. L sketched a graph of $f(|x|)$ by using the linear graph of $f(x)$ by means of a method that he developed. The graph was accurate but the interviewer tried to make L justify his method.

I: Do you think the graph is accurate?

L: The graph… (pauses for a while)

I: How could we know?

L: (he substitutes a couple of values of x into the equation) yes, I think it is

I: Sure?

L: You know the answer, don't you?

I: Yes, but you should find this out by yourself

L: Why don't you tell … right or wrong?

I: But you need to do this by yourself!
L: I did the drawing by myself... you already know if this is right or wrong... if it is right, it is ok, my way [method] is right, no need to waste time... If it is wrong tell me what and I'll work on it...

L was clearly frustrated and reluctant to provide an elaborate justification of the accuracy of the graph: he just wanted to know if the graph was right or wrong. The root of this reluctance lies, we believe, to a considerable extent, in common practices of Turkish educational culture which emphasises getting an answer as quickly as possible. This is considered important for students to be successful in the high-stakes university entrance examination. In many mathematics classes students are instructed how to get the 'job done' in the shortest possible time. Teacher-student expectations are that 'teachers lead and students follow'. In the case of L, he expected the interviewer to tell him if his method was right or not. If right then the 'job' is done; but if something is wrong then L expected to be told so that he can "work on it".

Returning to H and S we observe the interviewer playing a 'leading role' rather than adopting an 'open' approach. This was because the interviewer thought he recognised the students' frustration when employing a particularly 'open' approach. The interviewer's leading role could be detected in his utterances such as "I want to know which part of $f(x)$ is taken symmetrically? "(80I) and "I would like you to examine the graphs ... and discuss which part always change(s)" (116I) (see also 118I). H and S followed the interviewer whose interventions eventually led them to develop the reflecting method (episode 4). This form of dialogue is rooted in the participants' common understanding of social context, which, as is the case with L and H and S, has the potential to shape the individuals' expectations and the way in which they communicate with one another.

3.3 Tasks

Tasks are of central importance in education but this is not reflected, in the literature in English, in a large number of research papers on the matter. Hoyles (2001, p. 284) notes that the "design of activities and the design or choice of the tools introduced to foster mathematics learning ... bring knowledge and epistemology back into centre stage." How one views tasks is interrelated to one's theoretical perspective. Chevallard (1999), for example, sets forth an anthropological approach which views tasks as artefacts that are constructed and reconstructed in institutional settings; further to this tasks are not isolated objects of study but features of a way of viewing mathematical practices in terms of tasks, techniques, technology and theory. Researchers working outside of the anthropological approach often focus on qualities of tasks that facilitate specific mathematical actions or observations: Sahlberg and Berry (2003), for example, focus on the qualities of tasks that do and do not facilitate small group discussion and Monaghan and Ozmantar (2006) consider the qualities of tasks that facilitate the consolidation of newly formed constructions.

Research on tutoring (e.g. Chi, Siler, Jeong, Yamauchi, and Hausmann, 2001; Rogoff, Ellis, and Gardner, 1984) suggests different modes of interaction amongst the participants depending on the content of the task: highly verbal, e.g. a reading task or highly physical, e.g. completion of a puzzle; how structured a task is and the level of its difficulty. These studies suggest that the way in which tutors approach students' work is greatly influenced by the specificities of the task. Noss and Hoyles (1996) argue that even the phrasing of the questions in tasks is important and cite two different phrasing of the 'same question' which resulted in a considerable difference in student performance: "how many halves are there in 2 ½?" and "2 ½ ÷ ½"; the success rate for the former question was approximately 30% higher than the latter.

Task 2 influenced the way in which interaction amongst the participants unfolded, gave the students a direction and created a need for the students to develop a method. The task was designed by the interviewer to lead the student into the construction of the structure of $f(|x|)$ through a set of 5 questions. However, students' interpretation of the questions can be different to that of interviewer's and this can, in turn, shape the course of interaction. For example, in Q4, four graphs of $f(x)$ without equations were presented and the intention here was to lead the students to develop a (general) method to draw the graphs of $f(|x|)$ using the given graphs of $f(x)$ rather than using equations. Despite the fact that the interviewer tried to call attention to this (44I), the students interpreted this question differently (45H) and decided to find the equation first. Consequently, H and S first drew the graph of $f(|x|)$ by substitution and then tried to develop a general method.

Despite the differences between the interviewer's and H and S's interpretations, the task gave the students a particular direction by encouraging them to focus on symmetrical patterns and on the relationship between the graphs of $f(x)$ and $f(|x|)$ on the basis of these symmetries. H and S's realisation of the symmetric relationship in the graph of $f(|x|)$ provided them with a particular starting point in search of a method while transforming the graph of $f(x)$ into the graph of $f(|x|)$and hence played a crucial role in developing the reflecting method.

The task also created a need for H and S to develop a new method unavailable to them before. As noted, Hershkowitz et al.'s (2001) model of abstraction is an activity theoretic one which views the genesis of an abstraction as commencing with a need for a new structure. If there is no need of a new structure, then it appears extremely unlikely that one will attempt to create an abstraction. This need may occur in many different ways in the course of an activity as a result of, for instance, the context of a problem, the demands of the tasks and/or interaction with others during the activity. In our study, this need was partly created by the goal of the task. To achieve this goal it was essential for H and S to come up with a method to sketch the graphs of $f(|x|)$, given the graph of $f(x)$.

4. CONCLUSIONS

Even though 'situation' and 'context' are problematic terms, we conclude that mathematical abstractions are situated (arise and are applied in contexts). Of course it could be claimed that what we are talking about is a completely different type of abstraction than is meant in mathematics. Our response would be "we don't think so; we think the 'fairytale' account of abstraction is way off course with regard to what really happens when people construct big mathematical ideas".

'Situation' involves so many elements but they are here in H and S's construction: who you are, where you are from, what you know, what you respect, whom you get on with, how you get on with them, what you are doing, why you are doing it, what you are doing it with, how what you are doing is structured.

Situated abstraction[10] is messy but we can see patterns. It is like a tessellation of a quadrilateral in dynamic geometry, you drag a vertex of the central quadrilateral and the whole screen sways but there is beauty in the apparent chaos.

REFERENCES

Boero, P., Dreyfus, T., Gravemeijer, K., Gray, E., Hershkowitz, R., Schwarz, B., Sierpinska, A., & Tall, D. (2002). Research forum 1, abstraction: Theories about the emergence of knowledge structures. In A. D. Cockburn & E. Nardi (Eds.) *Proceedings of the 26th annual conference of the International Group for the Psychology of Mathematics Education* (Vol. I, pp. 113-138). Norwich, England: PME.

Chevallard, Y. (1999). L'analyse des pratiques enseignantes en théorie anthropologique du didactique. *Recherches en Didactique des Mathématiques, 19*(2), 221-266.

[10] Hoyles & Noss use this term for a specific construct but the argument of this chapter is that abstractions are situated.

Chi, M. T. H., Siler, S. A., Jeong, H., Yamauchi, T., & Hausmann, R. G. (2001). Learning from human tutoring. *Cognitive Science, 25*, 471-533.

Cole, M. (1996). *Cultural psychology: A once and future discipline*. Cambridge, MA: Harvard University Press.

Davydov, V. V. (1990). Soviet studies in mathematics education (J. Teller, Trans.). In J. Kilpatrick (Ed.), *Types of generalization in instruction: Logical and psychological problems in the structuring of school curricula* (Vol. 2). Reston, VA: National Council of Teachers of Mathematics.

Dienes, Z. P. (1963). *An experimental study of mathematics learning*. London: Hutchinson.

Engeström, Y., & Cole, M. (1997). Situated cognition in search of an agenda. In D. Kirshner & J. A. Whitson (Eds.), *Situated cognition: Social, semiotic and psychological perspectives* (pp. 301-310). New Jersey: Lawrence Erlbaum Associates.

Forman, E. A. (1989). The role of peer interaction in the social construction of mathematical knowledge. *International Journal of Educational Research, 13*, 55-70.

Gee, J. P. (1997). Thinking, learning, and reading: The situated sociocultural mind. In D. Kirshner & J. A. Whitson (Eds.), *Situated cognition: Social, semiotic and psychological perspectives* (pp. 235-260). New Jersey: Lawrence Erlbaum Associates.

Goldstein, L. S. (1999). The relational zone: The role of caring relationships in the co-construction of mind. *American Educational Research Journal, 36*(3), 647-673.

Hershkowitz, R., Schwarz, B. B., & Dreyfus, T. (2001). Abstraction in context: Epistemic actions. *Journal for Research in Mathematics Education, 32*(2), 195-222.

Hoyles, C. (2001). From describing to designing mathematical activity: The next step in developing a social approach to research in mathematics education? *Educational Studies in Mathematics, 46*(1), 273-286.

Lave, J. (1988). *Cognition in practice: Mind, mathematics and culture in everyday life*. Cambridge: Cambridge University Press.

Leontiev, A. N. (1981). The problem of activity in psychology (J. V. Wertsch, Trans.). In J. V. Wertsch (Ed.), *The concept of activity in soviet psychology* (pp. 37-71). Armonk, NY: M.E. Sharpe. Inc.

Locke, J. (1689/1964). *An essay concerning human understanding*. London: Fontana.

Monaghan, J., & Ozmantar, M. F. (2006). Abstraction and consolidation. *Educational Studies in Mathematics, 62*(3), 233-258.

Noss, R., & Hoyles, C. (1996). *Windows on mathematical meanings: Learning cultures and computers*. Dordrecht: Kluwer.

Ohlsson, S., & Lehtinen, E. (1997). Abstraction and the acquisition of complex ideas. *International Journal of Educational Research, 27*, 37-48.

Ozmantar, M. F. (2005). *An investigation of the formation of mathematical abstractions through scaffolding*: Unpublished PhD Thesis, University of Leeds.

Ozmantar, M. F., & Monaghan, J. (2005). Voices in scaffolding mathematical constructions, *Fourth Congress of the European Society for Research in Mathematics Education*. Available through http://cerme4.crm.es/, see Working Group 8. Accessed 13 March, 2006.

Piaget, J. (1970). *Genetic epistemology*. New York: W.W. Norton.

Piaget, J. (2001). *Studies in reflecting abstraction* (R. L. Campbell, Trans.). Philadelphia, PA: Psychology Press.

Rogoff, B., Ellis, S., & Gardner, W. (1984). The adjustment of adult-child instruction according to child's age and task. *Developmental Psychology, 20*(2), 193-199.

Sahlberg, P., & Berry, J. (2003). *Small group learning in mathematics: Teachers' and pupils' ideas about groupwork in school*. Painosalama Oy: Finnish Educational Research Association.

Skemp, R. (1986). *The psychology of learning mathematics* (2nd ed.). Harmondsworth: Penguin.
van Oers, B. (2001). Contextualisation for abstraction. *Cognitive Science Quarterly, 1*(3), 279-305.
Wertsch, J. V. (1998). *Mind as action*. New York/Oxford: Oxford University Press.

Chapter 7

'We Do It A Different Way At My School'

Mathematics homework as a site for tension and conflict

Martin Hughes and Pamela Greenhough
Graduate School of Education, University of Bristol

Abstract: This chapter draws on Wenger's (1998) account of communities of practice to provide insights into the relationship between home and school mathematics practices and identities. The chapter presents and analyses an interaction between a 9-year-old boy and his mother as she attempts to help him with a mathematics homework task, consisting of a sheet of two-digit subtraction problems. The analysis reveals considerable tension and conflict at the boundary between home and school practices, as the different identities of mother and child negotiate with and challenge each other. These conflicts are exemplified by arguments about the appropriate methods for carrying out the subtractions, in which both participants justify their positions in terms of power and legitimacy instead of the underlying mathematical principles. One implication is that schools need to reconceptualise their approach to homework and parents' role in supporting homework if such interactions are to be more supportive of children's mathematics learning.

Key words: communities of practice, boundaries, identities, mathematics homework

1. INTRODUCTION

In the late 1980s and early 1990s many mathematics educators were drawn to the novel ideas about situated cognition and situated learning emanating from writers such as Brown, Collins, and Duguid (1989), Lave (1988) and Lave and Wenger (1991). These ideas were attractive to mathematics educators as they challenged the traditional view embodied in much educational thinking that knowledge can be separated from the situations in which it is acquired and used. Instead, Lave and her colleagues argued that knowing and learning are essentially situated in social practices, and that in order to understand the nature of knowing and learning we need therefore to understand the nature of these practices. This meant that attention was drawn to the use of mathematics in everyday settings such as supermarkets, workplaces and homes, as well as to the acquisition of mathematics in the classroom (e.g. Watson, 1998).

Our own particular and longstanding interest is with the different worlds which young (pre-school and primary school) children inhabit as they move between home and school – and other places beside (e.g. Greenhough and Hughes, 1998; Hughes, 1986 and 2001; Tizard and Hughes, 1984). We are interested in the ways in which these different worlds are present and interpenetrate – or create obstacles between – each other in events and practices. We are also interested in what happens to individual children as they move between these different worlds – how they present themselves in each world, whether they experience them as similar or dissimilar, and how they make sense of any dissimilarities or discontinuities which they may experience. While our focus here is on mathematics, we are interested in these issues across the school curriculum and beyond.

In some of the early writing of situated theorists these kinds of issues were only sketchily addressed. For example, the practices studied by Lave and Wenger are considered primarily in isolation from other practices, and there is little sense of participants moving between a number of different practices. As others have pointed out (e.g. Walkerdine, 2007) a somewhat static and singular view of practice can come across from these writings. More recently, though, Wenger (1998) has given greater recognition to the plurality and dynamic nature of practice, and the ways in which individuals move between multiple communities of practice. For example, he suggests that organisations such as factories and schools might be more productively viewed as *constellations* of communities of practice, which can be linked together in various ways. He also pays particular attention to the *boundaries* between different communities of practice, and looks at ways in which continuities across these boundaries can be maintained. One way is through *boundary objects*, a term originally used by Star and Griesemer (1989) to describe "objects that serve to coordinate the perspective of different constituencies for some purpose" (Wenger, p. 106). A second way of maintaining continuity is through the practice of *brokering*, which occurs when individuals use their membership of multiple communities of practice "to transfer some element of one practice into another" (ibid., p. 109). Wenger points out that "the job of brokering is complex. It involves processes of translation, coordination and alignment between perspectives" (ibid., p. 109).

The multiple membership of different communities of practice is also central to Wenger's conceptualisation of identity. He argues that an identity should not be regarded as a static or singular entity, but instead should be viewed as *'a nexus of multimembership'*. This notion of identity as a nexus means that work frequently has to be done to reconcile the different forms of membership forming the nexus. Indeed, Wenger proposes that:

> The work of reconciliation may be the most significant challenge faced by learners who move from one community of practice to another. For instance, *when a child moves from a family to a classroom*, when an immigrant moves from one culture to another, or when an employee moves from the ranks to a management position, learning involves more than appropriating new pieces of information. Learners must often deal with conflicting forms of individuality and competence as defined in different communities (p. 160, emphasis added).

Wenger suggests that this process of reconciliation may not be easy, and that membership of multiple communities of practice may involve tensions and conflicts that are never fully resolved. At the same time, he makes clear that in his view "multimembership and the work of reconciliation are intrinsic to the very concept of identity" (p. 161).

While Wenger's work provides an important conceptual backdrop to this chapter, we will also draw on more recent work by Street, Baker and Tomlin (2005). This work represents one of the most far-reaching attempts to date to analyse the nature of home and school mathematics. Here, we will briefly describe some of the key constructs used by these authors.

Like Wenger, Street et al. see themselves as developing a 'social approach' to learning, although in their case the focus is specifically on numeracy. They argue for a perspective "which sees the social in terms of context, values and beliefs, social and institutional relations" (p. 17). They also refer to this as an 'ideological' model of numeracy:

> From this perspective social relations refer to positions, roles and identities of individuals in relation to others in terms of numeracy. Social institutions and procedures we see as constitutive of control, legitimacy, status and the privileging of some practices over others in mathematics… (ibid., p. 17).

Street et al. also make an important distinction between *numeracy events* and *numeracy practices*. Drawing on an earlier definition of a literacy event by Heath (1983), they define numeracy events as "occasions in which a numeracy activity is integral to the nature of the participants' interactions and their interpretative processes" (ibid., p. 20). Numeracy practices, in contrast, are said to focus on "the conceptualisations, the discourse, the values and beliefs, and the social relations that surround numeracy events as well as the contexts in which they are located" (ibid., p. 20). Numeracy practices are also said to be "broad notions about the ways numeracy is dealt with in different contexts and settings" (ibid., p. 21).

In addition, Street et al. make an important distinction between *domain* and *site*. Drawing again on previous work in literacy, this time by Barton and Hamilton (1998), they distinguish between 'sites' – as the actual places where the activities take place – and 'domains' – as areas of activity not located in specific places. Applying this to the distinction between home and school provides the 2 × 2 grid shown in Table 1 below:

Table 7-1. Sites and domains of numeracy practices (Street et al., 2005, p. 33)

	Domain: schooled numeracy practices	Domain: out-of-school numeracy practices
School site	Working on number bonds, counting, calculating. Numbers of children away and in class.	Dates, birthdays, aspects of data and measuring, Pokemon cards, money, playground games
Home site	Homework, commercially marketed texts, counting up and down stairs, patterns on car number plates, door numbers	Pocket money, time, laying the table, shopping, setting the video, home discipline, 'symbolic' uses of number systems, 'finger counting', door numbers, jigsaws and calendars

Like Street et al., we are interested in the relationship between home and school mathematics practices, and what happens when children move between them. In an earlier study (Hughes and Greenhough, 1998) we approached these issues by looking at children aged 5-7 years playing a similar mathematical game in two settings, with a parent at home and with a teacher at school. As well as being interested in what this told us about the boundaries between home and school, we were also interested in the ways in which children might or might not make connections across these boundaries. We observed that the children spontaneously made connections between the two settings, for example assuming that the rules of the game were the same in each setting. We also noticed examples of where the adult's lack of awareness of what had happened in the other setting had a significant effect on how the game was played. For example, one child used a measuring ruler as a number line in the school setting, but when she suggested this at home her mother refused on the grounds that it was irrelevant to the activity.

In this chapter we explore these issues further by looking at a 9-year-old boy carrying out a piece of mathematics homework at home. The data takes the form of a transcript of the conversation which ensues when the boy's mother attempts to help him. We will focus in particular on the different worlds which are present in the conversation, and the different ways in which these worlds relate to each other, in an attempt to increase our understanding of the different practices of home and school mathematics, and of the boundaries between them. In so doing, we are explicitly following a suggestion made by Lave[11] that homework can provide an interesting perspective on these issues, "because it moves back and forth between home and school, and actually to the bowling alley, burger bar and so on". In other words, by studying an object such as homework which crosses the boundaries between different communities of practice, we can learn something about those communities in particular and something about boundary crossing more generally.

2. RYAN, HIS MOTHER AND HIS HOMEWORK

In this part of the chapter we present a description of a numeracy event, as defined by Street et al., 2005, involving a 9-year-old boy called Ryan (a pseudonym) and his mother. The event occurs in the living room of the family home while Ryan is doing his mathematics homework. We will first provide a verbatim account of the event as it occurred, and then present an analysis of the event in terms of the different practices and identities involved.

The event was captured on video by Ryan's mother as part of her involvement in the numeracy strand of the Home School Knowledge Exchange project. The overall aim of the project was to develop and implement programmes of home school knowledge exchange activities and look at their impact on children, teachers and parents. The numeracy strand of the project involved children in Years 4 and 5 from four contrasting primary schools in Bristol and Cardiff. In each school six children were chosen for more intensive study, on the basis of gender and attainment, and in-depth interviews were carried out with these children, their teachers and their parents. Ryan was one of these 'target' children, selected at random from a group of low-attaining boys (see Winter, Salway, Yee, and Hughes, 2004, for more details of the numeracy strand of the project).

As part of the family's involvement in the project, Ryan's mother was loaned a video camera and asked to record mathematics events which took

[11] Situated cognition in mathematics, *Seminar held at Oxford University, Department of Educational Studies May 3rd*, 1996.

place in the home. This request was made after a long interview in which the kinds of mathematics taking place at home had been explored. When Ryan's mother returned the camera the tape was mostly filled with the homework event, although it also contained some footage of Ryan and his brother playing games outside.

At the start of the event Ryan is doing his homework on a box file balanced on top of a pouffe. He does not look happy. His mother is sitting on the floor next to him peering over his work. The work is in the form of a sheet headed 'takeaway revision work'

As can be seen from Figure 1, the worksheet consists of a number of subtraction calculations involving two-digit numbers. On the worksheet these calculations are printed in horizontal form, with an empty box in which to place the answer (e.g. 33 - 16 = □). However, Ryan's teacher has also written each calculation in a vertical form

e.g.

$$\begin{array}{r} 33 \\ -\underline{16} \\ \text{—} \end{array}$$

next to the horizontal form. In addition, next to each calculation is an empty number line with the number which has to be subtracted from (the *minuend*) printed at the right-hand end. For the first calculation, Ryan's teacher has added 16 dots and numbers to the number line, counting back from the minuend. These dots and numbers represent the number which has to be subtracted (the *subtrahend*). The answer to the calculation (17) can therefore be read off from the left-hand end of the number line.

The homework sheet thus affords a number of ways of carrying out the calculation. This is consistent with current teaching methods in primary mathematics in England, as laid out in the National Numeracy Strategy (DfEE, 1999). In particular, children are encouraged to develop a range of mental and informal written methods for addition and subtraction calculations before they are introduced to standard written procedures.

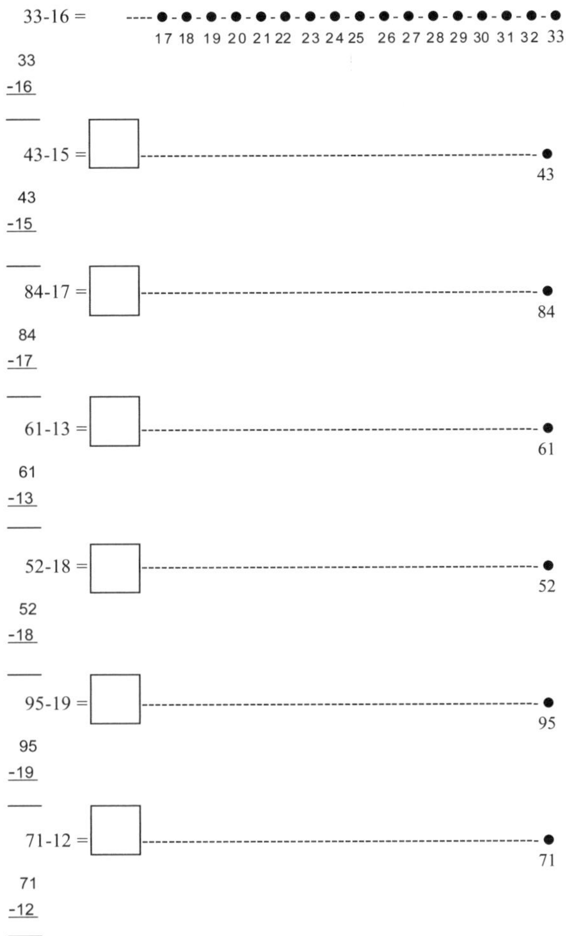

Figure 7-1. Ryan's homework sheet

For subtraction calculations such as these, where the number in the units column for the subtrahend is greater than that for the minuend, the currently favoured standard procedure is one of *decomposition*. This means that 1 is taken from the tens column of the minuend and 10 is added to the units column, as shown below:

3 3		2 [1]3
-1 6	becomes	- 1 6
___		___

However there is an alternative method which was favoured in the past, called *equal addition*. Here 10 is added to the units column in the minuend, while at the same time 1 is added to the tens column in the subtrahend (see below)

```
 3 3                  3 ¹3
-1 6     becomes    - 2 6
 ___                 ___
```

It is not clear which procedure Ryan's teacher wants him to use for these calculations, and there are no instructions on the sheet to provide guidance. Nevertheless, the fact that there are several different ways of carrying out these calculations is crucial for understanding the conversation which follows.

The conversation starts as Ryan is working on the calculation 84 – 17. He has already attempted the first two calculations.

1. M What's that you're doing?
2. C My work (sounds defensive)
3. M What's that? Let's see
4. C It's my work (He uses his arm to cover the part of his sheet he is working on. His body language generally suggests "get out of my face".) (Looks at the video camera.)
   It's on record, mum (defensive and accusatory)
5. M {What's/it's?} take away 15, take away 43
   {{You've} just dropped off one right?}[12]
   No because I just wanted to know if that was the way you were doing it, if it was the same as what I was doing
6. C I do it a different way from you
   (He has now gone back to the first calculation 33 – 16.)
   3 take away 6, I can't do that
7. M (Takes camera off the tripod to get closer to the work.)
8. C (Closes eyes and sighs.)
   {I keep doing them wrong}
   (Puts head on arm.)
9. M Well go on to the next {one} then
10. C Can you stop holding it too close

[12] We use the following conventions in this transcript:
() contains a description of non-verbal behaviour or our comment
{word} shows some uncertainty about what was said
[
[simultaneous speech
.. a slight hesitation or change of direction in what is said
… omission

(Mum has taken the camera off the tripod so that the sheet can be seen more clearly.)
That's why I hate it (presumably referring to the camera/filming)

11. M Go on to the next one then
12. C I *am* (with emphasis and an element of accusation)
13. M Right
14. C (Appears to write a number at the end of the number line next to the calculation 61–13)
15. M Have you no[13] to do this? (pointing to the filled in number line next to the first calculation 33 – 16) Put the same as what.. across {t}here at the top, no?
16. C It's there already for me, Miss done it
17. M Oh that's what it's there for, right
18. C {Mum, you're speaking}
19. M I know
20. C I'm just doing all that, why is that there
21. M I know, because I don't.. I don't understand why you've no put it there, here, there and there (pointing to the empty lines below)
22. C I don't have to put it all down there (argumentatively and upset)
23. M Oh right
24. C It's going to waste all my time.. Miss said
25. M But you're no in any hurry.
26. C (Sort of tuts and puts his arm down.)
Mum, I just want to play out
27. M Well, Ryan, you've got to do your homework first
28. C Can you stop speaking, I can't concentrate
29. M Right, sorry
30. C (By this point he has written 63 next to 61 – 13=)
(Works on the remaining calculations in the vertical format, then transfers the answers to the horizontal format, whispering to self.)
(Seems to finish with a slight bang of the hand holding the pencil.)
(Returns to the second calculation where he earlier completed the vertical format but did not transfer the answer to the horizontal format.)
31. M Right, can I check them?
32. C (rubbing out) I haven't done one (Writes 32 next to 43 – 15 =)
Right
(Bangs fist down on the work, as if to indicate he has finished.)

---

[13] Ryan's mother was partly educated in Scotland.

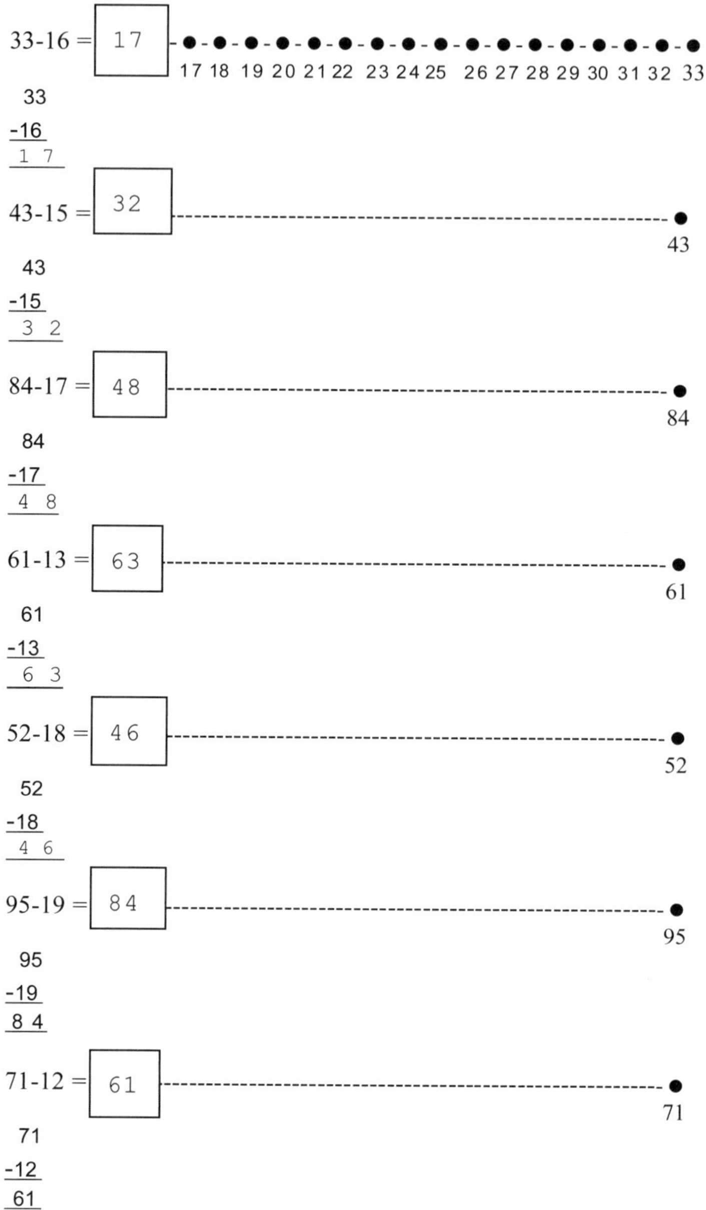

Figure 7-2. The homework sheet after Ryan's first attempt to complete it

Figure 2 shows the answers which Ryan has given to each calculation at this point. As can be seen, only the first one is correct. His most common mistake is simply to subtract the smaller number from the larger number in the units column, instead of using one of the standard methods described above. For example, for 43 – 15 he has subtracted 3 from 5, followed by subtracting 1 from 4, getting an incorrect answer of 32.

33.M That's it, finished?
34.C Yeh
35.M Right, all this.. see this here (Points to the vertical format of 61 – 13)
36.C Yeh
37.M It says 61 take away 13 (Points to the horizontal format.)
38.C Miss put it there for me (Points to the vertical format.)
39.M Oh she's put it there, right
40.C Yeh
41.M To make it easier for you, right
42.C Yeh
43.M Right, well let's have a look. That's.. I don't think that's right is it? That one there (pointing to 43 – 15 = 32 in vertical format) That's
44.C 4, 5, no 4 [{8} 2
45.M [I think you can't.. you can't take 5, you can't take..
46.C You have to take 3 away from 5 (rising intonation) 4, 3, 2. You don't get it, do you?
47.M No, because if I was doing a take away sum, I'd put
48.C (Raising voice, sounds indignant) It's the way I do it
49.M Stroke that, you say stroke that (pointing to 43 – 15 =) and take away one.. a 10
50.C It's the way I do it, we do it a different way
…
They're tens (pointing to the calculation 43 – 15 in vertical format)
51.M That's a 4 (points to the 4)
52.C Tens and units (pointing to the 3)
53.M A unit, so it's.. what.. take one unit away from 4 (rising intonation)
54.C That's a ten, the 4
55.M Yeh
56.C And there's the units, the 3
57.M To take em.. to be able to take 5 away frae 3 you have to put one unit off the 4 and put it onto the 3, do you not?
58.C No
59.M Well why.. you have to
60.C You don't, not in my school we don't, we do it a different way
61.M But it's no.. that's no your answer 32, 15 take away..
62.C I'll do it again then
63.M Let me see, I may be wrong, let me see right, em.. 43 right, take away 15, that's 33.. no, that's not
64.C (Rubs out.) {Let me do} {do a thing then} (truculently)
65.M Right, well that's all I'm doing, asking you to do it

66. C (Looks at the calculation with pencil poised above it.)
67. M The first time you done it right, you crossed off a unit, that's prop.. that's right (The first calculation had a line across the tens part of the upper number.)
68. C (Gets answer of 32 again.) I've got 33 again.. 32, that's the way I do it (Tone has softened somewhat.)
69. M But you stroke one unit off there, OK? (rising intonation, pointing to the 4 in 43)
70. C Oh I get it now
71. M And put one that you get there, yeh
72. C (Puts a line across 4 and writes 3. Puts 1 before the 3 units.) (Hesitates.)
73. M You're able to take 5 away from 13 now
74. C (Sigh) (After a while writes 8 in the units column of the answer, then 2 in the tens column.)
[{I'll have to do this again} (somewhat crossly)
75. M [That's right, 28, you had 48 the first time
Right what about the next one?
76. C (Writes 7 in units column and 6 in tens column of the answer to the vertical version of 84-17.)
77. M Right let's have a look, see if that's proper right
78. C (Rubbing out.)
79. M OK You've got to put.. There's a smaller number taking a larger number away and you're no able to do that, OK, do you understand now?
80. C Yeh (joylessly) (Rubs out the answers to the other calculations ready to redo them.)

This numeracy event might appear at first sight to be somewhat mundane. Ryan is doing his maths homework, his answers to the calculations are incorrect, his mother tries to help him, and as a result he starts using an alternative procedure which provides the correct answers. Yet beneath this mundane appearance the event reveals a good deal about the nature of mathematical practices, boundaries and identities.

2.1 The practice of homework

First, we note that the site of the event is the family home. At the time of the recording, this was not a particularly happy place. Ryan's mother and father were having difficulties in their relationship, and Ryan was undergoing counselling to help him cope with this. He was also having medical problems which seemed to be related to this. However, when we returned a year later the situation had improved considerably.

While the event is taking place in the home site, it does not belong to the home domain (Street et al., 2005). It serves no function within the family, either as a piece of domestic business or as a leisure activity. Instead, the event is a homework task, part of a practice by which an element of school can legitimately enter the home and demand the child's attention. This privileged status of homework is evident in the interchange which takes place on turns 26 and 27, when Ryan says "Mum, I just want to play out" and his mother replies "Well, Ryan, you've got to do your homework first". Here we see a home norm relating to homework within which the mathematics interchanges are embedded. The mother has the power to insist that the homework is done even though she cannot necessarily create a scenario wherein the task is done well. However, her insistence that the homework is done may itself be embedded in interchanges with the school that demand that parents see to it that homework gets done. There is also the society view of what constitutes a good parent, which despite the difficulties in her life Ryan's mother would like to be. For example, in her interview she said about his homework "I do make sure he'll sit and finish it".

While the school expects parents to make sure that homework gets done, it does not seem to encourage parental help or support. There are no instructions on the homework sheet, nor is there any information for potential helpers. Thus Ryan's mother has to infer what the task is, as she tries to do on turn 15. This lack of support (or dialogue with parents) implies that although the task has been sent home, the way in which it is done is still being circumscribed by the school. The ownership and control of the task remain with the school – and specifically with Ryan's teacher – who determines what is to be done and how it is to be done. It is the interactions which have already taken place at school between teacher and child which are intended to count, not those which might take place between parent and child. Thus we can see that the homework task comes into the home with strong boundaries around it which are intended to keep it firmly under the control of the school. However, as we shall see, these boundaries are challenged and renegotiated as the event unfolds.

The strong influence of the teacher on how the task is carried out can also be seen in the interchange which takes place at turns 24 and 25, concerning time. Ryan's mother has suggested that he uses the number line method which the teacher has completed for the first calculation, but Ryan seemingly repeats his teacher's view that this would take too much time. In practice, time is a key aspect when it comes to homework. School homework policies usually focus on time (in terms of how long homework should take for each year group) rather than the actual content of the homework. The teacher therefore has to judge and get right the amount of time the task will take. Filling in the number lines may help provide a way to access the answers but they will be time consuming and have therefore probably been discouraged. The teacher has to operate within a school policy framework and does not want parents complaining to the headteacher that their children spend far too long on their homework.

In fact, the reply given by Ryan's mother on turn 25 – "but you're no in any hurry" – suggests that she is unlikely to subscribe to this view. Her view of time reflects a more out-of-school perspective on time, in which taking/wasting time is only important if you are short of time or are in a hurry or have other things to do. Ryan's mother clearly thinks it is more important that Ryan spends time getting his homework correct than that he should do it quickly and badly.

2.2 Ryan's school and home identities

Bringing the school into the home also means that Ryan's identity in relation to school work becomes visible. At school, Ryan was a low-attainer. According to his class teacher, he had SEN[14] support in class but still found it hard to listen and concentrate. His reading was particularly poor and this spilled over into other subjects. His teacher described him as being a "loveable rogue" who was "very active, likes sport, but doesn't enjoy school work". Another teacher who had taught Ryan for some maths lessons said that he was "not into all this work, he does it against the grain… I like Ryan but there's not a lot there, maybe".

This picture of Ryan struggling with school work was supported by observations of him in class. During a lesson on percentages Ryan was seen to be having difficulty understanding throughout the lesson, and there was little evidence by the end that he had grasped the basic ideas. However he tried to be helpful to the teacher, for example by sorting out a problem with the lead for the OHP projector.

As part of the project, Ryan had a few months before the video completed a self-report questionnaire on his attitude to mathematics. On a five-point scale, he gave the most negative response to over half the questions. For example, he said that he "hated maths", found it "really hard" and thought he was "really bad" at it. However there were some areas where he was more positive, such as working out money problems and measuring.

When interviewed a year after the homework event took place Ryan was asked whether he thought he was different at home compared with school. He replied "loads". He went on "(at home) I forget about everything, I just forget about school and play". He thought that "in school I'm one person but

[14] Special Educational Needs

when I come home I'm another person…naughtier at home than around the school". He didn't think his teacher knew what he was like at home and didn't want her to know more about his home life.

In engaging with a school task at home, then, it is likely that Ryan was bringing with him an identity as someone who was struggling at school. Certainly he gave no indication of getting any enjoyment from the homework task; rather, it was an unpleasant chore to be completed before he could go off and play. His comments at turns 4 and 10 also suggest that he was not enjoying being filmed, unlike other children in the study who welcomed the opportunity to be the centre of attention. Ryan, in short, was a reluctant participant in this particular numeracy event.

2.3 Ryan's mother and mathematics

We also need to consider Ryan's mother's identity in relationship to maths. When interviewed she made clear that her view of herself and maths is not singular – it depends on which aspect of maths is being considered. She says that at school she was good at her tables but she could not get long division into her head. She is not good at measuring or fractions. She is, however, good at budgeting and this includes the decision-making about which bills to pay as well as the mathematics.

Ryan's mother reported that while she tried to help Ryan with his maths homework, she was often unable to do so and felt frustrated and 'thick' as a result: "I don't know if it's just the way they pronounce some things and he's explaining it to me and I just hav'na a clue and I just can't help him". She felt that much of this was due to her being taught mathematical procedures differently from Ryan:

Mother: I can read it out to him but he always says I'm wrong because I'm not doing it properly.. so.. and we end up at loggerheads and I just.. I think well you need to just take it back to your teacher and say you can't do it… "oh" she says, "I've showed him and I've showed him and I've showed him, but he just doesnae seem to take it in".

Interviewer: So do you think that you *are* doing it a different way?

Mother: Oh, definitely. I had.. see that's when I went to a meeting, the other week about the maths and everything, it's like you'll do your take away sum.. we used to do 10 to the top, 10 to the bottom, and she showed me, the teacher, you take 1 off the 8s it was and it came as 7 and you put that on there, the others. It was entirely different. But yet his dad does it the same.

These comments make clear that Ryan's mother was taught the process of equal addition when she was at school, although she had recently learnt Ryan's decomposition method from his teacher. They also suggest that Ryan is not slow to point out to her when he thinks she is using methods which are different from those of his school.

2.4 Tensions and conflicts during the homework event

We can now return to the homework event in the light of the above remarks on practices and identities. Throughout the event we can see tensions and conflicts emerging as Ryan's mother tries to help and Ryan responds in various ways to her attempts. Thus right at the start of the event (turns 1 - 4) we can see Ryan's initial defensive response to her interest, suggesting he does not find it welcome. On turn 5 Ryan's mother justifies her interest in terms of wanting to see whether they were both using the same methods, which we now know was an ongoing issue between them. Ryan responds on turn 6 by emphasising this difference, suggesting that he is using the difference to try to keep his mother at bay. However he is ambivalent here, as he recognises that he is stuck ("I keep doing them wrong" on turn 8) and will have to allow his mother into the domain of his homework. This is not easy: as we have already noted, it is not at all clear how the homework task is meant to be tackled, or how a parent might help, and Ryan is clearly reluctant – or maybe unable – to provide an adequate explanation for his mother.

After Ryan has completed (incorrectly) the calculations for the first time (see Figure 2) his mother takes on a new role, that of checking his answers are correct (she says "can I check them" on turn 31 and "let's have a look" on turn 43). This leads to further tension and conflict. Thus on turn 43 she somewhat hesitantly suggests that Ryan's answer of 43 – 15 = 32 may not be correct, and says "I think you can't.. you can't take 5.. you can't take". Here we can possibly hear a voice from the time when she herself was a child in the maths classroom: part of the mantra for the take away calculation decision making is the recognition of 'can't' if the number of units in the subtrahend is greater than in the other number, the minuend. Ryan's response to this ("you have to take 3 away from 5") has something everyday or matter of fact about it: if you can't do something one way, find another way to do it. At the same time he accompanies this with a derogatory accusation of his mother's ability to understand – maybe reflecting times when she has admitted not understanding the mathematics in his homework. He also calls on the authority of his school to emphasise the difference and justify his position ("It's the way I do it, we do it a different way" on turn 50).

The sense of conflict here may also be heightened by the rather unusual language which Ryan's mother is using to describe her method – she says "you say stroke that" on turn 49 (and again on turn 69) using a phrase with which Ryan is probably unfamiliar and which he may see as coming from another world. (It is interesting that she refers here to the physical action of putting a 'stroke' through a number, rather than seeing it as a mental process.) There is also an imprecision about her language which might well add to Ryan's confusion. For example, on turns 53 and 57 she talks about taking a 'unit' from the 4 in 43, although in fact it is a 'ten'. Indeed, Ryan corrects his mother at this point (turns 54 and 56) pointing out that the 4 is a 'ten' and the 3 is a 'unit'. This may explain why he thinks she does not understand his decomposition method, although it is becoming clearer around turn 57 that she is in fact suggesting the same method as used in Ryan's school. Nevertheless Ryan still resists this, and again calls on the authority of his school to justify his position. The nub of the conflict is revealed in stark terms in the following interchange:

59. M Well why.. you have to

60. C You don't, not in my school we don't, we do it a different way

Ryan's mother persists with her belief that Ryan's answer is incorrect and on turn 63 tries a different approach. She is somewhat hesitant here – "I may be wrong" – but perhaps surprisingly Ryan accepts her judgement that he has got the answer wrong and starts to rub out his answer. It is noteworthy that the method she uses to check accuracy is actually a mental calculation which starts by taking 10 of the 15 from 43. At this point she can see that the child's answer is incorrect since she is already just about at the same number as his answer (33 compared with his 32) and she still has more to subtract. What is interesting here is that she does not use the method talked about earlier involving 'stroking' tens and so on. Rather she uses a more informal method involving a mental calculation of the kind which is encouraged within the National Numeracy Strategy, although she is presumably unaware of this.

Despite the confusion and conflict, something has been communicated to Ryan and on turn 70 he says "Oh I get it now". This comment is justified by his subsequent behaviour, when he uses the decomposition method to complete correctly the calculation 43 – 15 = 28. However, his negative mood is not improved by this success. He states crossly on turn 74 that he has to repeat the rest of his work and on turn 80 joylessly admits that he now understands what he is doing. Perhaps he is more aware that not only did he fail to keep his mother out of his homework world but that he now has to repeat all his work – thus delaying even further the moment when he can go off and play.

3. DISCUSSION

Our analysis suggests that, beneath the surface of this particular homework event, the presence of a number of different worlds can be detected. Thus the event is an exemplar of the wider practice of homework, a practice which allows the school domain to legitimately enter and occupy the home site. With the practice comes a range of identities and presences. From the direction of school, we have Ryan's school identity as a low-attaining pupil with strong negative feelings towards mathematics; there are also the presences of his class teacher, the architects of the school homework policy, the publishers of the homework sheet and even the writers of the mathematics curriculum being used at the time. From the direction of home there is Ryan's home identity as someone who wants to forget about school and just play; there is also Ryan's mother and the different identities she brings – as helper, checker and enforcer of homework – and as someone with her own strong and ambivalent feelings about maths. We can even detect the presence of her own experiences of learning mathematics despite their taking place at least 20 years previously. In addition, we should not forget the presence of the research team, represented through the video camera which records the event with an unforgiving detachment.

As we have seen, these identities and presences do not co-exist harmoniously. There is a great deal of conflict and tension, as the various identities negotiate with and challenge each other. Moreover, this challenge is not present in every aspect of the interaction. For example, Ryan does not challenge his mother's insistence that he has to finish his homework before he can go out to play, possibly because he knows from experience that when his mother and the school are lining up on the same side he has ultimately little option. Instead, he vigorously challenges his mother's understanding of mathematics, calling on the legitimacy of his school to justify his own incorrect methods and to overrule his mother's attempts to persuade him otherwise. Thus we can see the clear presence of what Street et al. call issues of "control, legitimacy, status and the privileging of some practices over others in mathematics" (p. 17).

Unfortunately, it seems that the conflict and tension identified in this particular homework event are not atypical – either of Ryan or of homework more generally. As we saw earlier, Ryan's mother reported that they were frequently 'at loggerheads' over homework, as he regularly challenged her understanding of his school mathematics. In a wider study of homework (Hughes and Greenhough, 2002) we also found that homework frequently engendered heightened emotions between parents and children, as parents tried to make sure homework was completed or struggled to find ways of helping their children: as one parent commented "we often end up at screaming pitch". Similar tensions around homework have also been reported by Solomon, Warin, and Lewis (2002).

To what extent does our analysis of what is going on in this event relate to Wenger's (1998) framework for discussing communities of practice? We would suggest there are several fruitful areas of interplay.

First, the event can be seen as taking place at what Wenger terms a 'boundary' – in this case between home and school. At the same time, the event shows that this boundary is not a static or straightforward entity, but one which is dynamic and constantly being negotiated and renegotiated. A key factor in this negotiation is Ryan's ambivalence between wanting to keep his mother out of the world of his school work, and wanting her in so that she can help him get the correct answers. He thus oscillates between having the boundary drawn tightly around him and his work – indeed at more than one point he creates a physical barrier with this arm between his mother and his homework sheet – and opening it up to allow his mother entry into the school domain.

If Ryan and his mother are operating at the boundary between home and school, then is it appropriate to describe the homework sheet as some kind of 'boundary object'? In some ways it is. The homework sheet appears to play a similar role in this event to the claims processing form described by Wenger in his study. It is a physical object – in Wenger's terms, the product of 'reification' – which has the potential to connect up different practices by moving in time and space between them. At the same time, the potential of this particular sheet to connect up home and school is very limited. As we have already observed, there are no instructions on the sheet or suggestions of ways in which parents might help. There is no attempt to translate the decontextualised mathematics of the subtraction calculations into an activity more familiar from the home domain (e.g. turning the subtractions into problems about shopping and money). Again, our previous research on homework suggests this is not atypical: homework has the potential to link home and school but for the most part this potential is not realised (Hughes and Greenhough, 2002).

In addition to boundary objects, Wenger describes the process of 'brokering' as another means by which connections can be made between communities of practice. As we saw earlier, a broker is essentially someone who is a member of two (or more) practices and uses this multimembership to make positive connections between the practices. In the homework event, Ryan is clearly a member of both the home and school practices, and potentially could use this – as other children might do – to create links between them. In reality, as we have seen, Ryan has little desire to do this. He would prefer the practices to be kept separate, and so his role is more often one of 'blocker' than 'broker'.

In contrast, it is Ryan's mother who is trying to play the role of broker in this event. She wants to bring whatever understandings she has about mathematics to help Ryan with his school work. Her problem, however, is that she is not a member of the school community and so lacks valuable information about how the school expects the calculations to be done. As she admitted in the interview, she had tried to overcome this lack of knowledge by attending a meeting at the school about the methods used to teach mathematics, but her knowledge was still patchy. This, together with her own lack of confidence and Ryan's low opinion of her understanding, meant that her attempts at brokering frequently foundered.

It is also interesting to look at the homework event in the light of Wenger's ideas about identity, and in particular his view that identity should be seen as a 'nexus of multimembership' which involves the important work of 'reconciliation'. As we indicated earlier, both Ryan and his mother bring several facets of their identities to the homework event. For Ryan, though, there is little sign that the process of reconciliation has made much headway, if any. His interview comments make clear that he thinks he is very different at home and at school, and that when he is at home "I just forget about school and play". In contrast, Ryan's mother is more complex. Again there are several facets of her identity in evidence, such as her role as 'good parent', and her lack of confidence around maths, but these are not always working harmoniously together. Moreover, although she reports in interview that she has contemplated taking courses to improve her ability with mathematics, she has been inhibited from doing so by her perception that everyone in the class would be 'more intelligent' than her. Thus while Ryan's mother has considered taking action that would help to reconcile aspects of her identity, her lack of self-confidence has prevented her from doing so.

Finally, we turn to the implications for mathematics education. No doubt there will be many mathematics educators who will find the content of this homework event somewhat depressing. The child is unhappy, and has a negative attitude towards many aspects of mathematics. The task is mundane, and makes no connection to real-life contexts or to his out-of-school life. The interaction between mother and child, although ultimately leading to the child adopting a correct procedure, is negative and bad-tempered. There is little appeal to mathematical principles to resolve disagreements, but instead regular references to power and legitimacy to decide which procedure should be adopted.

How might such a situation be improved? One suggestion would be for a fuller implementation of the principle, embodied in the National Numeracy Strategy, that children should be made aware that there are a range of different methods – all equally appropriate – for carrying out particular calculations. We do not know enough about Ryan's classroom to say whether or not he had been properly introduced to this principle, but if he had then he had clearly not internalised it. As we have seen, much of his difficulty with the homework stems from his reluctance to accept that there might be more than one way of doing it.

We would also suggest two further areas where practical steps could be taken to improve the interaction around mathematics which takes place between children and parents at home. First, there is much which can be done to improve the nature and quality of homework tasks. This would, however, require some fundamental rethinking about the purposes of homework and the role which parents – as well as family and peers – might be expected to play in the process. Thus if homework continues to be seen as a practice whose main purpose is to reinforce and extend the school curriculum, with the assumption that it will be carried out independently, then unstimulating and opaque worksheets such as Ryan's will continue to be sent home. If on the other hand, homework is seen as a genuine way of making connections across home and school practices, involving other family members and peers in collaborative problem-solving, then it will lead to very different homework tasks and interactions around homework. For example, in our previous research on homework (Hughes and Greenhough, 2002) one class of students was set a mathematics assignment which required them to locate a number of items (like cosmetics) which were still in their original packaging. The students were asked to construct a chart showing the overall volume of the goods purchased as a percentage of the overall volume of the package. The students found this task quite engaging and commented afterwards on how revealing it had been. In particular, it had enabled them to see how mathematics might be relevant to an out-of-school practice such as shopping.

In addition to rethinking the nature and purposes of homework, schools can also do much to reconceptualise their relationships with parents and the ways in which parents can support their children's learning. Many – if not most – parents share Ryan's mother's desire to help their children with their school work, in mathematics as well as other areas of the curriculum. At the same time, many parents may lack the knowledge and/or confidence to provide the most appropriate forms of support. In the numeracy strand of the Home School Knowledge Exchange Project we worked with schools to develop ways in which information about teaching methods and mathematics topics could be shared with parents. At the same time we developed activities where the exchange of knowledge between home and school was in the opposite direction, from home to school. For example, children were given disposable cameras and asked to take photographs of activities involving 'everyday mathematics' – such as card games, cooking or shopping – in which they had been involved outside of school. A full account of these activities and their impact on children, parents and teachers can be found in Winter, Andrews, Greenhough, Hughes, Salway, and Yee (forthcoming).

In conclusion, we have attempted in this chapter to show how mathematics homework can be the source of tension and conflict, and that this tension and conflict tells us something important about the various practices and identities which are present in the homework event. At the same time, we have tried to demonstrate the value of looking at the relationship between home and school in terms of Wenger's ideas about boundaries, boundary objects, brokering and the need to reconcile different aspects of identity. More generally, we have tried to show the importance of seeing the learning of mathematics as a social activity embedded in various practices which are not always in harmony. While we may not welcome such lack of harmony, we need to recognise it and learn from it.

ACKNOWLEDGEMENTS

This chapter draws on data collected as part of the Home School Knowledge Exchange Project (ref no L139 25 1078) which was funded by the UK Economic and Social Research Council as part of its Teaching and Learning Research Programme. The HSKE project team consists of Martin Hughes (project director), Jane Andrews, Anthony Feiler, Pamela Greenhough, David Johnson, Elizabeth McNess, Marilyn Osborn, Andrew Pollard, Mary Scanlan, Leida Salway, Vicki Stinchcombe, Jan Winter and Wan Ching Yee. We are particularly grateful to Jane Andrews for the observations of Ryan at school. The chapter was written while the authors were supported by an ESRC professorial fellowship (ref no RES 051 27 0092) awarded to Martin Hughes.

REFERENCES

Barton, D., & Hamilton, M. (1998). *Local literacies: Reading and writing in one community*. London: Routledge.

Brown, J. S., Collins, A., & Duguid, P. (1989). Situated cognition and the culture of learning. *Educational Researcher, 18*(1), 32-42.

Department for Education and Employment. (1999). *The National Numeracy Strategy*. London: DfEE.

Greenhough, P., & Hughes, M. (1998). Parents' and teachers' interventions in children's reading. *British Educational Research Journal, 24*(4), 383-398.

Heath, S. B. (1983). *Ways with words: Language, life and work in communities and classrooms*. Cambridge: Cambridge University Press.

Hughes, M. (1986). *Children and number*. Oxford: Basil Blackwell.

Hughes, M. (2001). Linking home and school mathematics. In M. van den Heuvel-Panhuizen (Ed.), *Proceedings of the 25th International Group for the Psychology of Mathematics Education* (pp. 5-8, Vol. 1). Utrecht, Netherlands: PME.

Hughes, M., & Greenhough, P. (1998). Moving between communities of practice: Children linking mathematical activities at home and school. In A. Watson (Ed.), *Situated cognition and the learning of mathematics* (pp. 127-141). Oxford: University of Oxford, Centre for Mathematics Education Research.

Hughes, M., & Greenhough, P. (2002). *Homework and its contribution to learning*. Final report to the ESRC, http://www.esrcsocietytoday.ac.uk/ESRCInfoCentre/index.aspx

Lave, J. (1988). *Cognition in practice*. Cambridge: Cambridge University Press.

Lave, J., & Wenger, E. (1991). *Situated learning: Legitimate peripheral participation*. Cambridge: Cambridge University Press.

Solomon, Y., Warin, J., & Lewis, C. (2002). Helping with homework? Homework as a site of tension for parents and teenagers. *British Educational Research Journal, 28*(4), 603-622.

Star, S. L., & Griesemer, J. (1989). Institutional ecology, 'translations' and boundary objects: Amateurs and professionals in Berkeley's museum of vertebrate zoology, 1907-1939. *Social Studies of Science, 19*, 387-420.

Street, B., Baker, D., & Tomlin, A. (2005). *Navigating numeracies: Home/school numeracy practices*. Dordrecht: Springer.

Tizard, B., & Hughes, M. (1984). *Young children learning*. London: Fontana.

Walkerdine, V. (2007). *Children, gender, video games: Towards a relational approach to multimedia*. Basingstoke, UK: Palgrave Macmillan.

Watson, A. (Ed.). (1998). *Situated cognition and the learning of mathematics*. Oxford: University of Oxford, Centre for Mathematics Education Research.

Wenger, E. (1998). *Communities of practice: Learning, meaning and identity*. Cambridge: Cambridge University Press.

Winter, J., Andrews, J., Greenhough, P., Hughes, M., Salway, L., & Yee, W. C. (in press). *Improving primary mathematics linking home and school*. London: Routledge.

Winter, J., Salway, L., Yee, W. C., & Hughes, M. (2004). Linking home and school mathematics: The home school knowledge exchange project. *Research In Mathematics Education, 6*, 59-75.

Chapter 8

Situated Intuition And Activity Theory Fill The Gap

The cases of integers and two-digit subtraction algorithms

Julian Williams, Liora Linchevski, and Bilha Kutscher
University of Manchester, Hebrew University of Jerusalem, The David Yellin Teachers' College, Jerusalem.

Abstract: We report here an instructional method designed to address the cognitive gaps in children's mathematical development where operational conceptions give rise to structural conceptions, such as when the subtraction process leads to the negative number concept. The method involves the linking of process and object conceptions through semiotic activity with models which first record intuitive processes on objects in situations outside school mathematics – invoking situated intuition – and subsequently mediate new mathematical activity, with mathematical signs, in the mathematical voice. We ground this in teaching experiments focused on (i) the negative integers and (ii) algorithms for two-digit subtraction. We conceptualise modelling as the transformation of outside-school knowledge into school mathematics, and discuss the opportunities and difficulties involved.

Key words: modelling (RME), situated intuition, primary pedagogy

1. INTRODUCTION

While much of the work on situated intuition and the case of the integers in this paper was done during the late 1990s (Linchevski and Williams, 1996, 1997; Williams and Linchevski, 1997) within a situated cognition and activity theory perspective, the work since then has broadened to include further significant empirical studies using the same instructional method, by the authors with others (Koukouffis and Williams, 2005; Kutscher, Linchevski, and Eisenman, 2002; Linchevski and Williams, 1999). In addition, however, new theoretical work has developed activity theory in ways that help us to better understand the semiotics involved: in particular we have come to view learning in the zone of proximal development as 'modelling' at the boundary between familiar practices, including outside-school practices invoking situated intuitions, and the new mathematics to be learnt (Koukouffis and Williams, 2006; Ryan and Williams, 2007; Williams, 2005; Williams and Wake, 2007a, 2007b). The presentation of this chapter subsumes the work in Linchevski and Williams (1997), but is modified, with extra empirical results, and a discussion of this new theoretical position.

At the core of our work is an instructional method aimed at helping to overcome the cognitive gaps in children's extensions of their number schema. Sfard (1991) identified major intuitive gaps in children's cognitive development with hurdles she observed in the historical development of mathematics, where mathematical processes had to be transformed into mathematical objects. This transformation historically involved long periods in which processes were constructed, encapsulated and finally reified; the final stage was a relatively sudden ontological shift which occurs when the familiar processes are finally understood to be mathematical objects in their own right. This transformation involves an unavoidable paradox; it requires that mathematicians manipulate the processes instrumentally as objects before they are able to mentally grasp them as such. (See also Gray and Tall, 1994; Sfard and Linchevski, 1994.) Our approach addresses this paradox; we claim that with the benefit of hindsight, one may find learning pathways which do not recapitulate the historical or logical development of mathematics.

In our instructional method, we build on Realistic Mathematics Education (RME) developed by Fruedenthal's followers at the Freudenthal Institute (Gravemeier, 1994; Treffers, 1987). We appeal to children's everyday common sense and intuition. Meaning is given to the manipulation of objects through the siting of the objects within familiar contexts. The children model with their existing number concept, but the context supports the development of new concepts and strategies. We see the emergence of the new numbers in the children's active organisation of the phenomena which "beg to be organized" (Freudenthal, 1983, p. 32). We help the learner to organise the phenomena presented by providing tools and models, and by introducing mathematical signs. This provokes a transformation of the children's discourse as their everyday conceptions are structured by the mathematical language. Finally, the children operate on the mathematical signs as such, in the 'mathematical voice' (in the sense of Wertsch, 1991).

It will be observed that such a strategy involves two related transformations: one of knowledge and one of language. First let us consider the transformation of knowledge. We expect the children to bring knowledge constructed in situations outside school, and to make use of this intuitively in classroom tasks. As with Realistic Mathematics Education and ethnomathematics we select contexts which are designed to recall these outside-school situations. The contexts we select are considered experientially real, or 'authentic' if they relate directly to the children's culture, they are essentially true to life and make sense to the child (Barton, 1996). Moreover, we look for situations that are likely to provide new affordances for fostering computational strategies (Linchevski and Schwartz, 2001; Williams and Linchevksi, 1998).

However we wonder if the authenticity of the outside-school situation often survives the transfer into the classroom situation. The transfer of knowledge across social situations is known to be problematic (Lave, 1988). When we introduce 'realism' into the classroom we cannot recreate exactly the social situation in which the children experienced the 'reality' outside school. Consequently, we might fail to evoke the essential intuitive knowledge of the real experience. So we look for conditions in which essential congruences exist, i.e. when productive intuitions might transfer or be appropriately transformed through classroom practice.

Situated learning provides one perspective on the authenticity of activity. Lave and her coworkers (Chaiklin and Lave, 1993; Lave, 1996; Lave and Wenger, 1991) locate authenticity exclusively in communities of practice, mostly outside school. In these outside-school communities of practice learning is implicit, picked up along the way, 'stolen' from old hands on the job without much explicit instruction. Some, inspired by this perspective, have tried to generate authentic classroom activity by replicating examples of such practices in the classroom (for instance those described by Brown, Collins, and Duguid, 1989; Heckman and Weissglass, 1994). We too design activities which engage the children in simulations or replications of familiar practices from outside school which seem to us to have mathematical potential. But we believe that authentic classroom activity can and should be developed in which mathematical goals are, at a certain point, *explicit*. Why?

We believe that classroom activity can only be understood in the context of the social system in which it is accomplished, in the tradition of the Vygotskian school and modern sociocultural theorists (Engestrom, 1991; Cole, and Engestrom, 1993; Leontiev, 1981; Wertsch, 1991). Thus children, teachers, and parents understand that the purpose of classroom activity is that children learn: the goals of learning are explicit and so not usually congruent with outside-school practices. Indeed, the authenticity of much of the activity of the mathematics classroom *requires* that the children learn new mathematics. The use of contexts which refer to outside-school situations must be validated by their success in supporting 'authentic' schooling activity in the classroom, in which the tasks presented, the pedagogical tools, and the teacher can and usually does play a critical role.

Let us now consider the second transformation, i.e. that involving 'language'. We must expect shifts in meaning when children call on everyday knowledge and language in new situations such as the classroom (Walkerdine, 1988). To facilitate *appropriate* shifts in meaning, we introduce models, mathematical language and signs to help the children to organise and transform their everyday, intuitive knowledge. In our approach, models and the children's already existing mathematical language act first as mediating tools for children to make sense of the classroom activity. Later the models are used as tools for mediating actions on new mathematical signs. Thus they form the essential links in a chain of signification from the children's intuitive knowledge and existing mathematical knowledge to *new* mathematics. Both the model and the mathematics first mediate other activity, and become later the *object* of conscious action (Leontiev, 1981). This shift in attention is caused by a deliberate shift in the role of the model in different tasks, which in the RME literature is referred to as shifting from model 'of' to model 'for' (Gravemeijer, Cobb, Bowers, and Whitenack, 2000).

To summarise: we developed our instructional method from RME by a) focusing on the intuitive gaps in children's mathematical development which we think call for the use of extra-mathematical knowledge to support reification, b) questioning the authenticity of 'realistic' contexts put forward in the classroom situation, and c) identifying the need for shifts in meaning in the classroom tasks put forward, from intuitive meanings to mathematical meanings. We seek to improve on the use of 'realistic' classroom contexts by the extent of the engagement of children's experience of familiar outside school practice involved. But we also depart from some situated learning approaches in that we want to engage the children in authentic learning in which the mathematical goals become explicit, whereby the 'authenticity' of mathematics is recognised for itself as a valid, scientific schooling activity.

2. IMPLEMENTING THE INSTRUCTIONAL METHOD

In the following sections we will describe two experiments in which this method was developed for teaching: one for the teaching of integers and the other for the teaching of the subtraction of two-digit numbers. We thereafter compare the two experiments. The first experiment originally involved year 6 and sixth graders in the UK and Israel and focuses on the concept of negative numbers, addition and subtraction, and the order-relation among them. The second experiment involved first graders and focuses on the subtraction of two-digit numbers. By choosing two very different cases we try to illustrate the scope of our approach and its potential.

2.1 The teaching of integers

Research on learning integers has an interesting history. Traditionally negative numbers introduce a new aspect into the study of mathematics: in Freudenthal's view reasoning in an algebraic frame of reference seems to be required for the first time. While counting numbers are constructed by abstraction from real objects and quantities, and operations performed on them are related to concrete manipulations, operations on negative numbers and the properties of these numbers are traditionally given meaning through formal mathematical reasoning. Moreover, some of these properties contradict intuitions for the counting numbers. This situation has led some in the mathematical community to look for an embodiment, a 'model' that will satisfy the need for the negative numbers, and will justify the arithmetical operations on them, and the relations between them. In this literature, reviewed in Linchevski and Williams (1996) (see particularly Semadeni (1984) and Liebeck, (1990)) the term 'model' sometimes implies a manipulative aid (such as a chart, or the double abacus) and at others refers to some situation in which 'intuitive' knowledge can be used (such as balloons with weights attached). In general it does not imply a 'real experience' and in fact the use of manipulatives and the rules for handling them are often ready-made and unjustified.

Fischbein (1987) argued against using the existing models for negative numbers. He said they lack 'comprehensiveness', are based on artificial conventions and so do not address the cognitive obstacles confronting the students. The purpose of a model is to add 'obviousness' and 'correctness' to mathematical concepts and operations on them, but this purpose is not achieved by them. He therefore concludes that the topic of negative number should be taught only when the students are ready to cope with intra-mathematical justifications. In our terms, then, he recommends that the teacher avoid any attempt to give 'out of school' meaning to the negative numbers.

While disputing the argument that any single model can or should be comprehensive, we accept the requirement that the models we put forward should be 'obvious'. Situations and models must describe a reality that is meaningful to the student, in which the extended world of negative numbers already 'exists' and the students' activities serve to discover it. This world must include the practical need for two sorts of numbers, and the relevant laws must be deducible without mental acrobatics. How? Following earlier research, our teaching was based on the neutralisation of equal amounts of opposites, and every integer had many physical representations on a double abacus (Dirks, 1984; Lytle, 1994). Clearly the double abacus affords representation of the two kinds of numbers, and allows addition and subtraction of the integers, though not multiplication and division: so the model cannot achieve 'comprehensiveness'. Also, the integers so modelled are based on an extension of the children's existing cardinal schemes.

However, in contrast to previous studies we wanted the integers and the operations on them to be constructed intuitively. The difference in heights of the stacks of beads on the two wires of the double abacus (see the figures below) is the feature to which attention must be directed, so that one blue and four yellow beads on the two wires, (1,4) say, must be seen as *obviously* the same as (2,5) and so on to other pairs in the equivalence class that essentially defines the integers as an extension of the natural numbers. The obviousness of the equivalence class needs then to be extended to the addition and subtraction operations on them, so that (2,5) subtract (0,2) is obviously just (2,3), because this subtraction just corresponds to 'taking away' 2 yellow beads from the 4 on the abacus. Finally the introduction of the 'mathematical voice' leads to a symbolization of this operation on the abacus as just (-3) subtract (-2) = (-1). Our method in the following experiment followed just this procedure.

2.1.1 Methodology of the integers game

The study involved teaching sequences, each of about 5 one-hour-long sessions, with groups (of three children at a time in the first experiment and four in the second experiment) of grade 6 pupils who had not yet received any instruction in operations on negative numbers. The final version was gradually developed: the early groups helped us to develop the tasks, which were refined in the light of the pupils' reactions. We researchers were both teachers and interviewers. The final sequence led the children to construct the integers and operations of addition and subtraction. It was repeated with several fresh groups of children (in the first experiment three groups, and in the second experiment, four groups, two in England and two in Israel). In a recent replication and extension of the study, a similar approach was developed with grade 5 children with 4 groups of four children each (see Koukouffis and Williams, 2005; Koukouffis and Williams, 2006). All the meetings were videotaped to allow further analysis.

Because we made it an aim that the teaching should construct the integers on an intuitive foundation, in Fischbein's sense, we always sought to identify intuitive 'gaps' in the children's progress, and for ways to overcome them. But we were also particularly looking for shifts in the use of language, of the abacus and of mathematical signs in the tasks presented, and how these facilitate mathematisation.

Figure 8-1. Playing the Integers game

2.1.2 The Integers game: making integers and the operations on them concrete

The first context we tried for developing intuitions for integers involved a context where disco-dancers arrive at a disco and are recorded at each gate (for a full description, see Linchevski and Williams, 1996). The total number of children entering and leaving the disco is important because, when full, no more children should be allowed in (something to do with fire regulations!) A simulation game then is played with cards that tell of the action at each of the disco gates: "3 in" "4 out" etc. A double abacus is set up at each gate and keeps record of what has happened at that gate, thus if "2 in" then "4 out" an abacus would record this with 2 blue beads and 4 yellow beads, say (2,4). Periodically, the children are required by a card turning up in the game to work out, by combining the scores on the abacus, what the total entry is, and if regulations have been broken. This requires the addition of the 'scores' on the different gates' abacuses, such as (2,4) + (10,3) = (12,7) which eventually can be transformed into the addition of integers.

The context was evocative of intuitions that helped the children to justify actions on their abacuses: the cancellation of equal numbers of the two colours of beads on an abacus was obvious because 'if three go in and three go out it's the same thing'. When the abacus filled up with beads this proved an essential strategy so that the game could continue, although some other solutions were proposed as alternatives, such as 'add it to the other gate's abacus' instead of this one, etc.

In addition, we identified an elision of the 'cancellation' strategy that we came to call the 'compensation' strategy: that is, when "3 in" had to be recorded, normally by adding 3 beads to the (full) blue column, say, then three would be taken off both columns (cancellation strategy) and the three blues put back on. Elision of the process led to the solution, 'instead of adding 3 blues just take away three yellows' and vice versa: this is justified because if you add "3 outs" that is the same as taking off "3 ins". We looked to this compensation strategy for the intuitive basis for a formulation that "instead of adding +3 you can take away -3" and vice versa. In this context however, we found two weaknesses that we could not resolve: first, the context seemed to us to be a useful simulation, but their 'outside experience' of a disco lent only little to their intuitive reasoning, and we wondered if other contexts might be more vivid for the children; second, and perhaps more important, we wanted crucially to provide an intuitive base context for subtraction of the integers. Beyond the 'taking away' of the beads on the abacus, we sought intuitive meaning for this in the context itself. Why would one 'take away' the nascent integer, represented as a score on a gate? In this case we did this in an indirect way by 'inverting' the addition of the cards from the score on the gate: what was the score before those "3 ins" and "4 outs". We therefore designed a new experiment based on 'dice games'.

This experiment involved groups of four children playing a series of simple dice games in teams of two, recording points scored on the throw of dice (Williams and Linchevski, 1997). In the *first game*, a pair of dice is thrown, one yellow and one blue, which score for the yellow and blue teams respectively, and are recorded on the abacuses. The abacus has two wires, one with yellow beads for scoring the points on the yellow die and the other with blue beads for the blue die. We gave each pair of children an abacus, so the two abacuses had to be combined to get the score for each team. Occasionally the children are asked to report their abacus 'score' to the group. The team which gets 8 points ahead wins (10 or 12 for a longer game).

The limitations of the abacus provoke a crisis when the beads on one of the wires are all used up before the game ends. This crisis pushes the pupils to develop strategies based on 'equivalent' abacuses. After Liebeck (1990), we planned to encourage the idea of 'fairness', and asked the teams to justify their strategies on these grounds. The teams in every case developed a cancellation strategy on the dice, e.g. a throw of two blue and three yellow is scored as one for the yellow team. However, in *every* group the *compensation* strategy was the strongest intuitive strategy for dealing with the abacus when one of the wires is full (i.e. take from the blues instead of add to the yellows and vice-versa). Typically this was justified by Lena: "We can have 5 yellows, or we can take away from the blues instead of adding to the yellows". When pressed by the teacher, the children justified it intuitively: "It doesn't make any difference... it's the same difference", and "Well, yes, it's fair,... we have been doing it to them."

In an attempt to make subtraction concrete, in *game two* we added a third die, with all its faces labelled either 'add' or 'sub'. We asked the children to first throw the blue and yellow dice, decide the score which resulted, and then throw the add/sub die and either add or subtract this score accordingly. Game 2 (and also game 4, see below) begins with the abacus showing six of each colour bead so that the teams don't run out of beads immediately; this ensures an opportunity for the subtraction to be established as the taking away of beads before a 'crisis' arises. (In fact because compensation arises so intuitively the children quickly abandoned this rule.)

In the *third game*, we discussed with the children replacing the blue and yellow dice with a single die, labelled with integer signs +3, +2, +1, -1, -2, -3 : we asked them how the rules for scoring could work. This necessitated arbitrarily deciding which team was to be plus and which minus. This has the advantage of symmetry lacking in the first experiment. But we thought it to be counter-intuitive for the team represented as minuses, because they would have to see a minus as a score to be added to them, the minuses! We examine here a resolution of the conflict which arose particularly strongly for one of the groups in the third game. They have just discussed various possible games with a single die: look particularly for the justification of the fairness of the game and the significance of compensation in the children's explanation.

Teacher: We will take Ari's idea (who suggested that if we only want to use one die it should have three blue and three yellow faces), but because it is mathematics we are going to use the symbols + and - instead of blue and yellow colours on the dice.

Ari: Oh, we forgot it was mathematics completely!

Teacher: So instead of colours we put these signs [Ari: How?] We take the viewpoint of the blues. We could take the view of the yellows, but from the blues, when we have +3 it means 3 points for the blues... just like you suggested... if we had -1, who gets points?

All: The yellows.

Blue Team Pupil to Yellow Team pupil: One for you.

Teacher: So if you get this minus 3, so?

Blue Team Pupil: From the viewpoint of the blues, so I take 3 from me?

Teacher: Yes, three from you or 3 points for him (a yellow team pupil)

Ari (Blue team): So yellows will always win!... Because if it's a minus, we lose..

Jon: (Blue team) But if it's plus, you will be winning.

Teacher: If minus comes up.. it means?

Ari: To take away.

Jon: No, to take away from the blues.

Lena: (Yellow team) Or to add to the yellows, it's the same.

Ari: How do you know if it's yellow or blue?

Jon: It's from the viewpoint of the blues,.. Don't you see, you have 3 and 3, 2 and 2 (shows the die) there are pluses for everyone and minuses for everyone, and minus is like it's plus for us (the yellow team) and for you it's minus, it's luck, it's not certain...

Ari: Aha! Got it: for us it's exactly what is on the dice, and for them it's the opposite.

Jon: OK, lets start playing.

Ari struggles for the first 2 rounds, the group shows her how it goes and explains why; it is an important feature of the game that she has to be brought to understand sufficiently so the game can proceed. But the need to proceed, and the 'right' of the player to choose the strategy when it's their turn, seems to encourage the children both to operate rules instrumentally and also to evaluate each other's moves from the point of view of fairness. Thus a further discussion occurs later when Ari is instructed by the group, and she is finally brought to "imagine it from the point of view of the yellows" as well as the blues, which actually requires an appreciation of compensation.

In introductions of other groups to the third game this discussion was less problematic, the children accepted that 'minus' records yellow and 'plus' records blue team points, and re-discovered the compensation rule in the course of the game. In *Game 4* they again use the 'add' and 'sub' die used in game 2, but now also the integer die used in game 3. In the following extract, this new group which had run smoothly through the previous game, now faces the crisis of negotiating the two meanings of the minus sign. Stella makes a mistake and is challenged: she threw 'sub -1' but took away a blue instead of a yellow. Dave protests, then Stella, in the yellow team, picks up the -1 die, and says:

Stella: If I had only this I should've added one to us.

Sera: But you have a subtract.

Stella: Look, minus one it's one for us or take away one from you. Now I have to subtract minus one, so I have to subtract one.

Dave: OK, but from whom do you subtract?

Stella: (She pauses for thought, then sees Dave's point) From the yellows.

Shortly, Stella gets 'sub +1' and comments to herself: "it means to subtract 1 from the blues, subtract -1 means to subtract from the yellows". After this first game, all the children have some confidence already and the next game is fluent. Finally all the discussion is in a mathematical voice. The teachers pushed the children to verbalize "subtract plus 2", "add minus 1" etc. when they write down their games 'for checking'. This encourages conflict with the known truths about number. For instance when Stella has zero on the abacus, she throws 'sub -2' and correctly manipulates and ends up with 2 blues. She says: "It doesn't make sense". She is persuaded it does make sense by Limon, and she agrees "it's OK in the game, but not in mathematics." Clearly there is now a long process of conflict-resolution to be completed, where intuitions must be revised to incorporate findings about these new numbers.

In a final evaluation session for each group a written test was set, of purely mathematical questions such as +3 add -2, +2 sub +4, -5 add -2, -5 sub -2, 0 sub -2 etc. We presented some of the groups calculations in fully symbolic form without causing any apparent difficulty: (+3) + (-2), etc. All but one of the children got 100% correct on this assessment. When asked to justify their written calculations, they intuitively appealed to the dice game, but were capable of losing the metaphor. For instance, asked to explain why +5 - +7 is -2, Dave comments: "*I am explaining from the viewpoint of the blues. Plus 5 is like 5 blues and we have to take 7 blues, so we take 5 blues and we add -2 which means 2 yellows....*" And then for +28- (-9) is 37: "*28 plus 9: it's like minus minus is like plus, that's the reason why the result is not 19 but 37*".

2.1.3 Comparing the two integers experiments and generalizing the method

In both experiments, the children saw the abacus as a useful recording instrument, relating the difference in the heights of the stacks of beads to their team's score. In contrast to the first experiment, however, it seems the children were able to operate concretely with subtraction both on the abacus (even when going through zero, because of the strength of the compensation intuition) *and* in the situation. On the other hand, the problem of the arbitrary assignment of plus and minus symbols to the teams now led to some new difficulties which had to be resolved through discussion of fairness in the game situation.

The intuitive strength of fairness in games was immediately accessible in this experiment because the social structure of 'playing games' was relived in the classroom. This was an *authentic* game, with teams and points and some real competition. We see the strength of this in the children's use of the game as a medium of justification for the compensation strategy, and in their use of the situation in their justification of their mathematical operations. It seems that 'minus, minus is plus' is, finally, a rule which makes sense to the children because they see it as fair in an authentic situation.

Whereas the first experiment led to an intuitive gap at the point where subtraction was introduced (it was a secondary concept, defined by inverse-addition), the second experiment introduces a gap earlier, when the signs are introduced to refer to teams in an arbitrary way. In this sense both have strengths and weaknesses. This seems consistent with Fischbein (1987) that no *single* model will be both intuitive and comprehensive. However, in a recent experiment replicating the dice games approach, Koukouffis and Williams (2005) have compared this with a different approach in which the arbitrariness is avoided. Instead of the pair of dice such as (3,4) invoking 3 points for the blues and 4 points for the yellows, in this modification (3,4) is given the meaning 3 points FOR your team and 4 points AWAY from your team. In this version of the experiment each team records 'plus points' as points for them: analysis is ongoing but promising.

In any event we do not think that our model will extend from addition and subtraction to multiplication and division. In our view a multiplicity of models will be needed for integer operations, as Behr., Lesh, Post, and Silver, (1984) argued is required for rational numbers. Indeed we acknowledge that multiplication and division *may* require a purely algebraic approach as some have argued, and that the concrete models we have put forward in this case would need to be left behind. But we believe that we have shown that at least we can avoid introducing integers and the operations of addition and subtraction purely algebraically from the beginning. Further we argue that even if a formal algebraic treatment of all the operations will come later, the early treatment of the concept and the operations through models is desirable, because they allow the children access to intuition.

The integer is identifiable in the children's activity first as a process on the numbers already understood by the children, then as a 'report' or score recorded (concretized by the abacus). The operations on the integers arise as actions on their abacus representations, recorded in mathematical signs. Finally, the operations on the mathematical signs are encountered in themselves, and justified by the abacus manipulations and games they represent. Thus the integers are encountered as objects in social activity, *before* they are symbolized mathematically, thus intuitively filling the gap formerly considered a major obstacle to reification.

As developmental researchers we begin with the mathematical concepts to 'be discovered', explore models and then select contexts which are evocative of appropriate situations. This design aims to engage the children with phenomena to be organized with the relevant models and targeted mathematics. We then look to the classroom situation to see how the task sequence might be constructed so that the outside-school experience might be authentically evoked. This is a process of concretization of our mathematical concepts through tasks related to the culture of the child as we understand it. But for the children the activity *begins* with recalling or re-living these outside-school experiences in the classroom, and the sequence *ends* with formal, written mathematics. In Williams and Linchevski (1997) we generalized this instructional strategy, 'Object-Process Linking and Embedding', as a sequence of tasks in which children a) build strategies in the situation, b) attach these to the new numbers to be discovered, and finally c) embed them in mathematics by introducing the mathematical voice and signs. We now turn to the second case.

2.2 The teaching of subtraction of two-digit numbers

Research on computational methods of first graders show that first grade students usually do not demonstrate ability to invent algorithms for the subtraction of two-digit numbers besides counting back or up, or going back on a number line (if they were introduced to this model). This is in contrast to addition (Carpenter, Fennema, Petersen, and Carey, and Kutscher et al., 2002) even though they had been previously provided with opportunities to solve both addition and subtraction word problems that involved two-digit numbers and solved them meaningfully. This situation has led some educators to postpone any experience with multi-digit numbers to upper grades and others to the introduction of base ten models followed by the teaching of the standard algorithms. However, when young children are taught to use the standard algorithms they tend to error. The main deficiencies related to the learning approaches that focus on the formal algorithms at this age are that they foster memorization of the calculation procedures, with an inadequate conceptual basis. Without 'intuitive' understanding one finds errors in calculation procedures that survive in the long-term (Fuson, 1992). This situation led us to try to design our instructional method and to look for an embodiment, a 'model', that will satisfy the need for the subtraction of two-digit numbers, would have the potential to provoke 'intuitively' an advanced computational method in first graders, and would survive the transfer into the classroom situation.

2.2.1 The Lotto Game: Making computational strategy for the subtraction of two-digit numbers concrete

Our hypothesis was that a 'buying-selling' situation could be a context for developing invented strategies for the subtraction of two-digit numbers in first-graders, and, since invented subtraction strategies generally start with the larger unit (Hiebert, 1996) we expected two possible invented strategies: (i) Decomposition (sometimes also followed by 'Opening-a-Ten'); (ii) Overshoot-and-Come-Back.

For example, if the children have to pay 28 cents out of 64 they possess (6 tens and 4 ones), in strategy i) they would first pay 2 tens out of the 6 tens, change another ten to 10 ones and then pay 8 ones out of the 14 ones, leaving 6 ones. Answer: 36. In strategy ii) the children would pay the 28 with 3 tens, get 2 ones back, so they are left with 34 + 2. Answer: 36. The first strategy we coined the 'Bank' method (sometimes referred to as Alternate Subtraction, and the second one the 'Change' method (sometimes referred to as 'Compensation', N10C, or 'Overshoot-and-Come-Back', (Carpenter et al., 1988)

The reason we have chosen to focus on situations that might afford the invention of advanced strategies like Overshoot-and-Come-Back or Decomposition is that in most problem-centred programmes these advanced strategies have not been invented by very young children. In addition, when they are eventually invented, it happens via secondary references, for example via the model introduced, rather than the primary context of the problem situation posed. The implication is that these strategies are based on fundamentally different grounds, and can only be constructed when the children have reached the relevant developmental level, generally not before the end of second grade or even later, depending on the strategy and previous concept exposure.

We will show that in our case, the strategies for 2-digit subtraction were elicited intuitively in the reference-context of 'buying-and-selling' that we offered.

2.2.2 Methodology for the 2-digit LOTTO subtraction games

The study pretest comprised of two episodes in small heterogeneous groups, followed by two periods of teaching in small heterogeneous groups in a cooperative learning environment: four children worked with the teacher-researcher in each group, and each session was about one-hour-long (Kutscher et al., 2002). Finally an individual one to one post-test of about 15 minutes took place.

The children were first-graders who had not yet received any instruction in adding or subtracting two-digit numbers, although they were acquainted with addition and subtraction of one-digit numbers. One of the researchers was both teacher and interviewer. All the meetings were videotaped to allow further analysis. The sequence of the teaching episodes was designed (a) to elicit intuitive, spontaneous, 'out-of-school' strategies for the subtraction of two-digit numbers, and (b) to allow the shift of the spontaneous strategies from intuitive meanings to mathematical meanings expressed in mathematical sentences (Linchevski and Williams, 1999). Students who participated in the study were those who on a pretest were able to read and order two-digit numbers, and understand that a two-digit number like 27 is 20 plus 7. Out of the 34 pupils of the class 20 students participated in our research. The 8 students that did not meet the above requirements and in addition 6 very able students of the class were excluded.

During the first teaching episode for each group a "buying-selling" game, a modification of the Lotto Game with a buying-selling requirement, was played by the groups. Each of the participants was given a different lotto-board that had 9 picture squares drawn on it. For each lotto-board there were 9 square picture-cards identical to the ones drawn on the board that the players had to accumulate in order to cover the pictures on their lotto-board and thus win. We modified the game by writing prices of two-digit numbers on the pictures; their values were between 11 to 45.

Each child started off with the same amount of money, 77 dollars ($): 7 'ten-strips' with division lines marking off each strip into ten squares, each strip representing a $10 note, and also 7 squares representing the seven ones. Each child had an empty 200-number-board on which he or she could store and arrange their money. There was also a communal bank of money where there were ample tens and units that the students could change from tens to units and vice versa, so that they could apply any strategy involving exchanging ten-strips for ten units etc. as they wished. The picture cards were distributed equally among the children, who took turns to try to collect a picture card. If a player did not have a picture-card for his or her lotto-board they had to buy it from another player in the group according to the "price" of the picture-card. This "buying" process laid the base for the subtraction and intuitive methods of payment and giving of change. Thus typically a child might want to buy a picture card for $47, by handing over $50 and getting $3 change; thus, the value of assets in their wallet might have gone down by $47 from $65 to just $18.

During the second teaching episode the children were asked to reflect on concrete cases that they experienced during their games in the first episode, and were offered ways of recording the transactions as, and hence 'transforming' their situated, intuitive strategies into, mathematical sentences (such as, in the above case, $50 - 47 = 3$, and $65 - 47 = 18$).

This transformation was done in four stages of translation:

1. A verbal formulation of the solving strategy in shared spoken-language terms for concrete cases from their games. The verbal formulation was based on the shared verbal communication they had developed during the games.
2. (i) A written account of the solving process for concrete cases using the verbal formulation; (ii) Solving expressions written in verbal formulation of hypothetical cases that could occur in a game.
3. (i) A written representation in mathematical symbols for concrete cases. (ii) Solving expressions written in mathematical symbols of hypothetical cases that could occur in a game.

4. Solving written abstract subtraction expressions (65 – 47 =?).

Finally a written test with abstract subtraction exercises was administrated about 4 weeks after the last meeting with the children.

2.2.3 Results from the Lotto game

When the children were faced with the problem of having to pay money for a picture-card (e.g. $47), having enough money in their wallet (e.g. $65), but not enough units, all the children who participated in this study used the 'change' method to solve their problem. The process of the invention of the 'change' strategy occurred in various ways, and can be analysed as two separate stages: first the identification of the problem and second, the joint construction of solutions of the problem.

Two patterns of identification were recognized. In some cases the problem emerged half way through trying to pay the exact amount, thus Yuri wishes to buy a flag for $17, she takes off 1 ten-strip and she counts her units. She takes off money of value $16, but has no more units on her board. She immediately takes off another ten-strip – the budding of the 'change' strategy. In other cases the identification of the problem occurred before paying by comparing the number of units on the board to the number of units that need to be paid. Kim wants to buy a car for $29 and says:

Kim: I don't have … [enough units]

Interviewer: What do you mean when you say 'I don't have'?

Tania (another player): That he doesn't have the number.

Interviewer: But you have even more (than you need).

Tania: That we don't have like these; [points at the units]

The solution generally emerges as a joint production. For instance Barry is interested in buying a sofa for $47. He removes 4 ten-strips and says:

Barry: I don't have.

Interviewer: Does anyone have an idea?

Angel: To give one like this. [she points at one of Barry's ten-strips]

Barry removes 1 ten-strip from his 200-board, and gives it to Vy, who hesitates:

Interviewer to Vy (the owner of the sofa-card): What will you do?

Vy: I'll give change.

Interviewer: How much change? You were supposed to get 47 and you got 50.

Vy and Angel announce a few numbers: 2, 5, 7.

And Angel shouts: "Three!"

This answer is approved by all. Indeed every member of this group took part in the construction of this Overshoot-and-Come-Back solution. Barry identified the problem, Angel suggested the need to "overshoot", Vy the

need to "come back" and Angel found the amount you "come back" with. In another typical case, Kev is interested in buying a trumpet for $12. He has 5 strips of 10, that is $50. He is puzzled.
Interviewer: Do you have 12?
Kev: No… I have 50…
Kev removes one strip and puts his hand on another strip, hesitating. The group is silent.
Interviewer: Do you want to give Danni (the trumpet –card's owner) 20?
Danni: But then I'll give him 10 back…
Kev: 8!

Danni follows Kev's suggestion, when Kev gives her the 2 tens she gives him 8 units back. With the progress of the game most children, either spontaneously or through group interactions, adopted the 'change' strategy and used it naturally and meaningfully. However, sometimes, some of them experienced difficulty with the calculation of the change to be given.

Elsi had $42, 4 tens and 2 units, and she wanted to buy a card that costs 36. She removed 4 strips from her wallet but before handing them to the seller she tried to calculate the amount of change she has to get. She placed the 4 tens on the table, counted 36 ones and thereafter declared the amount left: "4". In another case, Tania wants to buy a card that costs 28. She gives 3 tens to Kev. Kev takes one of the tens and by placing it on the wallet checks the amount of change he has to give.

Sometimes the difficulty at the beginning is with finding a strategy and the calculation of the change. During the first game played in group 2, after a round without a need for 'change', Yuri wants to buy a vase that costs 37 from Elsi. She starts to remove strips from her wallet counting "ten, twenty, thirty" she switches to the ones continuing counting "one, two, three, four, five, six…" and stops since she does not have more ones.
Yuri: I have 36
Interviewer: But you have money, (to the group) what can be done?
Elsi: She will give me 40 and I'll give her change.

Yuri takes another 10 and gives Elsi 4 tens.
Interviewer (to Yuri): You gave her 40. How much change are you going to get?
Yuri: 10
The Interviewer gives her 1 ten back asking: "If you get 1 back then how much did you actually pay? Has someone a suggestion?" He puts down the 4 tens: "She has to pay only 37".
Elsi: 3
Interviewer: How much change she has to get?
Yuri: 3

After a few games (not all of them were played in the same session) most groups played the game smoothly and cooperatively where the major help needed from the group was usually in determining the change. However, most children very quickly either were able to determine the change immediately without the help of any manipulative or developed some strategy based on the strips or fingers. The most common strategies were counting the required ones on a strip and then counting the complementary (the change), counting "imaginary" ones on the wallet and then counting on the wallet the complementary, and counting-on orally from the "price" to the next tenth (38, 39, 40…3!!).

As reported earlier, all children adopted the 'change' method in the game. But none found it necessary to find, after a transaction, the amount of money left on his or her 200-board: the answer to the multi-digit subtraction problem. Their behaviour corresponded to the situation that elicited their spontaneous strategy: in a buying-selling situation one does not usually calculate the amount left in the wallet – unless there is some concern that the amount will not suffice. Thus the children's disregard of the amount left on their 200-board is not surprising. The teacher would call their attention both to the initial amount on the 200-board and to the amount left after each transaction in the second set of teaching episodes, when the transactions would be translated to mathematical subtraction- expressions.

When encountering the problem of not having enough units to pay, in no instance did the pupils suggest that the buyer exchange a ten-strip for ten ones albeit the 'bank' was on the table. The decomposition of tens strategy did surface once, when the seller did not have enough change. Here Vinny is interested in buying a flute for \$16. She removes 2 ten-strips and gives Benni, who doesn't have change of 4 ones on his board and is stuck.

Interviewer: Does anyone have an idea? Vinny doesn't have 6 units and B doesn't have 4 units to give her change?

Abi: We'll change this (points to a ten-strip on Benni's board) to small ones.

Benni removes a ten-strip, puts it in the 'bank' and takes 10 unit squares and sticks them on his board on the row that was vacated when he removed the ten-strip. Only then does he give Vinny 4 square-units.

This last vignette suggests the powerful effect of situated intuition. Why is it that here 'decomposition of tens-and-units' was the spontaneous reaction and that in all the other instances only the 'change' strategy was evoked? It might be that in 'real' life experiences, it is usually the responsibility of the cashier to change the 'big' money into smaller denominations in order to be able to give change to the buyer. Furthermore, just as was observed in Barry's behaviour, in real life when the seller changes money into smaller denominations he places it in the till and only then gives the buyer his or her change. Thus, when both the 'buyer' and the 'seller' did not have the necessary units to complete the transaction, the solution that was offered in the game, mirrored the problem-solving strategy regarded as normative of a real-life buying-selling situation.

Halliday and Hasan (1985) have analysed the discourse genre involved in a market stall buyer-seller exchange and shown how the social expectations of the contextual configuration (field, tenor and mode) are realised in the grammar of the typical exchange. We assume that the children are re-enacting in play (as they do in the classroom 'shop' even from nursery age) precisely this social configuration. The strategy we made use of in this pedagogical design is therefore intuitive because it is imported from this situation as a cultural resource.

The specific decompose-tens strategy, although understood and approved by the group, was not used by the students later on and did not enter the mathematical voice. We hypothesise that this was because this intuition did not become the focus of the teacher-researchers expectations in the second lesson: thus the teacher plays a critical role in 'weeding' as well as 'sowing'.

2.3 Comparison of the 2-digit subtraction and the integer-operations cases

We now take the opportunity to compare and contrast the cases of the integers studies with that of 2-digit subtraction with much younger children. In such cases theory is strengthened by the diversity of the cases, e.g. the different topics, ages, countries, etc. (see Yin, 2002). What common threads can be drawn from these two cases then?

In both cases powerful intuitions from children's previous experiences were imported into the classroom activities. We noted in the case of varieties of contexts presented in our researches on the integers that quite small differences might have a significant impact on the children's intuitive strategies. When the outside-situation is invoked in the classroom as a simulation or a game it is not exactly the same as the outside situation being evoked: this worked to some extent against us in the context of the 'disco simulation', as we had no direct linkage of the notion of 'taking away 'ins'' with the actual case being simulated. On the other hand the 'taking away of points from the other team' was linked clearly with a concrete 'taking away' in the context, and could be equated with 'adding of points to us' in just the right way for making the integer operations concrete. But then the new integer-game situation led to a counter-intuitive model whereby one of the teams counted as 'minus'. The concern is that for this team 'more minuses' means a 'bigger score'. Further efforts to re-model this involved a game modification whereby plus points are for us and minus points are against, but like goals in a match, our minus points count 'for' them and vice versa. In this integer-game context we have found a significant improvement in effectiveness (with a small sample, see Koukouffis and Williams, 2006.).

To this integer-game experience we can now add the 2-digit subtraction case. The buying-selling situation evokes intuitions and expectations that seem very strong even at this very young age! Indeed the main problem experienced by these children is with the 'bonds to ten' on which the 2-digit work is building. It is also notable that the outside situation is in both cases transformed most naturally into a 'game': it seems that 'play' is a transitional activity for children, in that it carries across from outside school to inside school very readily. All the norms of this activity work well for these cases: every player has their turn, the rules must be 'fair' to all, and when the game gets stuck everyone has an interest in jointly fixing the game so it can proceed.

In addition, a most important feature of the game is that it has somewhat arbitrary rules: a new game can always be invented. 'Play' is understood as an activity in which 'real life' is explored, but objects are expected to represent things other than themselves, and strategies can be playfully tried out without real consequences. When play is situated within school, the teacher is allowed to make up some rules, because that is in the conventions of schooling: clearly we took advantage of that. In both experiments, the children were guided by the teacher and encouraged to explore solutions we were interested in helping them to mathematise later.

In both cases the concrete physical constraints and affordances of the models/manipulatives introduced were significant. The constraints included the inability to add beads to the abacus indefinitely, or to easily 'break' the ten-strip up. The latter can be easily motivated by the money-context (you can't physically break up a \$10 note to get 10 single dollar coins). But the marking of the ten-strips are a physical, visuo-spatial reminder that the value of the 10-strip is indeed ten units, thus helping to avoid the error of counting tens as ones and vice versa.

In semiotic analyses of children's reasoning with the double abacus, a similar aspect of the instrument became clear: the heights of the beads on the abacus strongly encourages an enactive communication. Thus for instance, when trying to explain why 5 blue and 3 yellow beads - (5,3) – represents 'two for us', there is a strong tendency for the children to actually handle the two 'extra blue beads', sometimes using their hand flat between the 3 blue that are 'level with' the 3 yellow to show the 'extra 2 blue' above their hand. The gesturing with the hand then is used to accompany the idea of 'level' by some children, and eventually with the verbalisation of 'score'. Thus the two extra beads become associated with a 'score of 2 for us'.

Thus the physical affordances in the model engage with the child's actions, and these actions presage gestures in communication, and the crucial verbalisation 'score' completes a semiotic contraction of the integer concept (Radford, 2003, 2005). Analysis of gesture-verbalisation dialectics in communication processes (especially of externalisation) suggest that (i) in learning, when gesture and verbalisation conflict, gesture is in advance of the verbal; (ii) in recall, the gesture precedes the verbalisation sought. These hints at how non-verbal communication might play a major role in learning encouraged us to see it as significant in our own theory (see Williams, 2005; and inspirational work in Goldin-Meadows, 2003; McNeill, 1992).

In the case of 2-digit subtraction, we similarly suggest the way a young child is able to hesitantly offer a ten-strip, which they know is 'too much', provides a physical, sensuous, semiotic expression of a partial, hypothetical solution to the problem of buying-with-change. Perhaps it only meant to the child something like "but look, I have this…" This effort by one child is then understood and built upon by the others, "but then … she will then give you change". We suggest that it is because partial problem solutions are visibly (or sometimes aurally, or both) represented that joint activity is possible, and the group is thus able to construct what the individual cannot manage alone.

For this reason we believe in the key significance of appropriate use of manipulatives for mediating joint activity in these two cases of the construction of a zone of proximal development. They afford a collective production of a semiotic chain in the form of an action – gesture – verbalisation sequence that allows mathematical generalisation to become manifest.

3. SITUATED INTUITION AND AUTHENTIC *CLASSROOM* LEARNING

Learning is structured by its social context and situation. Authentic activity in the classroom must involve learning mathematics in some authentic way. It is clear too that children bring cultural knowledge and language from outside school with them, and that this can help or hinder them in building mathematics. The introduction of outside school, 'real' or 'realistic' experiences into the classroom can be problematic. Generally the goals of the activity in which the knowledge was acquired outside school may be absent in the classroom, and the activity must be re-constituted in some way. We tried to reconstitute the children's knowledge through a 'simulation' and a 're-living' of a point-scoring game. In general, such classroom tasks demand 'transformation' of children's knowledge. The transformation may be facilitated if the relevant goals and intuitions from the outside-school activity are evoked in the classroom. We think this can be said, up to a point, of the integers experiment, in which the social framework of 'playing a team game' was imported almost wholesale into the classroom, and the intuition for 'fairness' supported important constructions.

Why almost, and why only up to a point? Even then, when it was not uncommon to hear the children cheering on their team, the children were aware that they were involved in a school-learning context, and that the mathematics would not be forgotten for long. The children's intuitive reasoning and sense-making is structured largely by the activity system of the classroom, where they expect to learn mathematics. Our method uses this to advantage. The classroom situation affords rule-bound play in which we as teachers have the right to introduce mathematical signs and tasks which provided a crucial element in the change of voice required to structure conceptions and actions mathematically. In simple terms the schooling activity which constrains us as teachers to link tasks to learning mathematics also afforded tasks which led the children to mathematisation!

In both experiments the use of manipulative as pedagogical tools and representations was critical. First the abacus links classroom tasks with the intuitive situation (the recording function which first models process with object). Then the abacus mediates the manipulation of mathematical signs (the embodiment function, which allows the abacus manipulations to be modelled by and transformed into mathematics). The actions on the abacus in the first function are, of course, identical with those in the second, and strategies developed in the first are transferred to the second. This is the sense in which the abacus provides the link in the chain of signification from phenomena-to-be-organised to mathematics.

Our instructional method involves accounting for significant shifts in meaning when children's knowledge is evoked in different social settings from that in which it was constructed. The classroom can relate to outside-school knowledge, but the children's activity has to reconstitute it and it may be thereby transformed. This transformation furthermore may be strongly affected by the way the classroom activity is enacted. Quite small differences in authenticity, for instance between simulation and what we called re-living outside school experience, might be influential in determining the quality of involvement and intuition evoked by activity. On the other hand formal school learning itself has authenticity in the school context, and involves a further transformation of knowledge which can sometimes be accomplished through the introduction of signs into the activity in ways which make sense.

3.1 Finally, what can we say about situated cognition and activity theory

All social practice theory has cognition as socially situated (if it has cognition as such at all). What distinguishes socio-cultural theory is that knowledge is mediated in social networks and systems by cultural artifacts (Bruner, 1996; Hutchins, 1990). Thus, for example, the observation instruments used in navigation (or come to that in a classroom, or in astronomy) are vital to the activity, just as much as the social division of labour and social customs and rites of the people that share in the activity (see Williams and Wake, 2007a).

Lave and Wenger (1991) situated learning as social, i.e. as the change of social context of an identity, and located it in changes of social participation in a community of practice. This approach can be criticised as being overly determinist (the individual learns as they move centripetally within a relatively static community controlled by old-timers), and simplistic (one is always a participant in many communities). Since then, these authors have helpfully developed the notion of multiple identities, especially involving identity at the interface between distinct communities, e.g. brokers (Wenger, 1998).

Nevertheless, a problem for these situated learning theorists of schooling has been to identify what activity theorists call the 'object' of the activity: is it 'learning' or 'schooling'? Lave seems not to resolve the issue in a way that affords a role for teaching within 'schooling'. Engestrom too resolved the issue by a call for schools to address their activity to 'life' instead of scholasticism (Engestrom, 1991).

However, 'third generation' activity theorists have argued for the notion of multiple activity systems as central to understanding system change: it is the interaction of distinct, contradictory activities (mediated by boundary persons and boundary objects) that introduce the contradictions in systems that provide it with a historic dynamic. Thus, in some versions of Cultural Historical Activity Theory (CHAT) the unit of analysis always contains *two* interacting activities or activity systems, and it is primarily the systems that learn, rather than people (Engestrom 2003).

The key theoretical points of value from CHAT for this paper are (i) the concept of dynamic contradiction which drives change coming from outside the community, through boundary systems and artifacts/objects and persons, and (ii) the concept of a boundary crosser as a person who brings knowledge across communities and practices, and thereby gives new meanings to their practices and cognitions in both. For the children playing our games, confronted with problems that demanded they resolve an everyday problem with some mathematics; this represented a conflict between everyday practices and schooling. Because they were enabled in the school situation to mathematise their commonplace, everyday knowledge, they built new mathematical knowledge which linked intuitively with their previous (mostly outside school) knowledge.

This, finally, is our view of 'modelling', and it is intimately tied to the notion of mathematics as communication, mediated by metaphorical use of 'language' and other semiotic tools such as manipulatives (Williams and Wake, 2007b). CHAT encourages us to consider boundary activities, boundary crossers and boundary objects as the natural place to find the cultural resources to help bring about change. In this case, theory provides a socio-cultural view of learning as the practice of construction of semiotic linkages between existing practices (that embed familiar understandings) and the new practices and discourses that we call 'new' mathematics.

REFERENCES

Barton, B. (1996). Making sense of ethnomathematics: Ethnomathematics is making sense. *Educational Studies in Mathematics, 31*(1-2), 201-233.

Behr, M., Lesh, R., Post, T., & Silver, E. (1983). Rational number concepts. In R. Lesh & M. Landau (Eds.), *Acquisition of mathematical concepts and processes* (pp. 91-125). New York: Academic Press.

Brown, J. S., Collins, A., & Duguid, P. (1989). Situated cognition and the culture of learning. *Educational Researcher, 18*(1), 32-42.

Bruner, J. S. (1996). *The culture of education.* Cambridge, Mass., London: Harvard University Press.

Carpenter, T. P., Fennema, E., Petersen, P. L., & Carey, D. (1988). Teachers' pedagogical content knowledge of students' problem solving in elementary arithmetic. *Journal for Research in Mathematics Education, 19*(5), 385-401.

Chaiklin, S., & Lave, J. (1993). *Understanding practice.* Cambridge: Cambridge University Press.

Cole, M., & Engeström, Y. (1993). A cultural-historical approach to distributed cognition. In G. Salomon (Ed.), *Distributed cognitions: Psychological and educational considerations* (pp. 1-46). Cambridge: Cambridge University Press.

Dirks, M. K. (1984). The integer abacus. *Arithmetic Teacher, 31*(7), 50-54.

Engestrom, Y. (1991). Non scolae sed vitae discimus: Toward overcoming the encapsulation of school learning. *Learning and Instruction, 1*(3), 243-259.

Engestrom, Y. (2003). Conceptualizing transfer: From standard notions to developmental perspectives. In T. Tuomi-Gröhn & Y. Engeström (Eds.), *Between school and work: New perspectives on transfer and boundary-crossing* (pp. 19-38). Amsterdam/Oxford: Pergamon.

Fischbein, E. (1987). *Intuition in science and mathematics.* Dordrecht, Netherlands: Reidel.

Freudenthal, H. (1983). *Didactical phenomenology of mathematical structures.* Dordrecht: Kluwer Academic Publishers.

Fuson, K. C. (1992). Research on whole number addition and subtraction. In D. A. Grouws (Ed.), *Handbook of research on mathematics teaching and learning: A project of the*

national council of teachers of mathematics (pp. 243-275). Reston, V.A: National Council of Teachers of Mathematics.
Goldin-Meadow, S. (2003). Hearing gesture. Ca, Mass: Harvard University Press.
Gravemeijer, K. (1994). Educational development and developmental research in mathematics education. *Journal for Research in Mathematics Education, 25*(5), 443-471.
Gravemeijer, K., Cobb, P., Bowers, J., & Whitenack, J. (2000). Symbolising, modelling and instructional design. Perspectives on discourse, tools and instructional design. In P. Cobb, E. Yackel & K. McClain (Eds.), *Symbolizing and communicating in mathematics classrooms* (pp. 225-273). Mahwah, NJ: Lawrence Erlbaum Associates.
Gray, E. M., & Tall, D. O. (1994). Duality, ambiguity, and flexibility: A 'proceptual' view of simple arithmetic. *Journal for Research in Mathematics Education, 25*(2), 116-140.
Halliday, M. A. K., & Hasan, R. (1985). Language, context, and text: Aspects of language in a social-semiotic perspective. Geelong, Australia: Deakin University.
Heckman, P. E., & Weissglass, J. (1994). Contextualised mathematics instruction: Moving beyond recent proposals. *For the Learning of Mathematics, 14*(1), 29-33.
Hiebert, J. (1986). *Conceptual and procedural knowledge: The case of mathematics*. Hillsdale, NJ: Lawrence Erlbaum Associates.
Hutchins, E. (1990). *Cognition in the wild*. Cambridge, Mass.: MIT Press.
Koukouffis, A., & Williams, J. S. (2005). Integer operations in the primary school: A semiotic analysis of 'factual generalisation', *BSRLM*. St Martin's Lancaster, UK.
Koukkoufis, A., & Williams, J. (2006). Semiotic objectifications of the compensation strategy: En route to the reification of integers. *Revista Latinoamericana de Investigación en Matemática Educativa, 9* (Numero Especial: Special issue on Semiotics, Culture and Mathematical Thinking), 157-175.
Kutscher, B., Linchevski, L., & Eisenman, T. (2002). From the lotto game to subtracting two-digit numbers in first-graders. In A. D. Cockburn & E. Nardi (Eds.), *Proceedings of the 26th Conference of the International Group for the Psychology of Mathematics Education* (Vol. 3, pp. 249-256). Norwich, UK: PME.
Lave, J. (1988). *Cognition in practice: Mind, mathematics and culture in everyday life*. Cambridge, Mass: Cambridge University Press.
Lave, J. (1996). Teaching, as learning, in practice. *Mind, Culture, and Activity, 3*(3), 149-164.
Lave, J., & Wenger, E. (1991). *Situated learning: Legitimate peripheral participation*. Cambridge: Cambridge University Press.
Leontiev, A. N. (1981). The problem of activity in soviet psychology (J. V. Wertsch, Trans.). In J. V. Wertsch (Ed.), *The concept of activity in soviet psychology* (pp. 37-71). Armonk, N.Y: M. E. Sharpe.
Liebeck, P. (1990). Scores and forfeits - an intuitive model for integer arithmetic. *Educational Studies in Mathematics, 21*(3), 221-239.
Linchevski, L., & Schwarz, B. (2001). Can interaction between inferior strategies lead to a superior one? The case of proportional thinking. *Proceedings of the European Society for Research in Mathematics Education*. Czech Republic
Linchevski, L., & Williams, J. S. (1996). Situated intuition, concrete manipulations and mathematical concepts: The case of integers. In *Proceedings of the 20th Conference of the International Group for the Psychology of Mathematics Education* (Vol. 3, pp. 265-272). University of Valencia, Valencia: PME.
Linchevski, L., & Williams, J. S. (1999). Using intuition from everyday life in 'filling' the gap in children's extension of their number concept to include the negative numbers. *Educational Studies in Mathematics, 39*(1-3), 131-147.
Lytle, P. A. (1994). Investigation of a model based on neutralization of opposites to teach integers. In L. Meira & D. Carraher (Eds.), *Proceedings of the 19th International Group*

for the Psychology of Mathematics Education (Vol. 3, pp. 192-199). Universidade Federal de Pernambuco, Recife, Brazil: PME.

McNeill, D. (1992). *Hand and mind: What gestures reveal about thought.* Chicago: University of Chicago Press.

Radford, L. (2002). The seen, the spoken and the written: A semiotic approach to the problem of objectification of mathematical knowledge. *For the Learning of Mathematics, 22*(2), 14-23.

Radford, L. (2003). Gestures, speech, and the sprouting of signs. *Mathematical Thinking and Learning, 5*(1), 37-70.

Radford, L. (2005). The semiotics of the schema: Kant, Piaget, and the calculator. In M. H. G. Hoffmann, J. Lenhard & F. Seeger (Eds.), *Activity and sign. Grounding mathematics education* (pp. 137-152). New York: Springer.

Ryan, J. T., & Williams, J. S. (2007). *Children's mathematics 4-15.* Maidenhead, UK: Open University Press.

Semadeni, Z. (1984). A principle of concretization permanence for the formation of arithmetical concepts. *Educational Studies in Mathematics, 15*(4), 379-395.

Sfard, A. (1991). The dual nature of mathematical conceptions: Reflections on processes and objects as different sides of the same coin. *Educational Studies in Mathematics, 22*(1), 1-36.

Sfard, A., & Linchevski, L. (1994). The gains and pitfalls of reification: The case of algebra. *Educational Studies in Mathematics, 26*, 191-228.

Treffers, A. (1987). *Three dimensions: A model of goal and theory description in mathematics instruction- the wiskobas project.* Dordrecht: Kluwer Academic Press.

Walkerdine, V. (1988). *The mastery of reason: Cognitive development and the production of rationality.* London: Routledge.

Wenger, E. (1998). *Communities of practice: Learning, meaning, and identity.* Cambridge: Cambridge University Press.

Wertsch, J. V. (1991). *Voices of the mind: A sociocultural approach to mediated action.* London: Harvester.

Wertsch, J. V. (1996). The primacy of mediated action in sociocultural theory. *Mind, Culture, and Activity, 1*(4), 202-208.

Williams, J. S. (2005). Gestures, signs and mathematisation. In H. L. Chick & J. L. Vincent (Eds.), *Proceedings of the 29th Conference of the International Group for the Psychology of Mathematics Education.* University of Melbourne, Australia: PME.

Williams, J. S., & Linchevski, L. (1997). Situated intuitions, concrete manipulations and the construction of the integers: Comparing two experiments, Paper presented to the annual conference of the American Educational Research Association. Chicago.

Williams, J. S., & Linchevski, L. (1998). Situating the activity of teaching and learning integers. In A. Watson (Ed.), *Situated cognition in the learning of mathematics* (pp. 143-160). Oxford: University of Oxford, Department of Educational Studies.

Williams, J. S., & Ryan, J. T. (2000). National testing and the improvement of classroom teaching: Can they coexist. *British Educational Research Journal, 26*(1), 49-73.

Williams, J. S., & Wake, G. D. (2007a). Black boxes in workplace mathematics. *Educational Studies in Mathematics, 64*(3), 317-343.

Williams, J. S., & Wake, G. D. (2007b). Metaphors and models in translation between college and workplace mathematics. *Educational Studies in Mathematics, 64*(3), 345-371.

Yin, R. K. (2002). *Case study research: Design and methods* (3rd ed.). London: Sage.

Chapter 9

The Role Of Artefacts In Mathematical Thinking: A Situated Learning Perspective

Madalena Pinto dos Santos and João Filipe Matos
Centro de Investigação em Educação, Universidade de Lisboa

Abstract: This chapter aims to explore and discuss the notion of learning as participation with artefacts in social practices. It uses insights and evidence from the empirical field of *ardinas*' practice in Cape Verde to support a discussion that combines a situated learning approach with elements of activity theory. Two artefacts are analysed – the calculator and the record table used by the *ardinas*. We claim that the regulation of participation made possible by these artefacts does not come from the artefacts themselves but from the way they become present in the everyday and the functions they have in the practice. The artefacts do not represent something fixed and external to the practice; their usefulness is not revealed in the characteristics identified outside of its use in the practice. Artefacts are artefacts in the practice, though they have to be read in the interaction with the forms of use that practitioners put into action. Our final discussion goes into two key concepts in situated learning that we connect with the notion of artefact and resource: technology of practice and shared repertoire. The two concepts are complementary: giving visibility to particular aspects: firstly, to the process of construction; secondly, to the history. In both the key idea of participation is present, and it is through participation that one contributes to construction and has access to history.

Key words: mathematics learning, communities of practice, activity theory, participation, artefact, shared repertoire, technology of practice

1. INTRODUCTION

Since the late 1990s, learning and participation in social practices are seen more and more as inseparable. Looking at what people do in the 'everyday' as participation in communities of practice offers a key to reversing the failure of some of the earlier psychology-based attempts to understand and foster learning. The original situated learning perspective has been used only partially by some scholars and many authors are now making efforts to use its full developments in researching education.

In this chapter we aim to explore and discuss the notion of learning as participation with artefacts in social practices, using insights coming from the empirical field studying *ardinas*' practice in Cape Verde. The theoretical background brings a situated learning approach (drawn from Lave and Wenger, 1991) together with elements of activity theory (following Engestrom, 1999a, 1999b).

2. RATIONALE FOR THE THEORETICAL OPTIONS AND THE FOCUS OF ANALYSIS

Emerging from research is a set of issues that frame the rationale for further development. This rationale is discussed here in two key questions.

2.1 Why bring activity theory into a situated perspective?

Firstly we note that there are common roots between the situated perspective of Lave and Wenger (1991) and activity theory as it is recognized by those authors and elaborated by, for example, Engestrom and Cole (1997) and Miettien (1999). Jean Lave finds in the socio-cultural approach, and in particular in activity theory, key ideas that meet her need to think about activity: (i) a way of conceptualizing activity that makes possible the analysis of its intrinsic organization through the definition of a categorization of levels (or segments) of activity, but that simultaneously recognizes and considers the holistic nature and the dynamics of that activity; (ii) the emphasis given to the relational nature of both the meaning (localized in the relations between the different levels of activity) and the system of activity (which operates between the levels of activity such as in the interface action-operation); and (iii) the dialectic analytical approach of activity and its meaning in the relations with the system of activities where it is integrated.

Secondly, we recognize the concept of social practice is explicit in activity theory. For example, Toulmin (1999) elaborates on the idea of knowledge and makes a comparative analysis of the epistemological ideas of Vygotsky and late Wittgenstein. He identifies in both a concern with the concept of practice, stating that for the future the key central notion in any new theory of knowledge should be 'practice'. In the foreword of their book

'Activity Theory and Social Practice', Hedegaard, Chaiklin, and Jensen (1999) give an account of how social practice is brought into discussion of activity theory, and the relations between 'social practice' and 'activity', indicating possible further research in the area. Although Hedegaard et al. underline the importance of the concept of practice as it "provides a way to characterize those aspects of social practice that are believed to provide the conditions for psychological development" (p. 19) they recognize the need to think more profoundly about the possibility of the existence of wider meanings of social practice that exceed the notion of activity. According to Jensen (1999) activity theorists "have not applied their insight about the situated nature of practice and the practice-situatedness of concepts reflexively, only rarely have activity theorists accounted for their own concepts and theories as embedded in activities and practices" (p. 84).

2.2 Why look deeper into the situated role of artefacts?

Artefacts gain relevance when we seek to understand learning as a phenomenon emergent from participation in social practices. In addition, the dimension of the relations between resources or artefacts and power is introduced by Giddens (1996); resources are in fact means through which social power is exerted. As we understand this, power is not presented as a resource but as something dependent on the use of resources. On the other hand, in the perspective of Lave and Wenger (1991), artefacts that, in their terminology, constitute the technology of the practice, or the resources, which in Wenger's terms include the repertoire of the practice, have a relevant role in the learning emerging from participation in a social practice. Therefore, artefacts, together with their social structure, are a part of the historical trace left by the reproduction cycles and they reveal the productive (not only the reproductive) character of those cycles and the contribution to the constitution and re-constitution of the practice over time. "Thus, understanding the technology of the practice is more than learning to use tools; it is a way to connect with the history of the practice and to participate more directly in its cultural life" (Lave and Wenger, 1991, p. 101).

3. CONCEPTS IN ACTION IN THE ANALYSIS OF THE STUDY

In the study reported in Santos and Matos (1998) two concepts were introduced: artefact and resource. The idea of artefact is quite strongly used in activity theory in parallel with the notion of *tool*. The concept of resource is explicitly used by Lave and Wenger (1991) in a situated approach. The study we describe here reflects the need we felt to go deeper into the discussion of the social nature of the human construction of mathematical artefacts.

3.1 Activity in activity theory

According to Davydov (1999), in the conceptual framework of dialectic materialism the notion of activity is an initial abstract. According to Bakhurst (1997), Ilyenkov elaborated on a theory of the ideal in which activity becomes literally part of the mind as the capacity to act in accordance with what is proper in a culturalized environment is constructed by thought, and therefore he identifies thinking as a kind of activity. Activity is not seen as an abstraction, however, but as the basic unit of analysis of consciousness.

On the other hand, Leontiev (1978) presents an approach of the concept of activity that draws from an idea of a structure with several components. Leontiev tried to establish basic categories of human activity that allow the possibility of researching the way individual consciousness is organized through particular and specific activities. One basic principle for Leontiev is the recognition of the constantly social and cooperative nature of human activity. Besides that, it is assumed that human individuality emerges from social activity, thus leading to a need to reflect on the relation between individual consciousness and the specific activities. For this purpose it is not enough to say that there is a relationship between consciousness and specific activities, we need to identify knots or particular units of activity that constitute a set in the personalities of the individuals (Axel, 1997) because it is on the basis of personalities that humans relate to and develop particular activities.

For Leontiev, activity is a molar unit, not an additive unit in the life of the person but a system with its own structure, its transitions and its internal transformations, its own development. Here we can identify non-adding elements linked to central concepts: activity (linked to a motive), action (linked to a goal) and operation (linked to conditions). The motive of the activity is intimately connected to the need felt by the individual: the form responding to that need. Activity may involve different processes, actions, that aim to produce certain results intimately related to the activity and in this way directing the activity. Action can be made concrete in different ways and forms, operations, according to the conditions available but always making sense in terms of the goal that is supposed to achieve.

In these terms, two methodological implications emerge: (i) activity cannot be reduced to a set of simpler stand-alone additive parts or processes, and (ii) its structural and functional unit can only be examined looking at the phenomenon in its active/live state. A particular activity is distinguished from another mainly by its goal and motives and this is what may help us as researchers to understand the development of the activity – in fact it is what makes the activity. Therefore, this perspective allows us, on the one hand, to identify elements of the activity and, on the other hand, to say that those elements have only a potential character, neither deterministic nor definitive, as activity can only be realized through a dynamic, transformative process of development.

3.2 Artefacts in activity theory

Taking the model of the structure of the system of activity proposed by Engestrom (1999b) we will concentrate on one of its elements, the artefacts, and in particular the links between the notion of artefact and the idea of mediation, one of the key concepts of the historical-cultural approaches. It is common to see people characterising artefacts in two ways: on the one hand people refer to tools and signs; and on the other hand, we find researchers considering external (or practical) artefacts and internal (or cognitive) artefacts. In both approaches it is often the internal character of the artefacts that is of interest, independent of the activity in which the artefacts are used.

Engestrom, speaking of the non-definitive nature of activity, considers that none of those dichotomised forms are useful. In activity, functions and uses of artefacts are in constant transformation and therefore elements that seem to be internal in certain moment can be externalized the next, for example through speech. Similarly, external processes can on occasion be internalized. Freezing or splitting these processes seems to be a poor basis for understanding artefacts (Engestrom, 1999b). Engestrom proposes a differentiation in regard to the uses of artefacts:

> The first type is *what* artefacts, used to identify and describe objects. The second type is *how* artefacts, used to guide and direct processes and procedures on, within or between objects. The third type is *why* artefacts, used to diagnose and explain the properties and behaviour of objects. Finally the fourth type is *where to* artefacts, used to envision the future state or potential development of objects, including institutions and social systems (Engestrom, 1999b p. 382, italics in the original).

An important implication of this classification is that artefacts should not be considered by themselves but always in relation to use within a system of activity. As Engestrom points out, the construction of objects mediated by artefacts is a process which is collaborative and dialectic in its core; one where different perspectives and voices meet, collide and mix.

3.3 Activity from a situated perspective

Lave, Murtaught, and Rocha (1984) asked what it is "about grocery shopping in supermarkets that might create the effective context for what is constructed by shoppers as 'problem solving activity'" (p. 68) where grocery shopping is seen "as an activity which occurs in a setting specialized to support it: the supermarket." (p. 68) and the supermarket as "an institution at the interface between consumers and suppliers of grocery commodities [...] The setting of grocery shopping activity is one way to conceptualize relations between these two kinds of structure. It may be thought of as the locus of articulation between the structured arena and the structured activity" (p. 74). Two activities are referred to here: grocery shopping and problem solving. It seems that one lives in the other as if they were two layers of activity; one of the activities (shopping) helps to shape the other (problem solving) through the setting that in a sense was created by the former.

Lave et al. see Barker's[15] conceptualization of the idea of setting as a "promising beginning to theorizing about activity, setting, and their interrelations" (Lave et al., 1984, p. 70) but they claim that "it assumes a unidirectional, setting-driven relation between activity and setting, which reduces activity to a passive response to the setting" (p. 70) precluding the analysis of the internal relationships of the activity. Lave recognizes as pertinent the way activity theorists conceptualize the idea of activity as system with structures, internal transformations, and self development as this view allows and creates a basis for the study of the intrinsic organization of activity. It is also recognized that the studies of Zinchenko about the holistic nature of activity, developed in the framework of activity theory, help to support the idea that to comprehend the nature, for example, of arithmetic activity as a whole requires a contextualized understanding of its role within that activity. This is a strong argument for the need to analyze any segment of activity in relation to the flow of activity of which it is a part.

Another relevant aspect of activity theory that receives Lave's attention is the parallelism found in the distinction made by Leontiev between personal sense and public, societal, meaning and the distinction Lave proposes between the constructs of personal setting and the public, non-negotiable, arena. In addition, the dialectic character of the analysis of activity is central to the situated perspective assumed by Lave et al.

From 1988 on, Lave uses the term 'ongoing activity' to refer to activity and this orientates our attention to the strongly fluid and dynamic character

---

[15] Barker (1968) sees the environment of behaviour as "relatively unstructured, passive, probabilistic arena of objects and events upon which man behaves in accordance with the programming he carries about within himself" (p. 4) in Lave et al., 1984, pp. 69-70.

of activity. It induces interrogation of the continuity of the activity, where it comes from, where it goes to. This is related to the holistic character of activity and it seems to call to the need to remember the local character of activity, developing here and now, with the resources and the constraints which are present in the situation for the actors in place. The ongoing character of activity seems consistent with Leontiev's view that activity should be analyzed in its active state.

Thus, Jean Lave's opting for an analytical focus of direct experience in a lived-in-world requires "reformulating the role of direct experience raising the question of how activity is made accountable while ongoing. An analytic focus on direct experience in the lived-in world leads to emphasis on a reflexive view of the constitution of goals in activity and the proposition that goals are constructed" (Lave, 1988, p. 183). This is not compatible with a linear view of action as directed towards established goals: "action is not 'goal directed' nor are goals a condition for action" (ibid., p. 183). Taking as support the idea from Wittgenstein and Giddens that it is through the recursive character of social life that it is possible to capture the nature of social practices as a continuous process of production and reproduction, Lave concludes that "the meaning of activity is constructed in action" (ibid., p. 184). Whence, therefore, comes the intentional character of activity?

> In this perspective, motivation is neither merely internal to the person nor to be found exclusively in the environment. That is, even as goals are not 'needs' (hunger or sexual desire are socially constituted in the world), they are not prefabricated by the person-acting or some other goal-giver as a precondition for action. And activity and its values are generated simultaneously, given that action is constituted in circumstances which both impel and give meaning to it. Motivation for activity thus appears to be a complex phenomenon deriving from constitutive order in relation with experience (Lave, 1988, p. 184).

More than adding a typical approach from activity theory, for which the external world is determinant, with a phenomenological reading that gives the 'power' to individuals there is a dialectic attempt to integrate aspects of the two theoretical fields that allow us to argue that setting and activity connect with the mind through its constitutive relations with the person-acting. Thus, instead of talking of goals (as in activity theory) a situated perspective claims that "expectations, dialectically constituted in gap-closing processes, enable activity while they change in the course of activity backward and forward *in* time at the *same* time" (Lave, 1988, p. 185, italic in the original). This is closely related to the way Jean Lave talks about intentions of actors in ongoing activity as they are "engaged in what they are doing. When that activity poses conflicts, difficulties, in short dilemmas, they engage in resolving them" (Lave, 1992, p. 80). The procedures adopted in solving them, which gain form and meaning in relation with those dilemmas, are finally what motivates their practices. It is the specific character of certain conflicts that shapes what are problems to be solved. What is seen as problematic in the activity emerges from and within that activity.

Finally, there is the notion that the ongoing character of activity associates the idea of transformation to the temporal dimension of the dynamic aspect of activity. If the context of the activity is not completely external to it, if activity is itself in movement and in interaction with the social world, context and activity structure each other and intervene in the transformations that occur.

3.4 Structuring resources and artefacts

For Giddens (1996), resources are ways through which transformative relations are incorporated in the production and reproduction of social practices. This means that resources are intimately connected to power, be it seen in a broad sense as an ability that transforms activity or in a more specific sense of domination or ability to intervene. Resources are means through which social power is implemented. Resources are the basis and the vehicles of power. Given that resources are equally structural components of social systems, they become also the means through which the structures of domination are reproduced. It is within this framework that Giddens considers that exerting power is not a type of action, power is instantiated in action while a regular and routine phenomenon. In this sense, power is not a resource but it depends on resources.

The way Restivo (1993) places mathematical objects in the framework of the social world where they are constructed and used can also be seen as part of the efforts to redefine artefacts in relation to people. For that redefinition it seems important that, in parallel to the material nature of the proposal of Restivo, we take into account the mediating and symbolic character of artefacts emphasized by socio-historical theorists and also recognized by situated learning perspectives. A strong claim of the mediating role of artefacts seems to be clear in the introduction of the book edited by Dorothy Holland and Jean Lave in 2001. These authors, assuming a theory of practice that emphasizes the processes of social formation and cultural production, look with particular attention to power of inscription of cultural forms[16].

---

[16] The notion of cultural form is close to Cole's conceptualization of cultural artefact; Holland et al. (2001) explicitly discusses "the materiality of cultural artefacts" (p. 63).

However, the terminology of artefacts offers and asks for relevant associations of ideas. On one hand, talking about resources raises two questions: whose resources? Resources for what? This forces us to think about the people who use the resources and supposes some intentionality or at least some space for initiative from those who use them. On the other hand, adopting a situated way of talking about resources, structuring resources, means recognizing them as elements of spaces of activity taking place in a social world which is structured[17] and where meanings are produced and reproduced. Thus, the notion of resources calls us to think about the users as persons who are part of communities and therefore are seen and thought of as social beings.

4. THE STUDY OF THE *ARDINAS*' PRACTICE

The study developed in Cape Verde addressed the key research question of understanding how learning can be conceptualized as "an integral part of generative social practices in the lived-in world" (Lave and Wenger, 1991, p. 35) and as "growing participation in communities of practice involving the whole person acting in the world" (p. 49). For the empirical field we decided to address a social practice that seemed to constitute a promising resource for the development of theory: the *ardinas*'[18] practice in Cape Verde.

The *ardinas* are young boys aged between 12 and 17 years, half of them in 5th to 7th grade at school, who sell newspapers in the streets of Praia, the capital of the Republic of Cape Verde in Africa. The two newspapers existing in Cape Verde during the time of the study (1998-1999) come out once a week and are written in Portuguese, the official language[19]. The number of *ardinas* who sold the two newspapers during the study varied from 19 to 32, all of them with no formal link to the institutions that owned the newspapers.

The artefacts assume an obvious and necessary material dimension and a conceptual or ideal aspect, an intentionality, whose substance is embedded in the world of its uses.

[17] We should underline here the dynamic nature of the structuring of the social world. Structuring is seen as a process or, as Holland and Lave (2001) put it, "produced in struggles or as struggles and never captured in global terms alone" (p. 6). When they use the word 'struggle' they try to avoid the notion of 'conflict' as something stable or self-contained and referring at the same time to the structure as "process, as a matter of relations in tension" (p. 23).

[18] *Ardina* is the Portuguese popular way to refer to the people who sell newspapers in the street.

[19] Although Portuguese is the official language, the actual language in use in everyday life is the Creole.

O Tempo, one of the newspapers, tried to implement a selling system based on the coffee shops and stationery shops but with very little success as the population did not adapt to this way of buying newspapers. Therefore, selling newspapers in the city of Praia was totally dependent upon the availability and interest of the *ardinas*.

There was no special external sign, such as a special t-shirt, a bag or a cap, that could help one to identify the *ardina* in the street, except for the fact that he was carrying a number of newspapers under his arm. However, the *ardinas* were careful in the way they dressed on the selling days as this represented a key issue to gain access to certain places of selling, for example, in state departments. Most of the *ardinas* were motivated by the idea of getting some extra money to help their families and a number of them were in this practice for six years. These boys lived both in the city of Praia and in the nearby small village of S. Martinho.

The work of the *ardinas* developed in three different phases: (i) receiving the newspapers from the agency, (ii) selling the newspapers on the street, and (iii) paying back the money to the newspaper agency. The organisation of these three phases was necessarily connected to the instructions of the owners of the newspapers but the *ardinas* positioned themselves in that organisation in their own way.

Every Friday morning, in the main building of *O Tempo*, the newspapers were delivered to Disidori, the man who was responsible for the whole process of selling and returning back the unsold newspapers. At the door of *O Tempo*, Disidori distributed a number of newspapers to each *ardina* (between 50 and 150 units) writing down a list of the names and the number of newspapers given to each one, this list being the reference document for the final phase of paying back.

The link of the *ardinas* to the newspaper agency was very informal and based on personal relationships with Disidori. He had a link to the administration of the newspaper that was made visible to all when he signed a document against the delivery of the newspapers (that made him responsible for paying back to the administration).

Immediately after receiving the newspapers the *ardinas* run very quickly to the usual places for selling in the city, some of them trying to maintain their own selling position in the street. However, those places varied during the day according to the rhythm of selling and the rhythm of the city (namely, in the street during rush hour, at workplaces during the work times, close to the restaurants at lunchtime).

The price for one newspaper for the customer was 100 *escudos*[20] and, when they finished selling, the *ardinas* had to pay back to Disidori 87.5

[20] 100 Escudos from Cape Verde corresponds to 1 Euro.

escudos per newspaper sold (plus the non-sold newspapers), these amounts being defined by the newspaper administration.

Besides the strategic role of interaction with the *ardinas* in the integration of the newspaper selling into the socio-economic life of the city, the central Square in Praia was the place where Disidori stayed for long periods during the day of selling. He also walked around to the different places where the *ardinas* were selling in order to check how the process was going. Some time after the distribution of the newspapers by the *ardinas*, Disidori used to go to the Square carrying with him a set of newspapers for those *ardinas* who were in school, and hence could not attend the distribution of the newspapers at the door of the agency. He could distribute also to those who sold out very quickly and asked for more newspapers. The Square was the main point of convergence of the boys at several moments during the day: (i) as a lunchtime resting place for those who did not sell in the restaurants and (ii) when they finished selling and came to pay back to Disidori.

As the *ardinas* finished selling the newspapers (or got tired of selling) they started showing up for payment. The *ardinas* approached Disidori, saying how many newspapers were left or how many they had sold. Disidori made the calculation with his pocket calculator; he showed the result on the calculator screen to the *ardina* who then gave him the money. Sometimes the more independent *ardinas* with made their own calculation with their calculator or Disidori's, but several operations were in progress: some of the *ardinas* were counting the number of newspapers left for returning; others were counting and organising the money according to the value of bills and coins; they deliver newspapers to Disidori, observe the calculation, give the money to Disidori. The environment could seem confusing at a first glance as there was a lot of money in sight changing from hand to hand but when we observed in detail we understood that everything was running in a certain order and that this allowed each *ardina* to see what was going on with the calculations – their own or those of a colleague.

This phase represented a very important step in the practice of selling newspapers. The *ardinas* exchanged stories of the day, and they had face-to-face discussions and they organized and prepared the moment of making the final account with Disidori. In the complexity of the different activities that were taking place simultaneously was a very rich opportunity to observe how the *ardinas* interpreted and solved their problems. It is worth noticing the total absence of any attempt to make their calculation strategies explicit either through verbal explanation, deliberately showing, checking processes or anything we could classify as some sort of mathematical conversation.

Although the first author (acting as the field researcher) was present during the whole process of selling, this final phase in the Square was one of the best settings to collect data for the study. There were plenty of opportunities to talk with the *ardinas* in a quite natural way videotaping informal interviews whose guidelines were mainly directed by the topics the *ardinas* wanted to talk about or the problems they were discussing among themselves.

In 1999 a second newspaper *O Espaço* came into the market and Disidori moved to a new position at this new newspaper. The majority of the *ardinas* were involved in selling both the newspapers. New rules were in action: for each newspaper sold the *ardinas* would receive 20 *escudos* (instead of the previous 12.5 *escudos*) but they had to go to the newspaper agency to receive the newspapers and go back there after selling in order to pay for the newspapers sold.

5. WHAT EMERGED FROM THE DATA ANALYSIS OF THE *ARDINAS'* PRACTICE

In the everyday practice of selling newspapers, the *ardinas* interact with several resources. This was one of the foci in our study and its importance comes from the fact that in order to understand and characterize the processes of calculating-in-action it was relevant to identify some of the artefacts used and to understand how those artefacts are constituted while structuring resources. We will concentrate on who uses or creates the artefacts, who has the power in using them and for what that power is used. In doing so, we will highlight the way each artefact, as a historical and mediating tool, is present in the ways of acting and thinking. The focus of analysis will be the presence of two artefacts, the table to record the selling of newspapers and the calculator, in the 'everyday' of the *ardinas* and their interactions as it became relevant to consider some aspects related to power.

5.1 First encounter: what is in a table and what does it tell us?

The appearance of the new journal (*O Espaço*) provoked profound modifications in the whole structure and organization. Manu, one of the *ardinas*, started helping Disidori in his relations to the *ardinas*, in particular in the distribution of the journal, in the support and control of what was going on in the Square and in the reception of payment at the end of the day. At the end of the day Disidori visited Manu in his home to make the final accounts noting the journals left and the money from selling.

Manu's responsibility was now greater. He felt the need to create a form of recording in writing that allowed him to keep control of a large set of data – amounts received, notes of the selling, payments from the *ardinas*, etc. Manu used an A4 sheet to record everything that had to do with selling the newspapers for the whole week. He considered this record fundamental to his ability to be accountable to Disidori. At the top of the sheet he wrote the date (e.g. 19/3/99) and the number of newspapers he received (e.g. 550+55=605 not visible in figure 1). On the same sheet he made some computations in writing and notes of events that would be useful in his dealings with Disidori at the end of selling.

This sheet had the role of a memo, "to not forget it" as he said, about the major responsibility of his activity. The central organizer of that record was a table with five columns labelled 'name, took, sold, rest, money'. This table was completely different to the one used by Disidori the previous year. Manu's table had the names of the several *ardinas* at the start of lines across the page, together with the number of newspapers each one got from him. As soon as the *ardinas* finished selling they came to Manu, gave him the money and showed the number of newspapers left. Manu used to count the number of newspapers and the money and fill in the table with that data. Finally at the end of the line he wrote the word "Paid". When the *ardina* took a new set of newspapers to sell, Manu inserted a new row as if a new *ardina* was coming. It became clear that in fact the design of the table referred not to the *ardinas* but to deliveries of newspapers.

A first reading of this table shows two facts that indicate the existence of a double point of view which we can connect to the duality of the status of Manu: on one hand he is an *ardina* and on the other he is coordinator of a group of *ardinas* – evidence of the ambiguity of his role in the selling. He uses the word 'took' as if referring to the *ardina* – the one who takes newspapers, suggesting that he is positioning himself as an *ardina* even if he is in a new role. He could use the word ‘delivered’ putting in the centre his activity and giving visibility to his role of distribution and control.

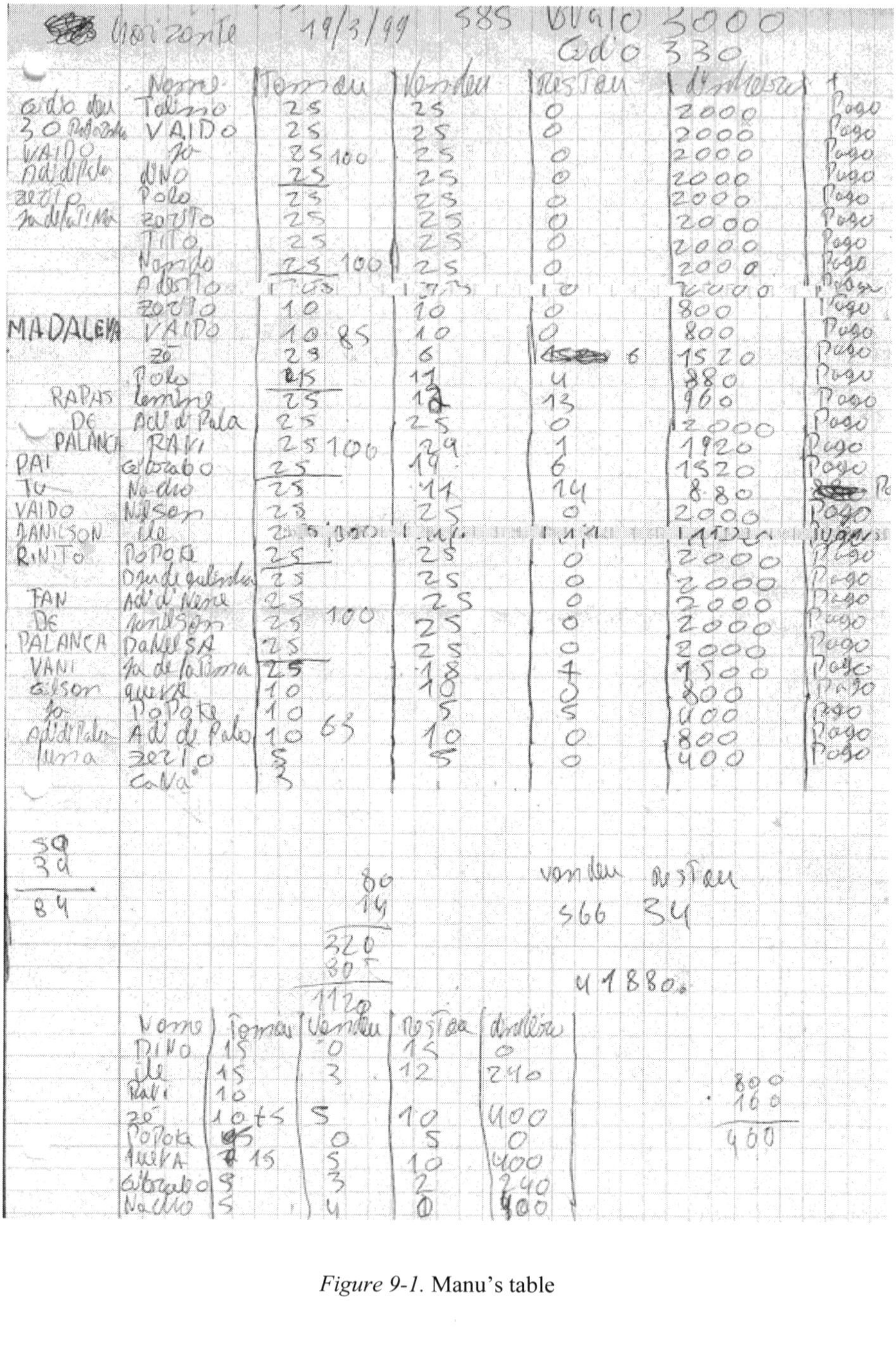

Figure 9-1. Manu’s table

But he enters a new row in his table for each new delivery of newspapers, even if it referred to the same *ardina* who already sold in that day, and this shows that he is careful in adopting a form of organizing the data that is simple to use during the selling when he is controlling in action. A 'clean' table, with no corrections, seems to be a tool for ease of recording, easy to read and transparent to everybody, including not only Disidori but also the *ardinas* themselves. Manu gives them access to the data inserted from the very beginning of the selling transactions.

One should refer to the discrepancies identified in the table (See Figure 1). Even if there are almost no errors, some entries are incoherent. For example, there is discrepancy between the number 605 on the top of the page in 550+55=605 (which Manu used as the number of newspapers he delivered to the *ardinas*) and the total number of newspapers inserted in column named 'took'. Is it the case that Manu had a margin of error in the number of newspapers in order to guarantee that he finally has the right amount to pay back to Disidori? Or is it the case that an approximate computation is part of the practice of selling in tune with a not-quite-well-planned operation from the very beginning of the printing of the newspapers?

The organization of the distribution of the newspapers reflected the articulation between the way the distribution was started by Disidori (the voice of the institution) and the emergent form of relationship between Manu and the *ardinas*. This introduced a 'breathing space' that allowed that the organization adapted to the rhythm of the city and the people, the specificity of the *ardinas*. The rigid structure of the table finally seems to be quite flexible in the hands of Manu as it was adjusted to the social world of the *ardinas'* practice and to the organization that was framing it.

We can see in the table that most of the *ardinas* received 25 newspapers at the beginning whereas in the previous year the *ardinas* used to receive from 25 to 100 or 150 newspapers. The reason behind this is the fact that selling was now more and more difficult given that there were two different newspapers in the market. On the other hand the number of *ardinas* increased and this led to a decision from Manu and Disidori to distribute fewer newspapers to each *ardina* allowing a greater number of them to make some money. Thus, it seems that the strategy implemented by Manu, according to the numbers introduced in the table, does not take as a key issue making quick sales to sell more newspapers than the other newspaper agency. On the contrary, the strategy followed shows an adjustment to the social world to which the *ardinas* belong – families with weak resources to whom the earnings of an *ardina* may be the only money coming in during the whole week. An institutional stance would suggest the strategy of delivering more newspapers to each *ardina* as leading to quicker selling on the street. Delivering an equal number of newspapers to each one allowed for the dynamic of the *ardinas* to control the quantities delivered, and not the institutional power.

The forms of calculating that Manu adopts for different categories of values brings in another relevant issue. In the column of the newspapers distributed ('took') the existence of a sub-column within that one (grouping the numbers in sets of 100 or close to this number) shows evidence of mental computation in action. The regularity of the amounts delivered (most of them 25) and the organization imposed by the process of putting together the pages of the newspapers at the agency (in groups of 25, thus sub-multiples of 100) can help to accept as obvious that the number 100 acquired the status of 'unity'. In the other columns, even for small amounts, we can discern the use of the pocket calculator – and Manu confirmed this. While explaining how easy it was to add the numbers in the column 'took', Manu showed the need to use the calculator in the other columns. However, he was not aware of the discrepancies of numbers in his table. This can be interpreted as a sign that in fact the record on the sheet was more a resource to give him a sense of security, to detect if something was radically wrong, and to give a message to the *ardinas* that he was in fact in control of the situation. The key issue was a happy end when the selling was over and the final account made.

Manu's record, in particular the table format, is an artefact with an historical trace of the practice of selling. The analysis of its use allowed us to understand how the history of the practice is relevant to the lived moments of other participants, making it present and revealing its ideal character and not only its material substance (Cole, 1997, 2001). This form of addressing the artefact makes knowledge visible while being part of the repertoire of a community that exists through time, and that is neither an isolated entity emergent in the mind of each individual, nor coming exclusively from the practice itself. This view of artefact is close to the perspectives of activity theory. On the other hand, although the Manu's table is constructed in a given organizational moment of the practice (distribution) and used for the recording of situations of distribution, its construction also addresses other uses during the practice: selling and payment. Besides that, the detail of the records made shows that the function of the table lasts longer than the strict time of selling as it stretches into the time period when Manu becomes accountable to Disidori.

In the case of Manu, the object which he shares with the other *ardinas* is selling during a given day. Within the global activity of selling newspapers the selling of a given day can be seen as an entity, the problem, that Manu must control, i.e. describe, re-construct and re-define in order to fulfill his need to be accountable to Disidori (the real motive for the construction and use of the table).

As an artefact, the table can be classified as an *artefact-what* (Engestrom, 1999b) as it describes the final situation of the distribution and selling for a day, and an *artefact-why* given that Manu identifies and explains with that table some behaviours that call for his attention during the selling both with the *ardinas* and with Disidori. This is apparent for example in the fact that Manu organizes the introduction of rows in the table for each delivery of newspapers and not according the names of the *ardinas* who take the newspapers. The form of use that Manu gives to the table is more framed by the activity than by the orientation that it could impose over the activity – thus not being recognized as *artefact-how*. Manu does not explore the potential for the table to become an *artefact-what-for* that predicts forth-coming situations or expands the development of the selling.

The power of Manu in the sale of newspapers, his status, does not come from the use of the table or of any other artefact, but from the way he uses the table and the way he uses artefacts as resources (Giddens, 1996). For example, it is Manu who defines which names are entered in the table, how he organizes it, how he uses it. It is Manu who makes the table accessible to the *ardinas* for reading but not for writing – they could consult, point, ask questions, but not add anything. It is also Manu who foresees uses of the table beyond the obvious, for example, when calling the attention of one *ardina* to his weak selling rate compared to others.

The table as it was created by Manu appears to answer his double need to control the *ardinas'* work and to answer Disidori's control over him. This seems to ensure Manu some security, and this is extended to other *ardinas* who consult the table and compare their performance in selling with others, even those they do not meet because they sell at different times and places. We see here a form of self-regulation that helps them to position themselves in the practice and towards the patterns described on the table. However, the transparency of the artefact, as Manu constructed it, involves visibility and invisibility (Lave and Wenger, 1991). The visibility of Manu's table allows a comparative perception of the behaviours of the several *ardinas* but only in relation to aspects relevant to the needs of the institutional power. With the table it is possible to control the amount of newspapers distributed and sold and the amounts paid, and to signal the *ardinas* who follow the rules for paying; but, for example, it gives no visibility to the earnings of each *ardina*. In its form and function, the table is an artefact of selling. Even if it was designed by Manu for himself as the key user (because it deals with his role of controlling the practice) the table has a presence in the everyday of the practice of selling of the *ardinas* as a collective entity. For example, the table contributes to the collective valuing of the practice of paying but it leaves the earnings of each one private.

The table reinforces institutional interests but it also contributes to the perception of belonging to a community among the *ardinas*. While making visible who is on the list (and thus showing who is not) and allowing some individuality to emerge, the use of the table contributes to the identification of who is who. The historical trace of the table seems to have the strong reproductive character of an institutional power and of the dominant culture that values being serious, but also reveals productive characteristics by contributing to the constitution of the meaning of the existence of a community. This double character comes in part from the fact that the artefact was created by someone who participates in both domains (as an *ardina* and as someone charged with controlling the *ardinas*) and in this frame constructs and uses an artefact that reifies aspects linked to the institutional conditions while maintaining an integral part of his identity as a participant in selling.

5.2 Second encounter: the calculator as artefact; what does it tell us?

In several moments of the selling practice a calculator was used by Disidori, Manu and other *ardinas*. One of the key issues here was that although the calculator was one of the few tools present in the practice of the *ardinas* that was associated to mathematics, in fact it was not a computation tool that they would have appropriated. In 1998 the use of the calculator was visible only while the *ardinas* made the final accounts with Disidori. The calculator was not an artefact present in the everyday life of most people and not used in school, although some people selling goods in shops used them (but not in the municipal market). It was a technological element with a quite restricted and limited impact in the daily practices of people and associated with specific domains outside the range of the mainstream of the population.

The data collection for this paper took place in a period of the life in Cape Verde when electronic technology (laptop computers and mobile phones) was seen by most people as 'magic' and 'automatic', something that people did not understand well but that carried a strong degree of power and rightness, something that people wanted to use as it had the social connotation of 'serious and important business'. The calculator is thus a technological tool within the category of desired and socially valued objects, although with a quite restricted access. Therefore, it was unnatural for *ardinas* to use the calculator when acting as *ardinas* as it was not a priority for them to have their own calculator. And in fact although the calculator was used everyday in the practice of selling, the interaction that the *ardinas* had with it was very limited both in frequency and in the nature of its use.

In 1998 the manipulation of the calculator was both associated with the phase of payment and with the need to control the selling situation. With regard to the forms of use adopted by Disidori we noticed a special focus on the role of the calculator in the interaction with the *ardinas* and the control of the selling process. When paying back to Disidori the *ardinas* were immersed in a routine dominated by the action of Disidori, and the calculator was an element always present and visible to the *ardina* he was addressing. In fact, it was through the calculator that he 'organized' his intervention with the *ardina*, an *artefact-how*, as it guided his procedure with the *ardinas* making it possible for each *ardina* to understand his particular situation. As he slowly typed in the numbers and operations, naming loudly each one, he transformed the calculator into a kind of 'guarantee', an *artefact-what*, because he described the situation of selling in detail. Giving access to the *ardina* to follow the whole process was a way not only of showing that he was not cheating but also of introducing a positioning of attention and honesty that Disidori found useful and important for the development of the autonomy of the *ardinas*. When finishing the process of computation he said out loud the amount that the *ardina* had to pay showing him the calculator screen. Here the calculator was used as medium for communication that allowed Disidori to offer to the *ardinas* different forms of representation of the amounts and this was important to those *ardinas* with less experience in arithmetic as they were learning about numbers and operations. In the frame of the social meanings associated to the technological tools (e.g. belief in the infallibility of the results produced) the calculator also presented the characteristics of *artefact-why*: if the numbers inserted were correct, the result would be correct also.

But the calculator was not used only as a resource for communication. When finishing the computations in order to make the final account with Disidori some of the older and more experienced *ardinas* effectively used a calculator for self-control. Thus the calculator had the status of a tool for confirmation of the final selling situation, an artefact-*why*, an affirmation of autonomy as it avoided a long interaction with Disidori. Our evidence for this is that newcomers were more dependent and talked longer with Disidori. This form of using the calculator can be seen as a non-structuring resource where the artefact does not play a major role in organizing the way the *ardinas* interact.

A key question is what do the *ardinas* learn while participating in these forms of interaction? We suggest: (i) a reinforcement of the hierarchical structure (not only dealing with age but also with experience and responsibility) underlying the social world where selling takes place; (ii) forms of talk and signs that become part of the repertoire of the social practice of calculating-in-selling – the decimal point, the operations (times, more), the big numbers; (iii) the place of profit in the sequence of actions that allow earnings to happen (what remains after paying back) and the place of the common *ardina* in the hierarchy of positions of power in the chain involved in the selling.

In 1999, the calculator became less of a tool for control and a marker of power. Manu gave less visibility and less organizing role to the calculator as his table occupied the major role. At the same time, the close and affective relationships of Manu with most of the other *ardinas* made it difficult to develop the formality and the rituals that were the norm when Disidori was in charge. An additional issue was the smaller need to use the calculator given that the amounts involved were easier to deal with: 20 escudos of profit (instead of 12,5 escudos in 1998) and 80 escudos to pay back (instead of 87,5). As most of the *ardinas* received 25 newspapers and sold them all there was less variation in the situations. The *ardinas* were now more disconnected, the field of selling was less adapted to ritual moments involving the whole group as more *ardinas* were in school during part of the day. Here the calculator was just a tool for computation for those who needed it and there was no evidence of a mediating role as artefact as noticed in 1998.

The calculator was a useful tool with an inherent character of rarity – an object that we seldom see in the hands of people – and of infallibility as no errors are allowed or imagined, and it justifies the amounts to be paid by the *ardinas*. But the way Disidori had used the calculator gave rise to a variety of learning opportunities for the *ardinas*. For example, they enlarged their repertoire of forms of naming and representing numbers and operations and they had support in learning to respect hierarchy as the calculator was used by the *ardinas* with more and more autonomy according to their experience of selling.

The perception that each *ardina* had of their selling behaviour was mediated by subjective interpretation of the situations and the feedback that Disidori gave through several signs. On the other hand, the almost individual nature that framed the payment phase was reinforced by the fact that Disidori used the selling record for his own use and gave a central role to the calculator in the interaction with the *ardinas*. In addition, as the state of the account of each *ardina* was reconstructed for himself alone, and because the screen of the calculator was only visible to the two participants, it was not possible in this process to compare his selling to that of the other *ardinas*. The way Disidori used the calculator reinforced this framework emphasising the power of the institutional organization while keeping the *ardinas* 'dependent' on the way they are considered by the 'authority'.

In the practice of the *ardinas* the calculator was an artefact for which the transparency of use by Disidori during payment gave visibility to those who were involved in the relation during payment, but it kept invisible the forms of computation which gave the final account. Therefore, in relation to mathematical thinking inherent to the computations, the calculator was not transparent at all. The visibility was given only to the sequence of actions and this was obscuring the processes behind the results. It was apparent that the calculator had no impact in the ways the *ardinas* calculated, and it did not mediate their thinking in the computations. The *ardinas* who had almost no familiarity with the calculator, when challenged by the observer to solve a problem using the calculator, always started by calculating their profit and they were not able to explore any other type of manipulation.

In the hands of Disidori the calculator seemed to be an artefact that essentially assumed a character of reproduction of the existing social structure. In the hands of Manu not even that character seemed to be present, it was just a tool for computation in specific moments. The way the calculator was present in the 'everyday' of the *ardinas* in 1999 also had no role in inducing a new meaning beyond the one embodied in the global culture of Cape Verde. It was easier to do some computations but not all, it was more accessible and its use was more generalized and thus it was not seen as distant and 'magic'. But it had no relevant role in the form the *ardinas* calculated.

The historical trace that the use of the calculator as artefact introduces into the practice of calculating in selling seems restricted to being a tool for computation. However, its major contribution seems to be in the regulation of the participation of the *ardinas*. But this role of the calculator was localized in 1998 when the form in which it was used was consistent both with the social meaning to which it was associated and with the structure of the broader social world.

6. ARTEFACTS AND RESOURCES: HOW THE TECHNOLOGY OF THE PRACTICE PRODUCES A SHARED REPERTOIRE

We were able to identify and describe the distinct forms of using the record table and the calculator and realize that their presence in the 'everyday' of the *ardinas* had different roles. For Manu, the calculator was a tool for computation that he used mostly for himself when checking the final accounts. It was rarely used when each *ardina* was paying. But the table was much more present in the practice and organized in a way that guaranteed real access to it. The table was a resource for securing Manu against the newspaper agency, valuing the moment of payment and regulating the activity of the *ardinas* while offering visibility to their place as members of a community. Thus, in the practice of the *ardinas*, it had a reproductive role (of the institution) and also a productive one, contributing to the idea of community. As Manu created it for his new responsibilities, the table reified aspects of the practice that he valued in his position but in a way reflecting his full participation as an actual *ardina*. For Disidori, the calculator was the more systematic tool used in the moment of payment of all the *ardinas* while the table was an extension of his memory or a support for his own computations but with no real impact on the *ardinas*. However, the calculator was the best support for the dialogue with the *ardinas* reinforcing on one side the individual character that Disidori gave to participation of each *ardina*, and on the other side the visibility of his authority. Within this framework, the calculator assumed essentially a role of reproduction as it was used in order to regulate the participation of the *ardinas* but sustaining the established social order (mostly hierarchical and almost no argumentative element).

The artefacts, while mathematical, had a structuring role in the activity of computation in 'paying back' although there was a small impact in the processes of calculating-in-action. They reinforced the act of paying back and the role of authority and regulated the participation of the *ardinas*. However, the regulation that was made possible by these artefacts did not come from the artefacts themselves but from the way they became present in the everyday and the functions they had in the practice. Thus, such regulation was in accordance with aspects connected to the social world that framed the resources and the activity and the people who organize, manage and act in it. But if the calculator in the hands of Disidori had an effect mainly reproductive of the social world, the table in the hands of Manu (beyond the reproductive effect) had a major role in the production of the community of practice of the *ardinas*. Such a role was able to emerge because the table was conceptualized by a 'senior' *ardina* reifying aspects of the practice where he was a full participant. In contrast, the calculator was introduced as an object already reified (with associated strong social meanings) serving someone not part of the community of practice of the *ardinas* but an officer of the institutional power.

We can conclude that these artefacts, although not being totally appropriated by the *ardinas*, were resources that had some role in structuring their activity of computation-in-action. There was a shared repertoire of the practice of the *ardinas* that reflected the framework of the activities of computation in the social practice of being an *ardina* but the repertoire used to compute-in-action was a different one. It was based on elements: (i) that emerged both in the structure of the broader social world where the activity was developed and in how people intervening in the activity coordinated it; (ii) that reflected the motives that were behind their participation in the practice. What really was structuring the activity of the *ardina* was 'paying back' (being honest) while what was structuring the computation-in-action was the 'gain' (the need of the *ardina*).

It is the articulation of participation and reification within the practice that allows and orients the construction or re-construction of artefacts with potentialities of going on functioning as resources for new needs that could emerge in the evolution of the responsibilities of the participants in the activity of control of selling. As Engestrom (1999b) puts it, the functions and use of artefacts are in a constant fluidity and transformation that goes along with the development of the activity. In this sense, the artefacts are not something fixed and external to the practices but are in the development of the practices; their usefulness is not revealed in the characteristics identified independently of use in the practices where they are put in action. Artefacts are artefacts-in-the-practice; they should be understood in interaction with the forms of use that users develop in those practices. On the other hand, the objects in relation to which artefacts are considered should be framed in a broader sense. For example, the table is a general tool for Manu but the analysis of its use shows aspects of the practice that interface with the particular selling activity of one day.

Our final remarks are about two key concepts in situated learning that we articulate in relation to the artefact and resource: the technology of practice and shared repertoire.

The idea of technology of the practice (Lave and Wenger, 1991) introduces a set of elements, artefacts, which people act with, associating a practice with the existence of a particular technology. Lave and Wenger (1991) stress the cultural nature of artefacts which carry part of the cultural heritage of the practice and relate their use to matters of power and access in the context of the discussion of the problematic character of the reproduction of a community of practice. The notion of technology suggests a certain stability, accepting renovation and transformation while based in the history of the practice and its membership. Entry into a new space of participation is thus associated with learning about its history and its technology.

The idea of shared repertoire (Wenger, 1998) refers both to a set of elements and a group of people who share them as resources for action and communication. The very word 'repertoire' leads us to aspects different from those related to technology and closer to forms of talking, acting and doing, or to stories that people tell and share. This means a broader spectrum beyond action that presupposes an audience to whom and with whom one acts. Such a collective entity shares constraints and affordances which involve action and interaction. And it is in that process that meanings and positioning are negotiated, reproduced and constructed. The notion of shared repertoire directs the attention to the dynamics of using, constructing and sharing certain resources and calls for a view of people as collective constructers of something and thus collectors of their own constructions. In doing so, it localizes knowledge on the collective and on the circumstances where the collective produces knowledge, uses it and reproduces it. The shared repertoire, reflecting the coherence of the practice, emerges as a source of the coherence of the community of practice (Wenger, 1998).

Talking about shared repertoire of a community of practice is talking about something in construction *via* participation and side by side with reification. The technology of the practice consists of those reified aspects of the practice that almost shadow the role of the practitioners in its construction, while maintaining a memory of its development.

The two concepts are complementary. Focusing on each one gives visibility to particular aspects, in one way to the process of construction (e.g. what facilitates or restricts the access to participation the practice) and in another way related to the history (that which allows or restricts access to meanings, comprehension and to the practice itself). In both concepts the key idea of participation is present and it is through participation that one contributes to construction and has access to history.

Both the artefacts analysed in this chapter, the recording table and the calculator, were present in the everyday of the *ardinas* as reifications; they made part of the technology of the practice of selling newspapers in Praia that every newcomer faced. The ways they were used in the practice gave visibility and reinforced the institutional order inherent to the social world where the action of the *ardinas* was taking place. Through interaction with such artefacts the *ardinas* gained access to certain aspects of the practice of selling, sharing meanings of the social world where the selling took place. The artefacts constituted structuring resources although with less direct impact on the strategies of computations of the *ardinas* which were appropriate to their participation in the selling. Such a contribution was rather more visible in the forms of talking and in the social meanings developed by the *ardinas*. The mediating character of the two artefacts in the mathematical-thinking-in-action of the *ardinas* was revealed in association to social meanings.

REFERENCES

Axel, E. (1997). One developmental line in European activity theories. In M. Cole, Y. Engestrom & O. Vasquez (Eds.), *Mind, culture and activity: Seminal papers from the laboratory of comparative human cognition* (pp. 128-146). Cambridge: Cambridge University Press.

Bakhurst, D. (1997). Activity, consciousness, and communication. In M. Cole, Y. Engestrom & O. Vasquez (Eds.), *Mind, culture and activity: Seminal papers from the laboratory of comparative human cognition* (pp. 147-163). Cambridge: Cambridge University Press.

Barker, R. G. (1968). *Ecological Psychology: concepts and methods for studying the environment of human behaviour.* Palo Alto, CA: Stanford University Press.

Davydov, V. V. (1999). The content and unsolved problems of activity theory. In Y. Engestrom, R. Miettinen & R.-L. Punamaki (Eds.), *Perspectives on activity theory* (pp. 39-52). Cambridge: Cambridge University Press.

Engestrom, Y. (1999a). Activity theory and individual and social transformation. In Y. Engestrom, R. Miettinen & R.-L. Punamaki (Eds.), *Perspectives on activity theory* (pp. 19-38). Cambridge: Cambridge University Press.

Engestrom, Y. (1999b). Innovative learning in work teams: Analysing cycles of knowledge creation in practice. In Y. Engestrom, R. Miettinen & R.-L. Punamaki (Eds.), *Perspectives on activity theory* (pp. 377-404). Cambridge: Cambridge University Press.

Engestrom, Y., & Cole, M. (1997). Situated cognition in search of an agenda. In D. Kirshner & J. A. Whitson (Eds.), *Situated cognition* (pp. 301-309). New Jersey: Lawrence Erlbaum Associates.

Giddens, A. (1996). *Novas regras do método sociológico* (António Escobar Pires, Trad.) (2nd ed. Vol. 34). Lisboa: Gradiva.

Hedegaard, M., Chaiklin, S., & Jensen, U. J. (1999). Activity theory and social practice: An introduction. In S. Chaiklin, M. Hedegaard & U. J. Jensen (Eds.), *Activity theory and social practice* (pp. 12-30). Aarhus, DK: Aarhus University Press.

Holland, D., & Lave, J. (2001). History in person: An introduction. In D. Holland & J. Lave (Eds.), *History in person: Enduring struggles, contentious practice, intimate identities* (pp. 3-33). Santa Fe: School of American Research Press.

Jensen, U. J. (1999). Categories in activity theory: Marx's philosophy just-in-time. In S. Chaiklin, M. Hedegaard & U. J. Jensen (Eds.), *Activity theory and social practice* (pp. 79-99). Aarhus, DK: Aarhus University Press.

Lave, J. (1988). *Cognition in practice: Mind, mathematics and culture in everyday life.* Cambridge, USA: Cambridge University Press.

Lave, J. (1992). Word problems: A microcosm of theories of learning. In P. Light & G. Butterworth (Eds.), *Context and cognition: Ways of learning and knowing* (pp. 74-92). Hemel Hempstead: Harvester Wheatsheaf.

Lave, J., Murtaught, M., & Rocha, O. (1984). The dialectic of arithmetic in grocery shopping. In B. Rogoff & J. Lave (Eds.), *Everyday cognition: Its development in social context* (pp. 67-94). Cambridge: Harvard University Press.

Lave, J., & Wenger, E. (1991). *Situated learning: Legitimate peripheral participation.* Cambridge, USA: Cambridge University Press.

Leontiev, A. N. (1978). *Activity, consciousness and personality* (M. J. Hall, Trans.). Englewood Cliffs, NJ: Prentice-Hall, Inc.

Restivo, S. (1993). The social life of mathematics. In S. Restivo, J. P. V. Bendegem & R. Fischer (Eds.), *Math worlds: Philosophical and social studies of mathematics and mathematics education* (pp. 247-278). New York: State University of New York Press.

Santos, M. P. & Matos, J. F. (1998). School mathematics learning: Participation through appropriation of mathematical artefacts. In A. Watson (Ed.), *Situated cognition and the learning of mathematics* (pp. 104-125). Oxford: Centre for Mathematics Education Research, University of Oxford.

Toulmin, S. (1999). Knowledge as shared procedures. In Y. Engestrom, R. Miettinen & R.-L. Punamaki (Eds.), *Perspectives on activity theory* (pp. 53-64). Cambridge: Cambridge University Press.

Wenger, E. (1998). *Communities of practice: Learning, meaning and identity*. Cambridge, USA: Cambridge University Press.

Chapter 10

Exploring Connections Between Tacit Knowing And Situated Learning Perspectives In The Context Of Mathematics Education[21]

Cristina Frade* and Jorge Tarcísio da Rocha Falcão**
Universidade Federal de Minas Gerais, Universidade Federal de Pernambuco

Abstract: This chapter explores connections between theories of tacit knowing and theories of situated learning and communities of practice aiming at a better understanding of school mathematics as a socio-cultural practice. By contrasting both school and other socio-cultural mathematical contexts, we discuss the usefulness of a perspective from which the identification and circulation of tacit knowing within school mathematics practice is a central concern. Empirical data are presented to illustrate our ideas.

Key words: explicit-tacit knowing, competencies-in-activity, situated learning, school and out-of-school mathematical practices, mathematics education

1. INTRODUCTION

Our motivation in investigating the tacit-explicit dimension of mathematical activity, in particular mathematics teaching and learning in the light of situated cognition and communities of practice, comes from the convergence of two main experiences. The first experience refers to studies of

[21] This chapter draws extensively on ideas that have been published in Frade (2003), Frade and Borges (2006) and Da Rocha Falcão (2005).

* Supported by FAPEMIG and CNPq. ** Supported by CNPq

mathematical activity in culturally meaningful contexts outside school, developed in Brazil by Terezinha Nunes, Analúcia Schliemann and David Carraher during the years from 1980 to 1995. These studies shared a common theoretical and methodological characteristic in their ways of focusing on mathematical activity and interpreting data. This characteristic can be roughly summarized as follows: mathematical activity is not limited to school context only. The studies of Nunes and colleagues (1993) demonstrated that very poor children from some Brazilian villages can make complex calculations about money, commercial costs and profit, and so on, without being able to solve mathematically isomorphic problems in school. Mathematical knowing on the basis of this street-mathematics activity might be seen by other theoretical approaches (e.g. Piaget, 1974) as 'hierarchically inferior' to mathematical knowing developed at school. This assumption concerning the differences of 'power' between *school mathematics* and *street mathematics,* following the expression proposed by Nunes and colleagues, motivated us to develop new contexts of observation in which mathematical activity would not necessarily be related to school mathematics patterns.

One of these contexts is discussed by Da Rocha Falcão (2005) and refers to a specific community of Brazilian fishermen, the *jangadeiros* from Recife. Da Rocha Falcão argues that these fishermen are able to pilot their sailing boats according to vectorial principles of composition of the direction and intensity of the wind and the orientation of the sail and keel. Most of these fishermen are illiterate and possess no conceptual-vectorial schemes at all. This is not the case for amateur sailing apprentices who learn in sailing courses how to navigate; these people, most of them belonging to upper socioeconomic and school levels, are able to discuss sailing and to explain the central principles theoretically, alongside significant competence in piloting their sailing boats. Nevertheless, there is an impressive example of a practical principle used by Brazilian illiterate fishermen that does not have a correspondent among amateur sailing apprentices: most of the *jangadeiros* know that they can improve the velocity of their sailing boats, *jangadas,* by throwing water on the sail. The scientific reason why these fishermen are right in doing so is conceptually based: it makes the sails more efficient by closing small spaces in the sail material through humidification and because of this the difference in air pressure between the two sides of the sail is increased. The power developed by the sail depends on the difference in air pressure between the side which slows down and compresses the current of air, and the side where there is normal circulation of air and therefore a zone of lower air pressure. This difference, according to Bernoulli's Law, explains why a sailing boat is not only *pushed* by the current of air but also *pulled* by the difference of air pressure between the two sides of the sail. In this sense, a boat sail behaves physically like the wing of a plane. The *jangadeiros* do not know anything about the Bernoulli Effect, but they know, in the context and history of their practice, the effectiveness of wetting the sails in order to optimize the speed of their boats. This is one of a number of operations characteristic of the *jangadeiros*' community of practice that distinguishes them from their literate, amateur sailor counterparts. The *jangadeiros* may not know the Bernoulli Effect and its application in sailing, but they could be considered better, faster sailors than many of those who do know this.

Though Brazilian fishermen and amateur sailing apprentices show clear differences in their psychological sailing competences, Da Rocha Falcão (2005), adopting Leontiev's (1994) theoretical concept of *activity*, points out that both groups of competences are *semiotic* and *culturally mediated*. In other words, the classification of these Brazilian fishermen's sailing competences, proposed by Vergnaud (1990) as being *competences-in-action*, or else *savoir-faire* as proposed by Piaget (1974), suggests the possibility of non-semiotic, strictly practical human actions. Da Rocha Falcão observes that the very fact that many people might not explain or discuss their competences should not be taken as being evidence of the purely enactive character of these competences.

The second experience relates to the mathematics curricular reforms of the 1990's. In the middle of that decade one of us participated actively in projects of mathematics curricular reforms for the elementary and secondary school levels of teaching in some different Brazilian educational institutions. This participation involved access to curriculum documentation, for different levels of teaching and from several countries, such as the United Kingdom, the United States, France, Germany, Spain, Canada, Japan and Portugal. In examining these documents a surprising similarity among the curricular objectives proposed by these countries was detected: the emphasis on the learning of types of mathematical competences, which the teachers had not previously had to consider. Some of these competences were familiar to them, as for instance those related to the domains of concepts and some procedures, and other aspects were less well-known, such as those related to the social, cultural and affective domains in mathematics.

In an effort to interpret those objectives in the context of these curricular reforms the work of Polanyi (1962, 1969) and Ernest (1998a,b), among others, has offered valuable theoretical insights. While the first author called attention to a *tacit dimension of knowing*, the latter showed the extent to which the practice of the production of mathematical knowledge could be seen to extend in this dimension. In particular, the work of Ernest suggests that the students' conceptual and procedural developments are only two among other components of mathematical learning. Further, most of these components are *mainly tacit*; that is, competences built upon activities, experiences or practices which cannot be fully communicated explicitly by rules or words. Put in pedagogical terms: the implementation of the curricula proposed for the teaching of mathematics as a school discipline should be careful not to disregard the tacit aspects that characterize mathematical knowing.

Aiming at developing this idea further Frade and Borges (2002) analyzed some curricular goals in the light of Polanyi's (ibid.) theory on tacit knowledge and Ernest's (1998b) model of mathematical knowledge. Frade and Borges used Ernest's model components of 'mainly explicit' and 'mainly tacit'. From his perspective and interpreting these components in terms of competence and knowing, *mainly explicit mathematical competence* is related to those competences that can be communicated through propositional language or other symbolic representation, as for instance: 1) accepted propositions and statements (e.g. definitions, hypotheses, conjectures, axioms, theorems); 2) accepted reasoning and proofs (all types of proofs including the less formal ones, inductive and analogical reasoning, problem solution including all analysis and computing); 3) problems and questions relevant to be solved by the mathematicians (e.g. Hilbert's problems, Fermat's Last Theorem). Alternatively, *mainly tacit mathematical competence* is related to the ways in which mathematicians use their knowledge, as well as how they appropriate mathematical experiences, values, and beliefs through their participation in mathematics practice. This, says Ernest, cannot be fully communicated explicitly. As mainly tacit mathematical competences he cites: 4) *knowledge-use* of mathematical language and symbolism; 5) meta-mathematical views, that is, views of proof and definition, scope and structure of mathematics as a whole; 6) *knowledge-use* of a set of procedures, methods, techniques and strategies; 7) aesthetics and personal values regarding mathematics. The arguments used by Ernest (1998a) to classify the model's components as either mainly explicit or mainly tacit, though not reviewed here, are very insightful and worth consulting.

As noted above, the materials examined in Frade and Borges' study were suggested by curriculum documentation from several countries and at different levels of teaching. From the analysis of each goal they identified the prevailing components of Ernest's model that required construction so that such goals could be achieved. The analysis demonstrated: 1) a strong similarity among the curricular goals proposed by these countries; 2) the prevalence of the mainly tacit components over the mainly explicit in such curricular goals.

We will now use these theoretical ideas to address the relationship between the experiences of the *jangadeiros* from Recife and the mathematics classroom activities addressed by curricular reforms from the 1990's.

2. EXPLORING CONNECTIONS BETWEEN SCHOOL MATHEMATICAL PRACTICES AND OTHER SOCIO-CULTURAL 'MATHEMATICAL' PRACTICES

Though apparently quite different contexts, the two experiences described above suggest for us a connection between school mathematical practices and other socio-cultural 'mathematical' practices. Indeed, in discussing analytical tools for the study of mathematical activity, Araújo, Andrade, Hazin, Da Rocha Falcão, Nascimento and Lins Lessa (2003) propose, among other things, taking into account pre-conceptual competences which can be characterized essentially in two ways: firstly, by their *effectiveness* in culturally meaningful contexts and, secondly, by the fact that these competences are, by nature, quite difficult to express by means of whatever symbolic-explicit representation is used. These authors suggest that these two characteristics, effectiveness and tacit quality, are *invariants* of mathematical activity, no matter if we are considering school or out-of-school mathematical practices such as those performed by tailors, carpenters, *cambistas de jogo do bicho*[22], sailing fishermen or participants within other communities of practice.

The 'mathematical' way in which the *jangadeiros* pilot their boats according to vectorial principles, as described previously, and the manner in which mathematical strategies are chosen by students in problem-solving school activities are examples of these two invariants of mathematical activity. It has been pointed out that tacit aspects are in fact present in all formal, high-level contexts: by Latour and Woolgar (1979) when referring to activities and competences of high-level workers in scientific laboratories; by Vergnaud (2000) when analyzing the competences of mathematics teachers; and by Da Rocha Falcão (2005) when discussing the differences and convergences between practical (effective) and conceptually-based competences. Actually both *activity* and *language* share an important common aspect: what is effectively done or said is the result of a psychological dispute among many possibilities. What is *done* by an

---

[22] *Cambistas de jogo do bicho* are a sort of Brazilian bookmakers dealing with what we call 'the animal lottery' (*jogo do bicho*).

individual shows something in behavioural terms, but at the same time defers many other possibilities that *could be done*. In the same way, what is said represents the emergence of one among many discursive possibilities, a result of a dialogical discursive fight inside the individual (Clot, Faïta, Fernandez and Scheller, 2001). According to this theoretical approach, explicit and tacit qualities of activity and language should not be seen as strictly opposite aspects, but as circumstantial interconnected states related to a psychological process of decision making.

Searching for a better understanding of the connection between school mathematical practices and other socio-cultural 'mathematical' practices we have been working on the ways in which tacit knowing or competences-in-activity 'circulate[23]' within these practices. This concern with the school context follows from our studies of Polanyi's theory of tacit knowledge and Ernest's constructivist view of mathematical knowledge. In relation to the former (see Frade, 2003) one particular element drew our attention, namely, the concept of 'tradition'– a system of values that describes how knowing circulates within a social practice – as proposed by Polanyi (1962):

> An art which cannot be specified in detail cannot be transmitted by prescription, since no prescription for it exists. It can be passed on only by example from master to apprentice. This restricts the range of diffusion to that of personal contacts...[For example] while the articulate contents of science are successfully taught all over the world in hundreds of new universities, the unspecifiable art of scientific research has not yet penetrated to many of these...*To learn by example is to submit to authority. You follow your master because you trust his manner of doing things even when you cannot analyze and account in detail for its effectiveness. By watching the master and emulating his efforts in the presence of his example, the apprentice unconsciously picks up the rules of the art, including those which are not explicitly known to the master himself.* These hidden rules can be assimilated only by a person who surrenders himself to that extent uncritically to the imitation of another. A society which wants to preserve a fund of personal [tacit] knowledge must submit to tradition...we accept the verdict of our appraisal, be it at first hand by relying on our own judgment, or at second hand by submitting to the authority of a personal example as carrier of a tradition. (p. 53, italics added)

[23] Our use of the term 'circulating' is based on the idea that people's tacit knowing is always present in any interaction and thus is available to be apprehended in some way by the others. We call the movement of this type of knowledge between people 'circulating'.

We combined Polanyi's theoretical contribution on *tradition* with Jean Lave's and Etienne Wenger's reflections on cognition as being a socio-culturally situated activity, as developed in the books 'Cognition in Practice' (Lave, 1988) and 'Legitimate Peripheral Participation' (Lave and Wenger, 1991). Our reading of Lave and Wenger's ideas and research with some communities of practice encouraged us to direct our investigations towards situated learning perspectives. It is interesting to notice how Polanyi's concept of tradition is quite close to the current characterizations of communities of practice (for instance, see Winbourne and Watson, 1998 and also Wenger, 1998). We highlight here two central similarities between these two theoretical frameworks: 1) boundary delimitation; part of knowing cannot be detached from its context of origin to be employed or used (in the case of tradition, such knowing would correspond to what Polanyi calls tacit knowledge); 2) learning means changes of participation and formation of identities within communities of practice.

These two similarities are clearly perceived in the cases of the *street mathematics* (Nunes and colleagues, 1993), the *ardinas* (see Santos and Matos' chapter in this book), the *jangadeiros* from Recife and many other professional and amateur communities of practice such as those studied by Lave and Wenger in their above-mentioned books. In particular, both the 'knowledge' of piloting the *jangada* and the social representation of the *jangadeiros* are strongly connected to the activity in the context of which this navigation expertise is exercised. This observation is confirmed by the following historical-cultural fact: the *jangadeiros* from Recife have changed the profile of their professional activity radically in the last twenty years, since *jangadeiros-as-fishermen* have been changed into *jangadeiros-as-tourist-operators*. Because of this their activity as navigators became less dangerous, since they do not need to navigate as far from land as they used to, and also they do not need to navigate by night. These factors have changed not only their competences but also the social representation of the *jangadeiros*: most of them do not know anymore how to 'read the sky' by night in order to find the way back to land, and nowadays the social valuation of a *jangadeiro* is clearly inferior to that of twenty years ago.

Concerning mathematics in school contexts we are convinced that the boundary delimitation mentioned above does not mean that *authentic*[24] *mathematics experiences* cannot take place outside the community of mathematicians. A good illustration of this can be found, for example, in Ponte and Matos (1991). The authors carried out research in which the

---

[24] Experiences that require the participation of the subject in activities involving, for example, formulation and evaluation of problems, questions, examples, conjectures, conclusions, argumentation, conversation and negotiation.

students, using the computer as a tool, play the role of mathematicians to experience several aspects of a scientific mathematical investigation. These experiences can be also found in Winbourne and Watson's (1998) work. Here the authors show up how some mathematics lessons can be planned in order to encourage children to experience modes of participation in *local communities of practices*. Both the examples do not have the intention 'to form' future mathematicians (though this might happen with some students); rather their intention is to allow the students to have the opportunity to learn mathematics by experiencing somehow different social-cultural aspects of mathematics practice. On the other hand and despite this, learning viewed as change in participation and formation of identities within communities of practice still represents a real educational challenge in regard to school mathematics.

Let us now turn to some of the differences or limitations between Polanyi's ideas of tradition and Lave and Wenger's ideas on communities of practice and situated learning. In particular, we remark on two main limitations of Polanyi's concept of tradition when considering the dynamic nature of situated perspectives. The first limitation refers to the fact that Polanyi does not discuss the interactions between master and apprentice and among apprentices. In his description of tradition, the learning process seems to take place in only one direction: from master to apprentice. This leads to a second limitation: learning taking place in only this direction suggests a strongly reproductive view of learning. It is our understanding that an essential feature of situated perspectives is a view of learning as dynamic or transformative: cognition is seen as a process occurring in practices, and so always changing or transforming individuals, including teachers and learners, activities and practices (see, for example, Frade, Winbourne and Braga, 2006). This is why we use terms like competence and knowing instead of knowledge, a tricky word as Wenger says (1998), to talk about learning as participation and identity formation in practice. Besides, in this chapter we are not writing a philosophical treatise on the nature of knowledge and knowing. For our purposes – exploration of the links between theories of tacit knowledge and theories of situated cognition – we can talk in terms of knowing and competence, ideas that are appropriately dynamic and transformative; we are confident that 'translations' to account for a discourse of knowledge would be possible, but that is not our purpose here.

Having said this, we would say that Polanyi was one of the precursors of the concept of communities of practice, more specifically, in scientific practice in the middle of the 20th century. As far as we know Polanyi did not develop his concept of tradition beyond the aspects described above.

The central claim arising from Ernest's constructivist view of mathematics is that mathematics practice is a social practice in line with

Wittgenstein's concepts of *language games* and *forms of life* (Ernest, 1998a). Interpreting such a claim in terms of competence and knowing Ernest argues that mathematics practice requires not only the development of competences related to the domain of discourse (know-what) but, above all, the development of practical competences (know-how), attitude, disposition, which cannot be replaced by rules or words, though they can ultimately be represented as such. Ernest justifies this form of knowing as follows: as long as knowledge is said to be tacit, its justification, or at least a part of it, is also tacit, or else it becomes contradictory. This means that acknowledgement of any tacit knowledge or its evaluation cannot be entirely explicit. Ernest asserts that the validity of tacit knowledge is demonstrated during performance.

Concerning the ways in which knowing circulates within the mathematics community Ernest says that while explicit knowledge of different cultures might well be inter-translated, tacit knowledge is not by definition. It would first have to be made explicit, but as long as this can be done partially, 'there must always be residues of that knowledge that remain tacit and bound to the forms of life that give it meaning' (Ernest 1998a, p. 250-251).

If we assume that teaching and learning mathematics have a socio-cultural history and demand an enculturation in selected features, most of them *tacit* or related to *competences-in-activity* of mathematics practice, then we can also say that teaching and learning mathematics involve, in some way and to some extent, adopting the tradition or the system of values of that practice. The same can be said in relation to the concepts of *language games* and *forms of life*: teaching and learning mathematics involve *play* language games inserted in a form of life. If this assumption is true, as we think it is, what Polanyi and Ernest are indicating in the case of teaching and learning mathematics is that a part of those features can only be apprehended by 'practising' or 'doing' mathematics. Put another way: *mathematics classrooms should function as a kind of community of practice, or have some characteristics of that community, to encourage the learning of selected features of mathematics practice.* This idea can be taken as being the core of this chapter, for in our view it links theories of tacit knowing and theories of situated learning/communities of practices by offering a very useful insight for approaching tacit mathematical knowing or competences-in-activity in school contexts. It also sets up another important connection between school mathematical practices and other socio-cultural 'mathematical' practices, including *street mathematics* and the professional activity of the *jangadeiros* from Recife; all of these practices can be seen somehow as specific kinds of communities of practice.

In a similar sense, Brito Lima and Da Rocha Falcão (1997) have demonstrated the effectiveness of 'contracts' (agreements about activity) concerning the 'drawing' of mathematical problems designed to introduce very young children to algebraic activity (year-2). Children accepted this contract as an important aspect of *becoming mathematical.* The social character of the negotiation of this contract encouraged them 'to attack' algebraic problems effectively, as illustrated by the protocol transcription in table 1.

Table 10-1. Drawing mathematical problems

Problem: Last Sunday, children from P and BV [P and BV are two different beaches in Recife – Brazil] took part in a kite-flying contest. On Saturday morning, the children from BV made a certain number of kites by hand and they made three times this number in the afternoon. Children from P made 24 kites in all. Both groups of children made the same number of kites. How many kites did BV's children make on Saturday morning?	Subject R's drawing-representation of the problem 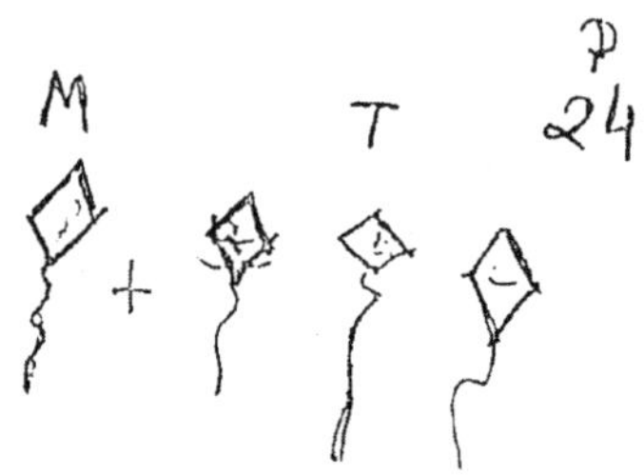

Drawing the main aspects of this mathematical problem allowed the student, R, to articulate a proto-equation through iconic representation of the unknowns; this led her not only to take on this very difficult problem, but to solve it. Through the earlier negotiation of a contract, she had become the kind of student who draws mathematical problems.

In the remaining part of this chapter we will try to explore further how the ideas we have presented can help us to improve our understanding of the identification and circulation of tacit knowing within mathematics school practices. This chapter does not have the character of a report of empirical research, with all its sequential phases. Rather, its aim is to explore both theoretical and empirical connections between theories of tacit knowledge and theories of situated learning and community of practices. Its sections should be seen as complementary in the sense that we explore in each of them different possibilities for these connections. Having said this, we will shift the focus of our line of thought a little from the one followed up to now. We attempt to provide analytical tools to question how tacit knowing, or competences-in-activity, manifest in mathematical activities as well as to discuss some pedagogical implications resulting from such an attempt. At the same time, we try to establish connections between school practices and culturally situated communities of practice. In doing this we also hope to contribute to a theoretical approach to tacit knowledge where both differences and similarities between tacit and explicit knowledge might be highlighted, without the suggestion of any ranking of forms of knowledge. Before we go on to develop our ideas, we want to clarify the concept of tacit/explicit knowledge (or knowing) and to consider some forms in which this concept has appeared in the mathematics education literature.

3. TACIT KNOWLEDGE (OR KNOWING) AND VARIATIONS

The concept of tacit knowledge (or knowing) does not have a single meaning[25]. Nor does it always have the same terminology when it appears in the literature of mathematics education. Nevertheless, whatever meaning and terminology are used, the authors who address this type of knowledge share Polanyi's (1969) epistemological thesis that all knowledge is tacit or constructed from tacit knowledge. Put another way: language alone is not enough to render knowledge explicit.

The concept of tacit knowledge comes from the idea that much of the competence that is relevant to performance is not openly expressed or stated by means of propositional language or even any other symbolic representation. 'Individuals often are not aware of the knowledge that underlies their action. Terms like *professional intuition* or *professional instinct* imply that some of the knowledge associated with successful [or non successful] performance has a tacit quality' (Grigorenko, Meier, Lipka, Mohatt, Yanez and Sternberg, 2001, p. 8). Wigner and Hodgkin (1997) add:

> the [tacit-explicit] threshold is moveable: when we are unselfconsciously engaged in action all relevant powers are integrated and are temporarily immune to reflective thought. When, however, we pause and monitor our action and even more if we discuss it, as we would if we were teaching a skill, then we make a conscious effort to lower the threshold so that some of the operational principles which were tacit may become explicit...(p. 431-432)

Tacit knowledge (or knowing) is then a concept strictly linked to activities or practices, either physical or mental, which take place in semiotically-mediated contexts such as cycling, sailing, reading or writing, teaching, recognizing a person, understanding someone else's attitude or solving a mathematical problem. Also, it has a strong situated character in

[25] See Frade and Borges (2006) for a detailed discussion.

the sense that what is taken as tacit varies from one situation to another; it depends on the context; the situation, the activity.

Several authors (e.g., Tirosh, 1994; Lins Lessa and Da Rocha Falcão, 2005; Frade and Borges, 2006) have been writing about the issue of types of knowledge that are more easily expressed by means of propositional language or other symbolic representations, and others which are not. We have already shown briefly how Ernest (1998b) addresses this issue from a philosophical perspective: he uses the words *tacit* and *explicit* as opposites to refer to different, but complementary dimensions of the same component of a certain practice.

Piaget, Vygotsky and Vergnaud, for example, approach the issue from a psychological perspective. Piaget (1974) distinguishes *savoir-faire* (knowing how, practical knowledge) from *savoir-dire* (knowing what, conceptual knowledge). He proposes that the relation between these two forms of knowledge be analyzed, at the same time, from a historical, ethnographic and psychogenetic perspective. Piaget focuses on the psychogenetic approach to the characterization and development of the conceptualization process. This can be seen in his emphasis on the operational-logical structures of the individual's cognitive functioning. Further, Piaget's view of conceptual knowledge includes and goes beyond practical knowledge; it is conceptual knowledge in a superior stage of development, since this type of knowledge emerges from empirical/practical knowledge through *abstraction réflechissante* (reflexive abstraction), leading to a *prise de conscience* (awareness) of conceptual relations (Piaget, 1974).

Vygotsky (1986) approaches the issue when studying adolescents' operations with concepts. He concludes that there is a surprising discrepancy between their abilities to form and use a concept and their abilities to define it by means of verbal language. For him, these abilities seem to develop as follows: the more complex the use of a concept, the more difficult it is to express it in words. Vygotsky says that such discrepancy also occurs with adults' thinking even at its more advanced levels (ibid., p. 141). Later, the author returns to the issue in a specific school learning context. Here, the study was focused on children's learning of 'scientific' concepts and their abilities to deal with these concepts as compared with their abilities to deal with concepts formed out-of-school. Vygotsky emphasizes the distinction between learning and development of these two kinds of concepts. He opposes Piaget's genetic epistemology within which formal-school conceptual knowledge would represent a superior level of development to that of practical-out-of-school conceptual knowledge. For Vygotsky, these two forms of knowledge, or children's abilities to deal with these two kinds of concepts, are in a complementary relation in which 'the strong side of one indicates the weak side of the other, and vice versa' (Vygotsky, p. 158).

Practical out-of-school conceptual knowledge is deeply rooted in children's real cultural experiences, but very difficult to transform into verbal language or to use as a basis for forming abstractions. On the other hand, formal-school conceptual knowledge reaches far beyond the immediate experience of children. It is rooted in verbalization under the conditions of systematic cooperation between the child and the teacher, but too abstract and detached from children's reality to be employed in a concrete situation.

Vergnaud (1990) has extended and redirected Piaget's psychogenetic approach to the study of the cognitive functioning of the 'individual-in-action'. Vergnaud's theory of conceptual fields is based both on the epistemological content of knowledge and the conceptual analysis of this knowledge domain. Here, the cognitive emphasis is not only on the conceptual aspect of the schemes (the great heritage brought from Piaget), but also, and more heavily, on the conceptual analysis of the situations, tasks, in which individuals are taken to develop their schemes either in or out of school. The importance attributed to situations in Vergnaud's theory leads him, inevitably, to distinguish between forms of knowledge such as explicit knowledge and implicit knowledge. The former refers to the several forms of symbolic representation in which the concepts, their properties, the situations, and the treatment of procedures can be expressed. The latter includes 'theorems-in-action' and 'concepts-in-action', based on schemes (similarly to conceptual-explicit knowledge) described as *invariant organization of conduct for a limited class of situations* (Vergnaud, 1990). These operational invariants are responsible for the recognition of the pertinent elements within a situation, and a great part of those types of knowledge is tacit in the sense discussed in this chapter. In fact, the operational invariants are those that decide how and which types of knowledge will be used in a subsidiary way (tacitly) to perform a task or a situation. What will be projected or become explicit within the task is, as Vergnaud says, only the visible part of the iceberg where the hidden part is formed by the operational invariants. It is our understanding that theorems-in-action and concepts-in-action can also be said to be tacit in the sense that individuals do not have access to the ways in which these types of knowledge operate within a scheme. On the other hand, as proposed by Vergnaud, it is not only possible but pedagogically important to offer situations through which students can migrate from concepts-in-action to explicit concepts. Language, for Vergnaud, would have a central role in this passage. The importance of symbolic representation as a landmark linking concepts-in-action and formal (school) concepts could be criticized from the perspectives of Vygotsky and Leontiev, since for these authors semiotic function is always present in human activities.

Despite theoretical divergences between Piaget's, Vygotsky's and Vergnaud's approaches on the issue, it is apparent that all these authors are referring in some way to that which has been called tacit and explicit knowledge (or knowing)[26].

4. CONNECTING TACIT MATHEMATICAL KNOWING, SITUATED PERSPECTIVES AND SCHOOL MATHEMATICS PRACTICES

Let us summarize what we have done so far. In the previous sections we have shown how Polanyi's theory of tacit knowledge and Ernest's ideas on the tacit-explicit dimension of mathematics practice and the theories of situated cognition are strongly linked. On the one hand, we have shown what these authors have to say about the circulation of tacit knowing in practices, thus how situated learning and communities of practice perspectives can be seen as a powerful framework to explain the situated character of tacit knowledge (or knowing). Those types of knowledge are to a great extent context-bound or situated within the communities of practice within which their meanings are constituted. Tacit knowledge is also deeply socially culturally situated in specific practices. To illustrate this we have cited the work of some researchers (Brito Lima and Da Rocha Falcão, 1997; Ponte and Matos, 1991; Winbourne and Watson, 1998) that shows how classroom activities can be organized in order to allow the students to experience aspects of 'authentic' mathematics practices – to become mathematical. In particular, we have claimed that for the students to experience the tacit aspects of these practices, classrooms should be designed to be as close as possible to communities of practice.

On the other hand, we have presented some psychological contributions concerning the ways in which tacit knowledge operates within activities. Ernest provides an insightful suggestion for elaborating a connection between the theoretical perspectives discussed, tacit-explicit knowing and

---

[26] In relation to Vygotsky we would like to remark that its main theoretical claim is that *thought* and *language* are indissoluble processes *via* meaning. Our interpretation is that Vygotsky recognizes the difference between these two forms of knowing, although assuming that this difference tends to disappear as long as thought 'realizes' in words (Vygotsky, 1986, p. 251): initially thought and speech are different processes, but as long as one is speaking the thought is modified in such a way that all is spoken when the utterance ends; the thought is modified along the speech; once this process is performed, *language* and *thought* are not 'different' any more; they were merged since there is no residue of thought left unspoken.

situated learning, and school mathematics practice. Other researchers have also focused on selected aspects of mathematics practice either to propose what 'school mathematics practice' should consist of, or to understand such practice (e.g. Burton, 1999, 2002; Ponte and Matos 1991; Romberg, 1992; Schoenfeld, 1992). In a sense, these authors are all saying that mathematics in school should be as close to a community of practice as possible. On the basis of what we have discussed so far, this implies that we cannot disregard the tacit dimension of mathematics teaching and learning.

Next we address some important questions with the purpose of offering analytical tools for the study of the tacit-explicit dimension of school mathematics practices. Here we present examples from the mathematics classroom not with the intention of showing how mathematics classrooms can be communities of practice – we think we have already done this. Instead, we use these examples to offer methodological possibilities for identifying tacit-explicit dimensions to, and aspects of, mathematics classroom activities.

4.1 What might a tacit-explicit dimension of school mathematics practice consist of?

Based on Ernest's tacit-explicit classification of scientific mathematical knowledge Frade carried out some research in which she proposed what might be called a tacit-explicit dimension of school mathematics practice (Frade, 2005). According to her, this proposal should be understood in the following sense: by the end of a period of learning, and for each level of teaching, learners are expected

- to have developed an appropriate a set of competences related to mathematical statements and propositions;
- to be able to use mathematical reasoning and justify it;
- to use mathematical language and symbolism;
- to have developed a certain view of the scope and structure of mathematics as a whole;
- to be able to decide which methods, strategies or procedures are more adequate to the resolution of problems and when to use those methods, strategies or procedures.

Moreover, and probably most importantly, learners are expected to have developed a favourable disposition towards participation in mathematics school practices.

Frade's research examined the stages of development of some Year 7 students in terms of the mainly tacit and mainly explicit components of area measurement. To this end, she proposed an adaptation of Ernest's tacit-explicit classification of scientific mathematical knowledge to the students' knowing[27]. In particular, the component 'aesthetics and values' was associated with the students' disposition, motivation and participation in classroom practices, in a process closely resembling those to be found in discussion of students' mathematical identity (see, for example, Winbourne in this volume). This component has a macro character in the sense that it is a necessary condition for the development of the remaining components[28]. Below we present some particular examples, without discussion, to illustrate briefly how this adaptation of Ernest's classification of the students' knowing was used in Frade's study (ME = mainly explicit-symbolic; MT = mainly tacit). In each case we first describe a classroom activity and then offer an example of the responses of a particular student-pair .

4.1.1 Propositions and statements (ME)

Activity: The teacher and the students are discussing the following proposition for K=3: if the sides of a rectangle are multiplied by K (K > 1), then its area grows K2 times.

Example (proposition):

Bruno: It's…, its side is 2 and D is 6. Then I saw that the area of the square C equals 4 and that of square D equals 36, which is 9 times bigger. Then after my mother gave me another example that it did not matter, that it was only that the number should be the triple, she put it here side 3 and the area 9. In the other 9, that is, if the side is 9 then it's 81, it's the same thing.

Activity: Answering a written questionnaire question: 'what do you mean by 'area' in mathematics?'

Example (statement):

Felipe: I remember that area represents a certain space or place. Based on that we can find that area is used to calculate the size of a space or place. Take the example of a piece of land, how many square metres it has. This is already a way to use area as measure.

---

[27] This adaptation took account of the fact that the students are learners, and that part of the learning process consists in a gradual improvement of their understanding and procedures, which in their initial manifestation may seem mistaken from the viewpoint of the discipline.

[28] The component 'problems and questions' was not investigated, as it was difficult to adapt it adequately.

4.1.2 Reasoning and proofs (ME)

Activity: A student is explaining to a classmate how he and his partner solved the following problem: a wall with height of 2.30 metres and length of 8.76 metres is tiled with square tiles having sides measuring 2 centimetres. Calculate the number of tiles on the wall.

Example (reasoning):

Bruno: We found out that the area of the wall is, we multiply length times width and to obtain the area of the tile we multiply side times side. When we find the result of the two, we divide the area of the wall by the area of the tile. The result was…

Activity: The students are working on exercises from their textbook. They are trying to calculate how many ceramic tiles are needed to cover the floor of a rectangular room.

Example (proofs):

Felipe: What is the result?

Bruno: 71 and 57.

Bruno: 178.

Felipe: 15, 1, 2, 3...15 times 12. 170? Because here, look 1, 2, 3 ... 16. [Felipe's multiplication is incorrect: $15 \times 12 = 180$]

Felipe: 1, 2, 3... 16. 16 times 12. 192.

Felipe: It's 16. 1, 2...16.

Felipe: The result is 192, isn't it? Because here, look, 16, we have to count the width of the tiles.

Bruno: 178

Felipe: 178?

Bruno: Then write it there: 178. (...) There are 178, 178 tiles on the floor.

Felipe: On the floor, on the floor.

4.1.3 Knowledge-use of language and symbolism (MT)

Activity 1: The students were asked to produce a written account of the subject under study, up to then.

Example (area and measure):

Bruno: The area is used to calculate forms like squares and rectangles; all area has its perimeter. The perimeter is the boundary of the square, the rectangle, etc... e.g.: 2cm $\times$ 2 = Area or 4 and the 2 $\times$ 4 = Perimeter. The area is used to calculate either a closed or an open space; it is used to compare sizes.

Activity 2: The same activity above, but at the end of the study with the subject in question.

Example (area):

Bruno: I think that area is a place or space, which can be used in many ways. The objects are bricks, tiles, floor and things which bear geometric forms, and also figures like squares, triangles, etc...

To obtain the area in cm^2 we have to multiply side × side, e.g.:

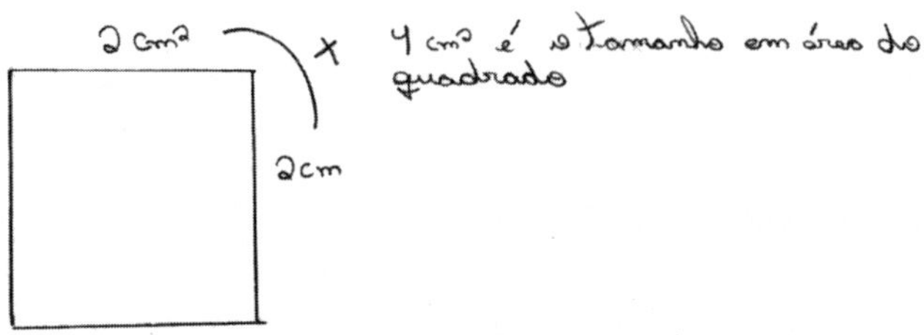

Area is used to calculate the size of a place or space.

Now perimeter is the boundary of the area. To obtain a perimeter you just add the sides, e.g.:

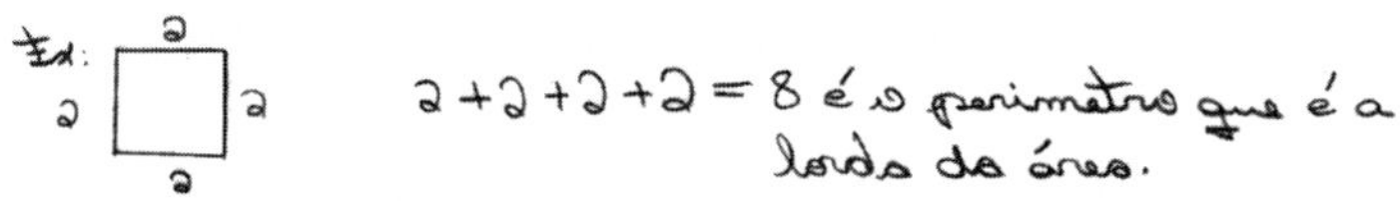

To obtain the area of an irregular figure you just have to divide by a geometric figure. e.g.:

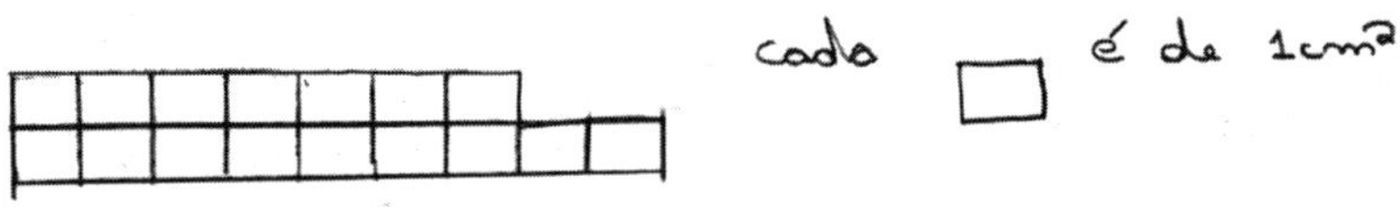

Volume is mass quantity.

The protocols above show that Bruno has improved his mathematical language and symbolism through the study of the subject. In fact, when comparing the two texts we can see that Bruno incorporates in the second text 'new' mathematical terms such as *figures*, *geometric forms* and *objects*, as well as drawings and symbols to express his understanding.

4.1.4 Meta-mathematics views (MT)

Activity: The students were asked to demonstrate their understanding about the differences between plane and spatial figures. To this end, they had to use a classifying table – flat, plane, volumeless forms *versus* spatial forms that can have a volume – from their text book. This table used only pictures.

Example:

Felipe: …the spatial figures that can have volume seem to be real.

Teacher: Okay, but what does this mean, why did you say that it seems real?

Felipe: This thing [spatial figure that can have a volume] here looks like an egg.

Teacher: Oh, yes.

Felipe: This thing [plane figure] here seems to be kind of a drawing.

Teacher: Oh, yes, this one here seems to be a concrete object; what about that one?

Felipe: No.

The protocol above indicates that Felipe has assigned a *status of entity*, an ontological status, to spatial and plane figures: his utterances suggest that spatial figures are real for him because they are concrete and tangible, '…looks like an egg', whereas plane figures are not real for him because they are representations, drawings. According to Frade (2005) such assigning of ontological status is an aspect of meta-mathematics views.

4.1.5 Knowledge-use of methods, procedures, techniques and strategies (MT)

Activity: Problem solving. For example, the students were asked to calculate the perimeter of a rectangular piece of land with an area of $450m^2$ and 25m in length.

Example:

Felipe: Now we have to find which number that is. The length is 25, we already know. And what about this one here? 25 times 20. Wait, I understood. 25 times 21. Oh, oh, no God, this is too much!

This example seems to capture a moment of strategy choice by Felipe: trying to fill the area with rectangles measuring 25 in length and variable width until he reached 450 for the area. We identify this strategy choice with a manifestation of the component *knowledge-use of a set of procedures, methods, techniques and strategies* by him. It is interesting to note that this component manifests itself when student strategy choice loses its 'tacitness', or else, when this strategy choice becomes explicit through the articulated computation.

4.1.6 Aesthetics and values (MT)

Activity: Problem solving. Whilst working on the wall tiling problem (see 4.1.2 above) the teacher told the students to observe that the measures of the units of measure of the wall and of the tiles were different; therefore they should work either with metres or with centimetres.

Example: [The students-pair spent a lot of time discussing the problem, making counts, erasing them and starting them again, doubting about the place of commas, asking the teacher for help …]

Felipe: Do you remember what the teacher said…?…She said that if you multiply one hundred times one hundred and removes the decimal point; you transform everything into centimetres, isn't that better? She told me so.

Bruno: What do we have to do?

Felipe: We did not use these 2 cm here.

Bruno: If we do it twice.

Felipe: We did not use centimetres, we skipped it. We only used these two here, look.

Bruno: Isn't it easier if we only do the last calculation? The last calculation only is the first thing.

Felipe: *Let's erase it and start it all again.*

Bruno: *That's it. It's better.*

Felipe: *Then we start it, it's better.*

As mentioned previously, the component *aesthetics and values* was observed in terms of students' curiosity, interest, motivation and participation in classroom practices. Through Frade's (2005) description of the performance of the students-pair's activity we can see a high degree of interaction between these students, and, in many cases, between them and the teacher. This was interpreted as a positive affective aspect in relation to mathematics. On the other hand, the utterances in italics show that both students have decided to put aside what they had done up to a certain point of the problem solving and started solving it again. According to Bishop (2006), values are beliefs-in-action. Our values are revealed when we make choices, this is when we express elements of our system of beliefs. Based on this and on the assumption that it is reasonable to say that at this level of teaching the students developed some sense of aesthetics and values regarding mathematics, one possible interpretation for the students' decision to leave out what they had done is that an element of the component aesthetics and values might have been triggered; their mathematical presentation of the problem was incorrect or not presentable, therefore they would have 'to erase it' and 'to start all again'. They possibly value the precision in a presentation of a mathematical task. Whether this is a personal value for them, or comes from the teacher is another question. Anyway, this decision making reveals their perseverance and disposition to do the task well, which are also positive affective aspects in relation to mathematics.

4.2 How does tacit knowledge (or knowing) manifest 'psychologically' in mathematical activities?

Activity: The students were engaged in the plane/spatial figures activity described in 4.1.4 above. After some time, the teacher engaged the students in conversation about their understanding of the task.

Example 1:

Jessica: …in some figures there are **some flat forms that make** a figure with a volume. Example, the **cylinder** has **two faces** with the **form** of a **circle**; the **prism** has two **faces** of one **hexagon** and two faces of a rectangle.

Example 2:

Lucas: The difference between the flat, the plane and without volume and the spatial figures that can have a volume is that the flat ones **cannot hold material inside** and the ones which have volume **can hold material inside**.

The words or expressions in bold above were used by Frade and Borges (2006) as clues about the knowledge the students were using in a 'subsidiary' or tacit way to elaborate an understanding of the difference between plane and spatial figures. Example 1 indicates that tacit knowledge related to surfaces of solids was deployed by Jessica. Example 2 indicates that Lucas was mobilizing tacit knowledge related to capacity. As seen in the second section of this chapter, it can be argued, in different ways, that when performing a task individuals often are not aware of the knowledge that underlies their action[29].

4.3 Characterising the site of learning for the jangadeiros

On the basis of our discussion so far we suggest that teachers would do well to work towards some kind of a *local community of practice* of the kind suggested by Winbourne and Watson (1998):

1. pupils see themselves as functioning mathematically and, for these pupils, it makes sense for them to see their 'being mathematical' as an essential part of who they are within the lesson
2. through the activities and roles assumed there is public recognition of developing competence within the lesson

[29] See Frade and Borges (2006) for discussion of other types of tacit knowledge.

3. learners see themselves as working purposefully together towards the achievement of a common understanding
4. there are shared ways of behaving, language, habits, values, and tool-use
5. the lesson is essentially constituted by the active participation of students and teacher
6. learners and teachers could, for a while, see themselves as engaged in the same activity. (p. 103)

The points listed above describe a context for effective interaction between teaching and learning that could equally be taken into account in order to describe what happens in the context of teaching and learning in other communities of practice, such as that of the Brazilian *jangadeiros*. These characteristics could be informative for more formal educational contexts.

1. *Jangadeiros* and their apprentices internalize a social representation of being or becoming a *jangadeiro*. The usual crew of a *jangada* is composed of a master-*jangadeiro* (the 'boss', owner of the jangada and pilot), a specialized fisherman, who is in charge of fishing activities and helping piloting the *jangada*, and an apprentice, a 12 to 16 year-old boy. Each of these persons knows very clearly their specific role, and also the common tasks in which they are engaged. At the present moment, as mentioned before, many *jangadeiros* have changed their professional profile and are now tourist-operators, and this change has excluded the specialized fisherman from the *jangada*s. Nevertheless, apprentices are still present, with a new function: helping with commercial activities during the tour-trip, like selling beverages, fried fish and other refreshment.

2. Through the activities and roles assumed there is public recognition of developing competence within the experience: apprentices are recognized as such by their social environment, and are expected to become a master-*jangadeiro* some day (even though this day is not institutionally established). On the other hand, an apprentice should not expect any introductory care from the other members of the crew. He is on board in order to help and contribute to a common task, not primarily to learn or be educated (even though he is expected to learn). Because of this, mistakes made by these apprentices in the context of navigation are not treated like pupils' wrong answers, but as the result of dangerous practice causing real trouble to the real activity. Inside this community of practice, then, there is not a clear separation between learning-contexts and 'real-life' contexts as is the case for regular pupils at school. Real-life contexts and learning contexts coincide for apprentice *jangadeiros*.

3. There are shared ways of behaving, language, habits, values, and tool-use: this is completely true for this community of practice. In fact, Polanyi's

propositions concerning *tradition* are particularly valid here. As pointed out above, the apprentice *jangadeiro* does not receive any introductory care, but he is expected to imitate his master, who is expected in turn to offer examples. In the context of the hierarchical organization of the activity on board, learning by example implies submitting to authority (see the quotation from Polanyi on page 6). In this specific context, there are no rigid, explicit rules regulating time and conditions by which an apprentice comes to 'graduate'. This up-grade emerges as a function of the set of competences developed by the apprentice, crossed with other circumstantial aspects like being offered an opportunity as a master *jangadeiro* or fisherman.

4. Despite some common points between Polanyi's concept of tradition and situated learning perspectives and communities of practice, it is our understanding that Lave and Wenger's (1991) notion of legitimate peripheral participation (LPP) elaborates more efficiently the issue of out-of-school learning contexts. Indeed, at the first moment of above-mentioned submission the learning can be a-critical. As Polanyi says, the apprentices rely on their master and surrender to his/her knowledge, without questioning, because they attribute to the master the legitimacy of his/her way of acting. However, at a later moment the apprentices are able to reconstruct the master's version of knowledge, as well as to judge his/her competence. Finally, when the apprentices are able to preserve the ideals of the tradition they are then liberated. The apprentice/master relation changes or is suspended.

5. The lesson is essentially constituted by the active participation of students and teacher; learners and teachers see themselves as engaged in the same activity. There are not 'lessons' on board, but real professional activities in the context of which apprentices and other members of the *jangada's* crew are expected to accomplish their tasks properly. From this point of view, apprentices and confirmed members of the crew are all submitted to the same professional engagements (even though important hierarchical differences of roles and responsibilities do exist).

4.4 What pedagogical implications result from connecting tacit mathematical knowing, situated perspectives and school mathematics practices?

Our response to this question comes from interconnections between mathematics school classrooms and *jangadeiro's* activities. Firstly, teachers can organize their classrooms as communities of practice, with rules, hierarchy, examples, negotiation, and other features of these communities. In our view this organization should not be restricted to some lessons only.

Classrooms can be organized as a kind of community of practice in a more general sense. Certainly there will be moments when students will learn by observing, listening and even imitating their teachers or classmates' learning to develop those actions as a part of mathematics practice. What is most important for the constitution of a community of practice in classrooms is that students feel their real engagement in this community, that they are participants in the practice and can share their doubts, understandings, meanings and experiences in it.

Secondly, classroom activities need to be relevant for students where relevance becomes identified with engagement, action, legitimate participation, and so on. Also, the students should understand what and why they are doing something even though certain tasks are abstract and not at all connected to identifiable daily-life contexts.

Finally, we have shown that mathematics learning involves a tacit-explicit dimension. Stressing the tacit does not mean that the explicit dimension should be ignored. Competences-in-activity benefit from discursive confrontation. It is through confrontation, demonstration, explanation and argumentation that students can conceptually develop, and become mathematical. Discursive dialogue (with others or with 'inside voices') is a crucial aspect of conceptual development. Lerman (2001), for example, argues that:

> Children become mathematical by getting used to what counts as being mathematical, which is constituted in the social practices of the classroom. This may be a more fruitful way of speaking about learning, in which learning is about speaking, about how to speak in the legitimated codes of school mathematics. (p. 50)

However, this element of mathematical knowledge, the social communication of mathematical knowledge, is basic for the production of mathematical knowledge in school if we understand it as social practice. So, teachers can play a crucial role in encouraging the student's development of that component by promoting conversational and representational practices.

5. CONCLUSION

Though situated in quite different contexts mathematics school practice and other socio-cultural 'mathematical' practices, as for example, that of the *jangadeiros* from Recife, are very similar in some fundamental ways: 1) both the effectiveness and tacit quality, what we have called *invariants* of any mathematical activity, are present in these practice-based activities; 2) all of these practices could be seen somehow as specific kinds of communities of practice. (In fact, most school mathematics practices are far from being communities of practices for their participants, but, as we have suggested above, they COULD be.) This second point emerges from our explorations of theories of tacit knowledge (or knowing) and theories of situated cognition and communities of practice and the connections we have made between them. These explorations show how situated learning and communities of practice perspectives can be seen as a powerful framework to explain the situated character of tacit knowledge (or knowing). This form of knowledge is deeply socio-culturally situated in specific practices. As a corollary, some important mathematical competencies acquired by the participants of these different practices cannot be evaluated in the same way. There is no hierarchy of tacit knowing or competencies-in-activities. We have explored different methodological possibilities to improve our understanding of the identification and circulation of tacit knowing within mathematics school practices. We have also discussed some pedagogical implications resulting from these. In doing so, we have drawn attention to similarities and differences between school practices and other culturally situated communities of practice, in particular that of the *jangadeiros*. We conclude this chapter with the observation that, if we look at school mathematics practice in the light of situated learning perspectives, we cannot disregard the tacit dimension of mathematics teaching and learning.

REFERENCES

Araújo, C. R., Andrade, F., Hazin, I., Da Rocha Falcão, J. T., Nascimento, J. C., & Lins Lessa, M. M. (2003). Affective aspects on mathematics conceptualization: From dichotomies to an integrated approach. In N. Pateman, B. Dougherty & J. Zilliox (Eds.), *Proceedings of the 27th Conference of the International Group for the Psychology of Mathematics Education* (Vol. 2, pp. 269-276). Honolulu, USA: PME.

Bishop, A. J. (2006). Values and mathematics education – a developing research field. Plenary lecture at culture and affect. Faculdade de Educação, Universidade Federal de Minas Gerais, Brazil.

Brito Lima, A. P., & Da Rocha Falcão, J. T. (1997). Early development of algebraic representation among 6-13 year-old children: The importance of didactic contract. In E. Pehkonen (Ed.), *Proceedings of the 21st Conference of the International Group for the Psychology of Mathematics Education* (Vol. 2, pp. 201-208). Helsinki, Finland: PME.

Burton, L. (1999). The practices of mathematicians: What do they tell us about coming to know mathematics? *Educational Studies in Mathematics, 37*(2), 121-143.

Burton, L. (2002). Recognizing commonalities and reconciling differences in mathematics education. *Educational Studies in Mathematics, 50*(2), 157-175.

Clot, Y., D., F., Fernandez, G., & Scheller, L. (2001). Les entretiens en auto-confrontation croisée: Une méthode en clinique de l'activité. *Education permanente, 146*, 17-27.

Da Rocha Falcão, J. T. (2005). *Conceptualisation en acte, conceptualisation explicite: Quels apports théoriques à offrir à la didactique des mathématiques et des sciences? Actes du*

Colloque Les processus de conceptualisation en debat-hommage a Gérard Vergnaud. Paris: Association pour la Recherche sur le Développement des Compétences.

Ernest, P. (1998a). *Social constructivism as a philosophy of mathematics.* Albany, NY: SUNY.

Ernest, P. (1998b). Mathematical knowledge and context. In A. Watson (Ed.), *Situated cognition and the learning of mathematics* (pp. 13-29). Oxford: Centre for Mathematics Education, University of Oxford.

Frade, C. (2003). Polanyi's social construction of personal knowledge and the theories of situated learning. *Philosophy Of Mathematics Education Journal, 17* (http://www.people.ex.ac.uk/PErnest/).

Frade, C. (2005). The tacit-explicit nature of students' knowledge: A case study on area measurement. In H. L. Chick & J. L. Vincent (Eds.), *Proceedings of the 29th Conference of the International Group for the Psychology of Mathematics Education* (Vol. 2, pp. 321). Melbourne, Australia: PME.

Frade, C., & Borges, O. (2002). Tacit knowledge in curricular goals in mathematics. *Proceedings of the 2nd International Conference on the Teaching of Mathematics* (p. 128). Crete, Greece: John Wiley & Sons.

Frade, C., & Borges, O. (2006). The tacit-explicit dimension of the learning of mathematics: An investigation report. *International Journal of Science and Mathematics Education, 4,* 293-317.

Frade, C., Winbourne, P., & Braga, S. M. (2006). Aline's and Julia's stories: Reconceptualizing transfer from a situated point of view. In J. Novotná, H. Moraová, M. Krátká & N. Stehlíková (Eds.), *Proceedings of the 30th Conference of the International Group for the Psychology of Mathematics Education* (Vol. 3, pp. 97-103). Prague, Czech Republic: PME.

Grigorenko, E., Meier, E., Lipka, J., Mohatt, G., Yanez, E., & Sternberg, R. (2001). *The relationship between academic and practical intelligence: A case study of the tacit knowledge of native American yup'ik people in Alaska.* ERIC: Educational Resources Information Center.

Latour, B., & Woolgar, S. (1979). *Laboratory life: The social construction of scientific facts.* Los Angeles, Londres: Sage.

Lave, J. (1988). *Cognition in practice.* Cambridge: Cambridge University Press.

Lave, J., & Wenger, E. (1991). *Situated learning: Legitimate peripheral participation.* New York: Cambridge University Press.

Leontiev, A. N. (1994). Uma contribuição à teoria do desenvolvimento da psique infantil. In L. S. Vygotsky, A. R. Luria & A. N. Leontiev (Eds.), *Linguagem, desenvolvimento e aprendizagem* (pp. 103-117). São Paulo: Ícone/EDUSP.

Lerman, S. (2001). Getting used to mathematics: Alternative ways of speaking about becoming mathematical. *Ways of Knowing, 1*(1), 47-52.

Lins Lessa, M. M., & Da Rocha Falcão, J. T. (2005). Pensamento e linguagem: Uma discussão no campo da psicologia da educação matemática. *Psicologia, reflexão e crítica, 18*(3), 315-322.

Nunes, T., Schliemann, A. D., & Carraher, D. W. (1993). *Street mathematics and school mathematics.* Cambridge: Cambridge University Press.

Piaget, J. (1974). *Réussir et comprendre.* Paris: Presses Universitaires de France.

Polanyi, M. (1962). *Personal knowledge.* London: Routledge & Kegan Paul.

Polanyi, M. (1969). *Knowing and being* (M. Grene, Ed.). Chicago: Chicago University Press.

Ponte, J. P., & Matos, J. F. (1991). Cognitive processes and social interactions in mathematical investigations. In J. P. Ponte, J. F. Matos, J. M. Matos & D. Fernandes

(Eds.), *Mathematical problem solving: Research in the context of practice* (pp. 239-254). Berlim: Springer-Verlag.

Romberg, T. A. (1992). Problematic features of the school mathematics curriculum. In P. W. Jackson (Ed.), *Handbook for research on curriculum* (pp. 749-788). New York: MacMillan.

Schoenfeld, A. H. (1992). Learning to think mathematically: Problem solving, metacognition, and sense making in mathematics. In D. Grouws (Ed.), *Handbook for research on mathematics teaching and learning* (pp. 334-370). New York: MacMillan.

Tirosh, D. (Ed.). (1994). *Implicit and explicit knowledge: An educational approach.* Norwood, NJ: Ablex Publishing Co.

Vergnaud, G. (1990). La théorie des champs conceptuels. *Recherches en Didactique des Mathématiques, 10*(23), 133-170.

Vergnaud, G. (1997). The nature of mathematical concepts. In T. Nunes & P. Bryant (Eds.), *Learning and teaching mathematics: An international perspective* (pp. 5-28). London: Psychology Press.

Vergnaud, G. (2000). Que peut apporter l'analyse de l'activité à la formation des enseignants et des formateurs? *Carrefours de l'éducation, 10*, 49-63.

Vygotsky, L. S. (1986). *Thought and language* (A. Kozulin, Trans. Revised ed.). Cambridge: MIT Press.

Watson, A. (Ed.). (1998). *Situated cognition and the learning of mathematics*. Oxford: Centre for Mathematics Education, University of Oxford.

Wenger, E. (1998). *Communities of practice: Learning, meaning and identity*. Cambridge: Cambridge University Press.

Wigner, E. P., & Hodgkin, R.A.. (1997). "Michael Polanyi" in *Biographical Memoirs of Fellows of the Royal Society*, (Vol. 23, pp. 413-438). London: The Royal Society.

Winbourne, P. (2002). Looking for learning in practice: How can this inform teaching. *Ways of Knowing, 2*(2), 3-18.

Winbourne, P., & Watson, A. (1998). Learning mathematics in local communities of practice. In A. Watson (Ed.), *Situated cognition in the learning of mathematics*. Oxford: Centre for Mathematics Education Research, University of Oxford.

Chapter 11

Cognition And Institutional Setting

Undergraduates' understandings of the derivative

Erhan Bingolbali and John Monaghan
University of Firat, University of Leeds

Abstract: This chapter examines Mechanical Engineering and Mathematics undergraduates' understanding of the derivative and addresses institutional issues in the social formation of knowledge. It summarises results from a study and examines student and lecturer data. Significant differences over the course of the first year are noted. It is claimed that these differences arise from their participation in different departments (institutions). The closing section examines students' developing conceptions of the derivative in institutional settings by addressing the question: what brought about these changes in students' conceptual development?

Key words: derivative, cognition, institutional setting, departmental affiliation, engineering and mathematics students

1. INTRODUCTION

We look at Mechanical Engineering and Mathematics undergraduates' understanding of the derivative in one Turkish university and find significant differences. We claim that these differences in their understandings arise from their participation in different departments. These differing forms of participation could be viewed in many ways, e.g. in terms of activity systems (Engeström, 1987) or identityy (Holland, Lachicotte, Skinner and Cain, 1998) or praxeologies (Chevallard, 1999) or communities of practice (Wenger, 1998). In this chapter we wish to use, rather than follow, constructs from these points of view and 'chip away at' what it means to learn about an important advanced mathematical concept, the derivative, in two different departments or, in terms of the main construct of this chapter, two different *institutional settings*.

Mathematics education research on students' understanding of the calculus was, until the mid 1970s, a relatively unknown activity. Since then research has flourished but the 'research norm' in undergraduate studies has been to ignore departmental affiliation. We illustrate this with two relatively recent, at the time of writing, studies. Asiala, Cottrill, Dubinsky and Schwingendorf (1997) investigated learning about the slope of a graph of a function of 41 engineering, science and mathematics students. They compared the performance of 17 students who took a reform-based calculus course with 24 students who took a traditional calculus course. Their focus was on the contribution of the different instructional treatments to students' conceptions of the derivative. They appeared to have ignored the possibility that the departmental affiliation of the students might have influenced their conceptions. Bezuidenhout (1998) explored 100 first year university students' understanding of rate of change from engineering, physical science and service calculus (the commercial and management sciences) courses. Again the departmental affiliation of the students was not considered.

Such studies appear to assume that departmental affiliation has no bearing on cognition. In this chapter we present a case that department affiliation matters a great deal: that two sets of undergraduate students from different departments, both taught by members of the Mathematics department, developed different conceptions of the derivative in their first year; that students' departmental affiliation impact upon their developing conceptions; and that the departments/institutions where learning and teaching take places need to be taken into account.

The chapter has the following structure. The next section considers a theoretical framework which helps us to understand the institutional dimension of learning of calculus in different departments. We then describe a research study that informs this chapter followed by a description of each department. We present summary results from this study and examine students and lecturers. The closing section looks at developing conceptions of the derivative in institutional settings by addressing the question: what brought about the changes in ME and M students' conceptual development?

2. THEORETICAL FRAMEWORK OF THE STUDY

We seek a theoretical framework that can help us understand departmental affiliation in the developing calculus conceptions of two groups of undergraduate students in different departments. There is little mathematics education literature to help us on our quest. We outline potentially relevant social theories, one with a particular interest in mathematics, and two mathematics education studies on mechanical engineering students and the constructs they offer.

At a general level we seek studies that can inform us about students' practices and their communities of practice. Engeström (1999) views these two foci as constituting 'weak' and 'strong' versions of situated cognition.

> The weak version argues that learning is situated in physical and social contexts … The strong version argues that learning is literally a by-product of participation in any social practice (p. 250)

Engeström regards most empirical work on situated learning in the 1990s as largely following the weak version of the agenda but notes that:

> This does not mean that the research is weak or ill-informed… it simply means that social practices and communities of practices are not very often taken as the starting point of analysis. (ibid., p. 250)

We think the research focus and its 'genetic domain' (Wertsch, 1998) will at least partially determine the 'strength' of the situated account: 'strong' for a socio-cultural study, 'weak' for a microgenetic study.

Daniels (2001, p. 135) notes that educational research has "insufficient empirical study of socio-institutional effects" and a "tendency to under-theorise differences between schools in terms of institutional effects on the social formation of mind". This, we hold, is the case in mathematics education research too. An exception is work following Chevallard's (1999) anthropological approach. Here 'praxeologies' (practices) are described in terms of tasks, techniques (used to solve tasks), technology (talk/discourse) and theory. Technology and theory concerns what is legitimised as knowledge *per se* and task and technique concern 'know-how'. Tasks are artefacts that are constructed (and reconstructed) in educational settings, institutions. 'Institution' here relates to Chevallard's notion of 'didactical transposition' where "mathematics in research and in school can be seen as a set of knowledge and practices in transposition between two institutions, the first one aiming at the production of knowledge and the other at its *study*." (Lagrange, 2005, p. 69). Techniques are institutionally privileged to the extent that only one, of many possible techniques, may be considered.

Castela (2004) works within the anthropological approach and notes that:

> When persons 'enter' an institution, their life in this institution is submitted to collective *constraints* and *expectations* that regulate their actions. These constraints and expectations specify their position as *subjects* of the institution. Several subject positions exist in a given institution: for example, students, lecturers and assistants at a university (pp. 41-42).

Castela goes on to say that institutions “give these persons opportunities to adapt themselves to the institutional customs – in other words to learn” (p. 42). We understand this to mean that what students value or not is related to the collective constraints and expectations of the institutions to which they belong and to their interpretations of institutional customs.

Work in the anthropological approach goes beyond adaption to institutional customs and addresses institutional influences on knowledge acquisition. Praslon (1999) examines a university entrance task on the continuity and differentiability of a function and stresses the important role of institutional values and norms in developing personal relationship with mathematical knowledge, emphasising that individual relationships with particular mathematical objects are shaped by institutional parameters. In the same vein Artigue, Assude, Grugeon, and Lenfant (2001) argue that:

> Mathematical knowledge cannot be considered as something absolute. It strongly depends on the institutions where it has to live, to be learnt, to be taught. Mathematical objects do not exist per se but emerge from practices which are different from one institution to another one. (p. 2)

Mechanical Engineering was not the only department we could have used, as a contrast to Mathematics, in order to explore undergraduates' conceptual development of the derivative concept in relation to their department/institution but it was a good choice – it uses advanced mathematics including calculus on a regular basis but it ‘uses’ mathematics rather than studies mathematics for its own sake. There are two mathematics education studies of mechanical engineering students that informed our work. Sazhin (1998) examined first and third year mechanical engineering students' understanding of physical and mathematical concepts and found that engineering students learnt physical concepts and concepts related to real life experiences much easier than they learnt mathematical concepts. Maull and Berry (2000) examined first and final year mechanical engineering and mathematics undergraduates alongside postgraduate students and professional engineers. They indicated that “the mathematical development of engineering students is different from that of mathematics students, particularly in the way in which they give engineering meaning to certain mathematical concepts” (ibid., p. 916). They noted that both groups of students demonstrated similar patterns of responses at entry but, by the final year, the groups' responses diverged. They did not, however, supply reasons for this divergence and called for further research.

Maull and Berry (2000) suggest that engineering students are socialised into ways of thinking and behaving, and ask whether the difference found stems from socialisation, from the interactions between students and their peers, lecturers and other professional contacts, or whether there is also a second acculturation process through their discovery of what is useful in the context of their study and work.

Although they did not explicitly mention the term 'institution', they called on researchers to carry out further research to explore whether the difference found between mathematics and engineering students were related to parameters of the department to which they belonged. Our work goes a little way towards addressing this issue.

3. THE RESEARCH

The study that informs this chapter set out to investigate first year Mechanical Engineering (ME) and Mathematics (M) students' conceptual development of the derivative with particular reference to rate of change and tangent aspects, and examined contextual influences of students' institutions/departments on their knowledge development (see Bingolbali, 2005). The study was conducted in a large university in Turkey. Data were collected by a variety of means: quantitative (pre-, post- and delayed post-tests), qualitative (questionnaires and interviews) and, to some extent, ethnographic (observations of calculus modules and 'coffee house' discussions). The study adopted a 'naturalistic' approach (Lincoln and Guba, 1985) – situations were not manipulated nor were outcomes presumed.

The study involved all first year university students majoring in ME and M from the same university. Both ME and M departments selected their students on the basis of the Turkish university entrance examination taken at the end of high school (typically in Year 11, students aged between 17-18). Differential and integral calculus is contained within the school curriculum but is not examined in the university entrance examination, so some students are not introduced to it at school. Both groups consisted of a similar attainment range on the basis of their university entrance examination results. In the academic year in which the study took place 85 students were admitted into the ME department and 65 students were admitted into the M department[30].

The study was conducted over one academic year and the first author was involved in testing, observations, interviews, document analysis and

[30] Both the ME and M departments, as is common in Turkish universities, offer the same course to two different groups of students: one group attends modules during the day whilst the other in the evening. This is done on the basis of students' attainment on the university entrance examination – lower attaining students generally attend evening classes. In this study we worked with day students.

informal 'coffee house' talks on a daily basis from October – January (semester 1) and April – May (half of semester 2).

We report on a number of test results in this chapter. This is important to establish the significant differences that emerged over the course of the year but, as befits a chapter in a book as opposed to a research report, we endeavour to keep tables, etc. to a minimum. An important set of tests were pre-, post- and delayed post-tests on the derivative. Prior to administering the pre-test students sat a basic test on the derivative – we deemed it unwise to ask students to do an extended test for which they could not attempt any questions[31]. Leaving aside these students and any who missed any one of the tests there remained 50 ME and 32 M students who completed all three tests. The three tests were administered in October, January and April/May. Test questions addressed 'rate of change' and 'tangent' aspects of the derivative in graphic, algebraic and application forms.

The pre-test showed no significant difference between ME and M students' performance. In the post-test both groups improved their performance, but in different ways. Overall, ME students did better than M students on all forms of rate of change questions whilst M students did better than ME students on all forms of tangent questions. In the period between these two tests, the calculus modules were observed and copies of students' notes were made to see how the topic 'the derivative' was taught in each department. A selection of lessons from students' other modules was also observed to find out more about the academic experience of each group of students.

We will dwell for a paragraph on this interpretation: overall, ME students did better than M students on all forms of rate of change questions whilst M students did better than ME students on all forms of tangent questions. When this interpretation appeared to us in February we were very excited. It suggests a fundamental shift in cognition over quite a short (one semester) period of time. It was not unexpected but we did not predict it and it was quite consistent, it did apply 'on all forms'. We called it the 'emergent theme' and will call it this in this chapter. As this theme emerged in the middle of data collection we were able to add additional data collection tools to explore this emergent theme when the first author returned to the university in April – May. We now describe aspects of February data analysis that impinged on how we collected extra data.

An initial analysis of the calculus module observations and students' notes revealed that the ME calculus lecturer, a member of the Mathematics department, privileged rate of change aspects of the derivative and the M

---

[31] As mentioned, the university entrance examination does not set calculus questions. Some teachers, a minority but a notable minority, do not teach it.

calculus lecturer, also a member of the Mathematics department, privileged tangent aspects. We thought that these different emphases in teaching were not just a matter of personal preference and that most lecturers would teach differently to students of different departments. To investigate this issue we interviewed, in addition to the ME and M calculus lecturers, four more lecturers, two mathematicians and two physicists, all of whom had taught 'service courses' to students of different departments. To obtain further insight into students' orientation to the derivative, tangent/rate of change, we designed the 'rate of change and tangent' test which had two questions (see the *Results* section for details).

4. INSTITUTIONAL CONTEXTS OF THE DEPARTMENTS

Educational practice has arisen through a long period of historical development. Resources, institutional structures and interpersonal rituals have *evolved* to their present forms, and researchers cannot take them for granted as if they simply provided a neutral backdrop to the interactional business of teaching and learning (Crook and Light, 1999).

Given that we focus on ways that departmental affiliation influences students' cognition, it is important that we describe salient, to us, features of each department.

4.1 The Mechanical Engineering department

The Mechanical Engineering department was one of six departments within the faculty of Engineering and Architecture. The department was founded in 1976 and aimed to "prepare/foster engineering students as those mechanical engineers who have a fundamental knowledge of technological development...who not only analyse but also synthesise..." (translation from the department's 'overarching goal' statement). The department had 62 academic stuff including 20 professors, 3 associate professors, 7 assistant professors, 8 lecturers and 24 research assistants. The department offered courses to 25 PhD students, 80 MSc and approximately 600 undergraduates.

In the first year of their degree ME students take the following modules:

First semester	Second semester
Calculus I	Calculus II
Physics I	Statics
Introduction to Mechanical Engineering	Chemistry
Introduction to computer programming	Computer-based technical drawing
Technical drawing	Computers in Mechanical Engineering

First semester	Second semester
Ethics of Mechanical Engineering (optional)	Introduction to engineering mechanics (optional)

But the vast majority of first year students did more than simply attend lectures, they mingled with students in other years and selectively involved themselves in departmental activities.

Research interests of academic staff were: construction and manufacturing, thermodynamics, energy, machine theory, dynamics, mechanics and automotive engineering. After their first year students could choose elective modules as well as following compulsory modules. In their fourth year students could specialise in the following areas: Design and Manufacturing, Heat Treatment and Installation and Automotive. The city in which the university is situated is a centre for the automotive and textile industries in Turkey and the department had close links with these sectors. ME students visited some local industrial factories organised by the department and voluntarily attended the activities of a local mechanical engineering chamber, an organisation of (generally) practising engineers.

4.2 The Mathematics department

The Mathematics department was one of 10 departments within the faculty of Arts and Science. The department was founded in 1983 and aimed to "foster mathematicians… to provide fundamental knowledge for those students who want to study mathematics and mathematics-related subjects". This statement also declares aims to prepare students to work as high school teachers and in the banking and financial sector. Many graduates choose to become teachers, which requires them to attend special teacher training modules. The department had 23 academic staff including 6 professors, 3 associate professors, 6 assistant professors, 2 lecturers and 6 research assistants who teach mathematics to approximately 500 undergraduate students. Mathematics lecturers taught courses in other departments, e.g. engineering, physics and chemistry, in which mathematics was a compulsory module.

In the first year of their degree M students take the following modules:

First semester	Second semester
Calculus I	Calculus II
Linear algebra I	Linear algebra II
Abstract mathematics I	Abstract mathematics II
Introduction to computer	Mathematical programming

The research interest of staff included: analysis, function theory, algebra, number theory, geometry, applied mathematics, logic and statistics. These interest areas were incorporated into the study programmes for undergraduate students. M students could choose elective modules from these areas starting from the second year of their studies. The department offered little by way of extra-curricula activities.

5. RESULTS

The results are presented in three sections: student tests, calculus modules and lecturer interviews.

5.1 Student tests

Student test results are presented in two parts: pre-, post- and delayed post-tests; and rate of change and tangent items.

5.1.1 Pre-, post- and delayed post-tests

We report on two tangent, Q1 and Q2, and two rate of change, Q3 and Q4, questions. These questions, displayed below, were included in all three tests. We do not report on every item in the tests but we have not 'hidden any data' which does not support the 'emergent trend'!

Question 1
Line L is a tangent to the graph of $y=f(x)$ at point $(5,3)$ as depicted in the graph below.

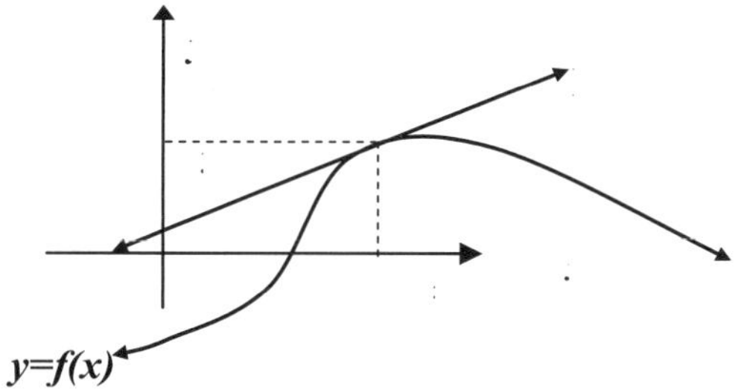

a) $f(5)=?$
b) $f'(5)=?$
c) What is the value of the function $f(x)$ at $x=5.08$? (be as accurate as possible) (Amit and Vinner,1990)

Question 2

a) Find the equation of the tangent to the curve $y=x^3-x^2+1$ at (1,1)
b) Find the equation of Line L by making use of the graphs given below (In administration of the question, this item presented the graph of $f(x)=-2x^3+x^2+x+1$ which, together with Line L, passed through the point (1,1).

Question 3

Find the rate of change with respect to the given variable of the following functions at the values indicated.
a) $f(x)=x^2-7x$, when $x=3$
b) $g(x)=(x^2-1).(x+1)$, when $x=1/3$

Question 4

For a certain period the population, Y, of a town after x years is given by the formula $Y=1000(50+2x-x^2/6)$. Find:
a) The initial population,
b) Its initial rate of increase,
c) The time at which the rate of increase is 1000 people per year,
d) The time at which the population stops growing and its value at this time.

Table 1 displays the test results. We display correct (✓) and incorrect (✗) responses. Student responses were initially coded under four categories: correct, partially correct, incorrect and not attempted. We collapsed the categories partially correct, incorrect and not attempted into the incorrect category to keep the presentation of data to a minimum in this chapter. The collapsed table hides the fact that there were many partially correct answers, e.g. in Q1-b 34% of ME and 25% of M provided partially correct answer in the pre-test. This table does not show responses to Q1-a which virtually everyone answered correctly on each test.

Table 11-1. Students' responses, percentages, to questions 1 – 4

		Pre-test		Post-test		Delayed post-test	
		ME (n=50)	M (n=32)	ME (n=50)	M (n=32)	ME (n=50)	M (n=32)
Q1-b	✓	22	22	32	42	42	56
	×	78	78	68	58	58	44
Q1-c	✓	10	18	4	34	10	34
	×	90	82	96	66	90	66
Q2-a	✓	28	22	50	63	50	69
	×	72	78	50	37	50	31
Q2-b	✓	28	13	34	56	38	56
	×	72	87	66	44	62	44
Q3-a	✓	24	22	84	53	74	53
	×	76	78	16	47	26	47
Q3-b	✓	18	22	64	53	76	50
	×	82	78	36	47	24	50
Q4-a	✓	56	41	70	44	78	44
	×	44	59	30	56	22	56
Q4-b	✓	6	6	22	13	36	22
	×	94	94	78	87	64	78
Q4-c	✓	8	6	30	22	60	19
	×	92	94	70	78	40	81
Q4-d	✓	8	3	28	9	52	13
	×	92	97	72	91	48	87

A careful inspection of Table 1 should reveal what we call the 'emergent trend': very similar response patterns in the pre-test but M students consistently performing better on tangent items (these have been highlighted) and ME consistently performing better on rate of change items (these have been highlighted). We remind the reader that this is mathematics performance, not student preferences, and this trend appears to indicate diverging cognition from a common base level over a relatively short period of time, one semester.

Due to the importance we ascribed to these results we applied an appropriate statistical test, a Mann-Whitney U test. The test was applied to the sum of students' improvement score on the rate of change and tangent items from the pre-test to the post-test and from pre-test to delayed-post test. The results showed a significant difference between the two groups of students on: the tangent items from the pre-test to the post-test ($p=0.003$) and from the pre-test to the delayed post-test ($p=0.002$); the rate of change items from the pre-test to the post-test ($p=0.027$) and from the pre-test to the delayed post-test ($p<0.001$). This, we feel, provides strong supporting evidence that the 'emergent trend' reflects a socio-cognitive phenomenon.

5.1.2 The rate of change and tangent test

The rate of change and tangent test, as mentioned above, was developed to further explore the 'emergent trend' and was applied in the second round of data collection. The test is shown below.

Item 1: If the following two questions (A and B) were set in an examination and you only had to solve one of them, which one would it be? Please tick just one option and explain why you chose that one.

A) At a certain time (*t*, seconds) the rate at which water flows (m^3 /sec) into a water tank is given by the formula:

$$f(t) = \frac{t^2}{4} + 24t + 125$$

Find:

a) The initial amount of water in the tank and its initial rate of change?
b) What is the rate of change of flowing water at any time, *t*?
c) The time at which the rate of change is $32\ m^3/sec^2$.

B) Find the solutions of the following questions:

a) Verify that the gradient of the tangent to the curve $y=x^2$ at a point $(x_1, x_1^2)=2x_1$.
b) Find the equation of the tangent to the curve $y=2x^2-x+3$ which is parallel to the line $y=3x-2$.
c) Show that the graph of $f(x)= x^{1/3}$ has a vertical tangent line at (0,0) and find an equation for it.

Item 2: Two university students from different departments are discussing the meaning of the derivative. They are trying to make sense of the concept in accordance with their departmental studies.

Ali says "the derivative tells us how quickly and at what rate something is changing since it is related to moving object. For example, it can be drawn on to explain the relationship between the acceleration and velocity of a moving object".

Banu, however, says "I think the derivative is a mathematical concept and it can be described as the slope of the tangent line of a graph of *y* against *x*".

a) Which one is closer to the way of your own derivative definition?

Please explain!

b) If you had to support just one student, which one would you support and why?

Item 1 provided students with one rate of change-oriented question (A) and one tangent-oriented question (B). Item 2 was intended to include core ideas on rate of change and tangent aspects of the derivative and presented these ideas as the words of two imaginary students, Ali and Banu. The question has two parts (2a and b). Note that Ali and Banu present reasons for their choices but no explicit indication is made of their department.

In the analysis which follows, numbers in brackets refer to alternative answer choices. In item 1, 60% (40%) of ME students (n=45) chose question A (B). 22% (78%) of M students (n=32) chose question A (B). The responses for item 2 are best displayed in a tabular format, see Table 2.

Table 11-2. Students' responses (percentages) to item 2

	ME		M	
	a	b	a	b
Ali (A)	51	49	19	13
Banu (B)	27	49	63	78
Both (A &B)	22	2	16	3
Not Attempted (NA)	0	0	3	6

We now turn to students' reasons for their responses. Repeated reading, and refinement of open codings, of students' responses produced six categories: 'real life and application', 'mathematical and scientific', 'department', 'practice', 'ease' and 'not categorised'[32]. We first explain what these categories stand for and how we allocated students' responses to these categories. We then give examples of students' response for each category. Table 3 presents a categorisation of students' reasons for their choices in items 1 and 2.

Real life and application

Student responses were assigned to this category when they referred to real life or application in support of Ali in item 2 and question A in item 1.

[32] Some responses fall under more than one category e.g., since ME 4 mentions both real life and engineering, his response was assigned to both 'real life & application' and 'department' category.

ME 1: Ali's one, because his definition is related to real life. It shows the area of application of derivative. I think that maths concepts are attractive in as much as they are applicable.

ME 2: I would support Ali. I am thinking with an engineer mentality. This makes me tend to be closer to the practicality and the concreteness.

Scientific and mathematical

Student responses were assigned to this category when they made reference to being mathematical or scientific in support of Banu in item 2 and question B in item 1.

M 1: Banu gives the definition while Ali gives the explanation. I would support B because Banu explains it in a scientific way.

ME 3: I would support Banu because the root of the derivative concept is mathematical and the derivative's other applications are based on its mathematical concept.

Department

Student responses were assigned to this category when they made reference to belonging to a Mathematics/Engineering department or being a mathematician/engineer.

ME 4: Calculating rates of change seems to me more real. On the other hand what Banu says is not far away … but since I am going to be an engineer, Ali's idea would be just different. Because I would be the one who makes mathematics concrete.

M 2: Banu interprets the derivative from a mathematician's perspective, and Ali interprets it from a physicist standpoint. At the end of the day, since I too am from mathematics department, I find Banu's explanation closer to myself.

Practice

Student responses were assigned to this category when they referred to their calculus module.

M 3: It is similar to what we are learning now and easier to answer.

Ease (ME and M)

Students' responses were placed in this category when they mentioned the 'ease' of this way of thinking about the derivative.

M 6: Because it is easier.

Not Categorised

Student responses were assigned to this category when they did not give a reason, i.e. they just repeated their choice.
M 4: I would support Banu because Banu's is closer to my understanding.

Table 11-3. Categorized responses (percentages) to items 1 and 2

	ME						M					
	Item 1		Item 2a		Item 2b		Item1		Item 2a		Item 2b	
	A	B	A	B	A	B	A	B	A	B	A	B
	60	40	51	27	49	49	22	78	19	63	13	78
Real life/application	22	0	29	0	29	0	3	0	9	0	6	0
Scientific/mathematical	0	0	0	13	0	29	0	9	0	25	0	28
Department	0	0	11	0	11	0	0	0	0	19	0	25
Practice	9	0	9	4	7	0	0	38	0	19	0	9
Ease	27	29	-	-	-	-	9	47	-	-	-	-
Not categorized	2	11	9	9	9	20	9	6	9	6	6	22

NB: Some responses in item 2 fall under more than one category

In item1 more ME students, 22%, than M students, 3%, cited 'real life/application' in explaining their reasons for choosing the rate of change-oriented question A, suggesting that application is seen as more important for ME students. For question A, 9% of ME students and no M students referred to the calculus practices and 27% of ME and 9% of M students cited 'ease'. Of students who chose question B, 29% of ME and 47% of M students cited 'ease'. 38% of M students and no ME students cited 'practice'. Very few, 9% of M and no ME students, chose question B because it was 'scientific/mathematical'.

In item 2a (2b), 29% (29%) of ME but only 9% (6%) of M students cited 'real life/application' in explaining their preferences for Ali's rate of change interpretation. These results are consistent with the 'emergent trend' and suggest, not surprisingly, that engineers are more concerned with the real world than mathematicians. 13-29% of ME and 25-28% of M students cited being scientific/mathematical and, when this was cited, it was only to support Banu's statement for item 2a (2b). The percentages in the 'department' category are not large but the pattern of responses, without exception, follow the 'emergent trend': no ME (M) student who chose Banu's (Ali's) interpretation cited department in explaining their choice.

5.2 The calculus modules

We present a brief analysis of both calculus modules regarding the emphasis placed on rate of change and tangent aspects, the number theorems, proofs, and definitions, and the examinations questions set in both modules.

The ME calculus module took 2 hours 15 minutes per week (three 45 minute lessons) and 15 hours (20 lessons) in total were devoted to teaching the derivative. The M calculus module took 4 hours per week (six 40 minute lessons) and 24 hours (36 lessons) in total were devoted to teaching the derivative. The calculus modules of both departments were observed and compared with students' notes to gain insights into which aspects of the derivative were 'privileged' (Kendal and Stacey, 1999). Table 4 presents the data with regard to rate of change and tangent, which clearly shows that rate of change aspects were privileged in ME calculus lectures and tangent aspects were privileged in M calculus lectures.

Table 11-4. Analysis of calculus modules with regard to rate of change and tangent

	Rate of change		Tangent	
	ME	M	ME	M
Duration	≈133 minutes	≈11 minutes	≈10 minutes	≈85 minutes
examples	(9 examples)	(no examples)	(no examples)	(7 examples)

With regard to theorems and proofs, both lecturers referred to 20 theorems. The ME lecturer proved 10 of these theorems and the M lecturer proved 17 of these. With regard to definitions the M lecturer provided 14 definitions and the ME lecturer provided 10 definitions. Analysis of ME and M departments' mid-term and end of semester 1 calculus examinations revealed: ME – one rate of change and no tangent question; M – one tangent and no rate of change question. M students were asked to prove two theorems in both their mid-term and final examinations but ME students were not asked to prove a theorem in any examination.

5.3 Lecturers' views regarding their teaching practices

We interviewed ME and M calculus course lecturers, hereafter referred to as L1 and L2, in the course of semester 1 data collection. As mentioned, to investigate different emphases in teaching (for further understanding of the 'emergent theme') we interviewed two more mathematics lecturers (L3, L4) and two physics lecturers (L5, L6), all of whom had taught 'service courses'. L5 and L6 were selected for interview to gain the views of lecturers from a department other than Mathematics on service teaching.

The data presented here were obtained through semi-structured interviews with L1 – L6. Lecturers were asked: (1) if they teach students of different departments in a different way; (2) if they set different types of questions on examinations for different students; and (3) if they use different textbooks for different departments but the interviews were free to pursue directions raised by the lecturers. We report on their views regarding emphases in teaching, examination questions and textbooks and their awareness of distinct departmental features.

5.3.1 Lecturers emphases in teaching, examination questions and textbooks

All lecturers stated that in teaching in different departments they emphasised different aspects of topics. Both the calculus module lecturers we observed stated that they taught engineering students differently from mathematics students.

L1: ...in engineering departments it is more application-oriented.... I give examples regarding objects in motion with regard to time, pressure and so forth... I focus on rate of change aspects in teaching engineering students. But in our department (*maths*) we focus more on concepts rather than on application aspects, for example, tangent aspects.

L2: If they *(maths students)* were physics department students, lots of examples regarding physical meanings of the derivative would be given. But in the mathematics department the derivative as a concept is prioritised. For maths students to see just how it can be applied is enough.

These views accord with what was observed in their calculus modules (Table 4) and the numbers of theorems and proofs given by each lecturer. When asked whether they reflected this difference with regard to concepts and examples in examination questions as well, they and the other four lecturers affirmed that they did so, e.g.:

L1: Maths students would be specialists in this area...they need to know this job's reason and logic. That is why you can ask them theorems in their examinations. This is their job. You can ask them some definitions as well. Nevertheless, if you do this in engineering departments, it would not do anything good to them but get them bored.

L2: I set at least two theorems in maths students' examinations to prove. But I don't set theorems in other departments' examinations. I only set application-oriented ones in other departments.

Again these views accord with what L1 and L2 actually set in calculus examinations. The accounts of L3 and L4 suggested that they did almost the same thing as L1 and L2. The Physics lecturers made similar remarks, e.g. L5 stated that he set more theoretical questions in physics examinations and "did the opposite" in engineering examinations.

As regards textbooks, both mathematics and physics lecturers stated that they used different textbooks or privileged different parts of the same textbook for different departments, e.g.:

L1: The books I use in the engineering department are generally application-oriented ones…. But in our department I would use those books which include more theorems and proofs…

5.3.2 Lecturers' awareness of distinct departmental features

We attend to lecturers' stated reasons for why there were differences in teaching, examination questions and textbooks. In responding to prompts of why they taught students in client departments in the way they did L1, L4 and L5 referred to departmental demands from 'higher authorities'.

L1: They demand from us some stuff. It is like we use maths here and there, we want our students to know this and that so that they can be successful in the coming years' modules.

L5:....we contact administrators of department and ask them what they want their students to get from physics and the physics we teach to their students is hence based on their demands as well.

All lecturers but L2 referred to 'students' needs' regarding mathematics. We cite from the interview with L3 with regard to the needs of ME and M students and from L5 and L6 with regard to the needs of Physics and service course students.

L3: The main aim is where maths and engineering students make use of maths. Maths students need to know everything but engineering students only need to know the parts which are useful for them.

L5: …in the physics department ...topics are given in detail and their 'hows' and 'whys' are investigated.... But in engineering or maths departments it is not possible get into detail much…

L6: (of client department students)…we need to think how the physics we teach is going to be useful to them. What we teach them should be useful to them.

All lecturers but L1 stated they taught or set questions the way they did due to 'departmental features', e.g.:

L5: In the maths department I tried to give examples concerned with the essence more while in the engineering department it was more towards the application aspects… I tried to choose some typical questions which are peculiar to this or that particular department.

L6: Topics are presented so that they are useful for the department's job..., are close to these departments' features.

We found L5's 'essence – mathematics; application – engineering' comment interesting but defer discussing this further until the next section.

6. INSTITUTIONAL SETTINGS: STUDENTS AND LECTURERS

We consider three sets of results – the 'emergent trend', students' stated proclivities towards forms of the derivative and lecturers' privileging – to address a fundamental question: how did these two groups of students come to think about the derivative in such different ways? Our attempt at an answer explores: the role of calculus modules; lecturers' calculus module practice; students' situated developing conceptions; and the role of the institutionsal setting, the department.

6.1 Calculus modules and students' developing conceptions

Our assumption is that 'teacher privileging' in the calculus modules contributed to students' differing conceptions – but how? The analysis of the calculus modules shows differential treatment with regard to: rate of change and tangent aspects; emphasis placed upon theorems and their proofs. With regard to theorems and proofs, although the same numbers of theorems were given there were many fewer proofs in the ME module. With regard to rate of change and tangents aspects, Table 4 shows ME students' 'greater exposure to' (time and number of examples) rate of change aspects and M students' greater exposure to tangent aspects.

These modules, which include lecturer privileging of aspects of the derivative, 'shaped' students' developing conceptions. These developing conceptions had, one might say, a cognitive dimension (right and wrong answers as shown in the pre-, post- and delayed post-test results) and an affective dimension as shown in the rate of change and tangent test. In explaining the reasons behind their preferences/proclivities students in both groups acknowledged the impact of their calculus lecturers' privileging on their preferred forms of the knowledge: Table 3 shows that students in both groups, but especially M students, referred to calculus practice in explaining the reasons behind their preferences for rate of change or tangent aspects. The common response given by students who referred to calculus practices was about the way calculus was being covered and used in their departments

– it was 'their calculus module' – and they recognised this when it was presented to them.

The category to which we ascribed the majority of students' reasons, however, was 'ease'. Statements like "because it is easier" can be interpreted in many ways, so we are careful not to overstate our case but we feel that one reason why a rate of change or a tangent derivative question or viewpoint is easier is because it is 'your way', i.e. the way you have been 'indentured' by your department. So we are arguing that even 'ease' can be traced back, at least in part, to departmental affiliation. This argument has some support, quite strong in the case of M students, in that 47% of M students whom we ascribed to the category 'ease' picked the tangent question.

6.2 But why did 'what you teach' differ?

Although ascribing differing conceptions to different calculus module practices is partially correct this only touches the surface of this complex phenomenon and might convey 'you get what you teach'. A question which might take us farther (though still does not tell the whole story) is 'why did 'what you teach differ?' Why did ME and M calculus lecturers privilege different aspects of the same concept? Can lecturers' privileging be reduced to their personal preferences for ways of teaching? We draw on lecturers' accounts of their teaching in different departments to answer these questions.

The interviews showed that both ME (L1) and M (L2) calculus lecturers were aware of their privileging and that this was intentional, i.e. that their views and practices were compatible. L3, L4, L5 and L6 also stated that they adapted their instruction to suit students of different departments. The data consistently points to the fact that lecturers privilege different aspects of the derivative, set different questions on examinations and use different textbooks in teaching modules in different departments. Analysis of the interviews generated three factors that 'lecturers suggest influence their teaching': departmental features (L2, L3, L4, L5 and L6); departmental demands (L1, L4 and L5); and students' needs (L1, L3, L4, L5 and L6).

When we say "lecturers suggest influence their teaching", we must point out that these three factors are clearly our constructs, though they are closely based on the lecturers' own words (in translation, of course, from Turkish). These 'factors' clearly overlap; there may be other factors and they do not 'influence' lecturers in isolation. We do not, in fact, view them causally at all, they are 'woven into the fabric' of lecturing to different groups of students and this 'weave' has a past, present and future.

With regard to 'departmental features', all but L1 referred to distinct departmental features in explaining why they taught students of departments other than their own differently. They reported that the way they taught students was compatible with the departmental perspectives and the content they selected was "peculiar" (L5) to particular departments. Lecturers' accounts suggested that they perceived the departments as having distinct goals not only in terms of preparing students in line with their professions but also in terms of the modules they were teaching. We thus argue that lecturers' interpretation of the goals/nature/features of the departments influenced how they taught to students of different departments.

With regard to 'departmental demands', L1, L4 and L5 referred to the 'demands of the departments' in explaining the changes they made in teaching students from different departments. They reported that they had consultations with the departmental administrators on what to teach to the students. The key point here is the extent to which the demands of the administrators impacted upon what lecturers taught. University lecturers are metaphorically 'free' to teach what they want but this freedom is really 'restricted movement'. As a lecturer you know the level of mathematics you expect to, and are expected to, teach but modules have written requirements and sometimes you have to ask departmental personnel about the subject matter and the focus. The way that L1, L4 and L5 reported that they taught other departments' students was interrelated to the way they perceived the departments, and this is related to the last factor we isolated, students' needs.

All lecturers but L2 referred to 'students' needs' in explaining the difference in their teaching. Lecturers' accounts suggest that they were quite conscious of teaching students from different departments as they mentioned different students' needs. This was mainly the case when lecturers explained why they taught mathematics and engineering students differently, see, e.g., L1 "Maths students will be specialists in this area ..." and L6 "we need to think how the physics we teach them is going to be useful to them". L3 clearly differentiates between engineering and mathematics students in terms of their mathematical needs. He considers that since mathematics students are going to be mathematicians, they "need to know everything" but engineering students "only need to know the parts which are useful for them". This perception of student need is clearly tied to departmental considerations, i.e. the construct may be more appropriately named 'students' needs with regard to departmental expectations'.

To return to the question 'why did 'what you teach' differ?', our considerations suggest that lecturers' perceptions of the departments for which they teach have strong impact on what they teach. Lecturer privileging of approaches to the derivative cannot be reduced to their personal preferences, and what lecturers privilege has a strong impact on students' knowledge development and attitudes.

6.3 Students' situated developing conceptions

We now turn to the developing conceptions of students situated in departments. We attend to cognition (their test performance) and affective matters (their stated preferences for forms of the derivative). Cognition and affect are, to us, interrelated and develop together over time.

These students' knowledge development has at least two sources: the calculus modules and what we have termed 'student proclivities'. We argue that 'departmental considerations' feature in both of these sources. The previous sub-section considered the first source and we argued that institutional settings influence calculus module practices, which influence knowledge development. But what evidence do we have: that student proclivities are interrelated with knowledge development; that student proclivities are influenced by institutional settings?

Before addressing this we note problems with the data and ways that we may interpret these data. A methodological problem is that it is very difficult, if not impossible, to obtain data that establishes if, and if so how, institutional setting influences student performance. A second problem, if 'problem' is the right word, is that the trend 'ME students to A and M students to B' in the 'rate of change and tangent' items is more pronounced in our categorisation of the M students' responses than it is in the ME students' responses. This may be a statistical quirk or a data collection anomaly but we suspect that this difference, even though the 'trend' is present in both groups of students, reflects something that we have not considered.

Table 3 has no absolute trend but interesting patterns can be seen. It might be expected that students would ascribe their reasoning to what they were taught in the calculus modules – the 'practice' category – but this, apart from M students' responses to item 1, is not particularly strong in their explanations for their choices. Some, all following the 'ME to A and M to B' trend, referred to their departments in explaining their choices. Others referred to 'real life/application' and 'scientific/mathematical'. Note the diagonal pattern in table 3:

	0		0	
0		0		0

in the distributions of the zeros in the rows for these two constructs – students as a whole appear to see mathematics/science and real life as incompatible or as poles. Note too that the 'ME to A and M to B' trend is present. ME students generally referred to 'practical', 'application' and 'real life' aspects in explaining their preferences. They appeared to associate engineering with 'application', 'practicality' and 'concreteness' and some of these ME students regarded Ali as an engineering student, even though this was not stated, because of his rate of change interpretation. We view this as evidence of a link between students' knowledge development, their proclivities to forms of the derivative and their 'sense of themselves in their departments'.

Of the M students who chose Banu's interpretation, 25% cited 'scientific/mathematical' and 19% cited 'department' for item 2-a and 28% cited 'scientific/mathematical' and 25% cited 'department' for item 2-b. These M students generally referred to 'definition-oriented', 'scientific/ mathematical', 'abstract' and 'being from mathematics department' aspects in explaining their preferences for a tangent-oriented interpretation. Some of these M students found Banu's tangent interpretation of the derivative more 'formal' and regarded Banu as being from a mathematics department. We take this as evidence for the way (most) M students perceive the features of their department.

The final piece of evidence to support our claim that student proclivities are influenced by institutional settings is that that none of the ME students who chose Banu's explanation and none of the M students who chose Ali's explanation cited 'department' in explaining their choices. This indicates that being a member of the ME department directed some ME students to choose Ali and being a member of the M department directed some M students to choose Banu. Further support for this is the fact that some ME and M students thought that Ali was an engineering or physics student and that Banu was a mathematics student.

There are clear patterns which appear to be strongly linked to departmental affiliation but this trend is not deterministic or causal. There are exceptions to the overall trend and these appear more pronounced in ME students. We do not have the data to explore these exceptions but we believe that one factor may be successful study, that academically successful students are more likely (and academically 'struggling' students are less likely) to form a bond with their department. In the next and final section of this chapter we explore what this 'bond' might be.

6.4 How do institutional settings influence lecturers and students?

So what is it about a university department, an institutional setting, that influences lecturers' teaching and students' conceptual development? Crook and Light (1999, p. 187) suggest that institutional settings are not a "neutral backdrop to the interactional business of teaching and learning" but that they have a 'directive role' in what their 'players' do and how they act. We accept this but what is this directive role? We think that this has to do with the nature of the departments themselves and how they are perceived by lecturers and students. In addressing this question we first attempt to exemplify what we mean by 'the nature of the department'.

Barab and Duffy (2000) argue that every community has a common cultural and historical heritage which may be manifested through many forms; each community has and develops its own goals, practices, norms, conventions, rituals and histories. This applies to both the ME and M departments, they have their own cultural forms, goals, practices, etc., which have developed over decades, and they continue to develop. Both departments have 'stated goals' and an 'overarching goal' – to foster future mechanical engineers (or mathematicians). Intertwined with these goals both departments have programmes and specific modules and particular features which they are often associated with, e.g. engineering is associated with 'practical', 'application', and 'real life' (Becher and Trowler, 2001) and is seen as an 'applied' discipline (Biglan, 1973) whilst mathematics is associated with 'abstract' and 'theoretical thought' and is seen as 'pure' discipline (ibid.).

As noted in the introduction to this chapter, there are a variety of complementary ways to view departments: as activity systems; as communities of practice; in terms of identity; with regard to praxeologies. In the Theoretical Framework section we cited from Castela (2004), that institutions have constraints and expectations which position students as subjects of an institution. Although our work is broadly consistent with this claim we would add that students do not position themselves in a uniform manner as subjects of a department.

Institutions have customs. Crook and Light (1999) speak of 'institutional structures' and related goals, rituals and resources. Castela states that "institutions give these persons opportunities to adapt themselves to the institutional customs—in other words to learn" (ibid., p. 42). Daniels (2001), with regard to work by Resnick and LeGall, posits that school cultures (departmental settings) may act to position learner and teacher beliefs. We think the 'positioning' metaphor is cogent and explore individuals' positioning in terms of their adaptations to the 'customs' of the ME and the M departments.

When lecturers teach in particular departments, they come to know norms, values, rituals as well as the overarching goals of these departments. These norms and values enable lecturers to make judgements as to what the students' (mathematical) needs are – this was evident in the interviews with all lecturers'. L2, for instance, stated that "if they were physics department students, lots of examples regarding physical meanings of the derivative would be given. But in the mathematics department the derivative as a concept is prioritised" (our emphasis). Similarly L1, speaking of M students, stated that "they need to know this job's reason and logic. That is why you can ask them theorems in their examinations" (our emphasis). Our emphases here point to socio-mathematical 'norms, values and rituals' with regard to an institution, and that lecturer-privileging of aspects of the derivative contributes to students positioning themselves in relation to the valued mathematical knowledge of their department.

Of course, lecturers' knowledge of different departments varies but we believe that the ways in which departments are historically and culturally perceived affords and constrains lecturers to perceive departments and to interpret students' needs in particular ways. This can be the case even if the lecturers have no experience of the department; the title of the course may suggest a need. The Physics lecturer L5 stated that physics for engineers should be application-oriented and that physics for mathematicians should be concerned with the 'essence' – was he thinking of Mathematical Physics, a separate degree course?

For the students the norms, values and rituals of the department come through their whole departmental experience, their 'lived-in world'. In semester 1 the ME students attended a module *Introduction to Mechanical Engineering* and most attended a module *Ethics of Mechanical Engineering.* The first author attended some of these lessons, the ethos was one of inducting novices into the customs of a special community of practice. M students had fewer semester 1 modules and these were all 'abstract' in nature instilling, we posit, a sense of a cerebral community of practice. In both departments, semester 1 activities contributed to students coming to know what their departments were about.

It is very difficult to give a definite answer as to the extent to which students' positioning influenced their developing conceptions on the test questions from pre-test to the delayed-post test. There are all sorts of data that we would, in retrospect, have liked to obtain, such as the Dewey numbers of the library mathematics books they used, that may have informed our understanding of the relationship between departmental positioning and knowledge development. Holland et al. (1998, p. 57) state "knowledge…cannot be divorced from position, and …position married to knowledge". Our data also suggest that there is a close relationship between position and cognition, but further research is required to find out how intimate this marriage is.

REFERENCES

Amit, M. & Vinner, S. (1990), 'Some Misconceptions in Calculus: Anecdotes or the tip of an iceberg?'. In G. Booker, P. Cobb & T. N. de Mendicuti (eds.) *Proceedings of the 14th*

Conference of the International Group for the Psychology of Mathematics Education, Oaxtepec, Mexico, 1, 3-10.

Artigue, M., Assude, T., Grugeon, G., & Lenfant, A. (2001). Teaching and learning algebra: Approaching complexity trough complementary perspectives. In H. Chick, K. Stacey & J. Vincent (Eds.), *The future of the teaching and learning of algebra, proceedings of 12th ICMI study conference*. Melbourne, Australia: The University of Melbourne.

Asiala, M., Cottrill, J., Dubinsky, E., & Schwingendorf, K. (1997). The development of student's graphical understanding of the derivative. *Journal of Mathematical Behavior, 16*(4), 399-431.

Barab, S. A., & Duffy, T. M. (2000). From practice fields to communities of practice. In D. H. Jonassen & S. M. Land (Eds.), *Theoretical foundations of learning*. Mahwah, NJ: Lawrence Erlbaum Associates.

Becher, T., & Trowler, P. (2001). *Academic tribes and territories; intellectual enquiry and the culture of disciplines* (2nd ed.). Buckingham: SRHE/Open University Press.

Bezuidenhout, J. (1998). First-year university students' understanding of rate of change. *International Journal of Mathematical Education in Science and Technology, 29*(3), 389-399.

Biglan, A. (1973). Relationships between subject matter characteristics and the structure and output of university departments. *Journal of Applied Psychology, 57*(3), 204-213.

Bingolbali, E. (2005). *Engineering and mathematics students' conceptual development of the derivative: An institutional perspective.* Unpublished PhD thesis, University of Leeds, UK.

Castela, C. (2004). Institutions influencing mathematics students' private work: A factor of academic achievement. *Educational Studies in Mathematics, 57*(1), 33-63.

Chevallard, Y. (1999). L'analyse des pratiques enseignantes en théorie anthropologique du didactique. *Recherches en Didactique des Mathématiques, 19*(2), 221-266.

Crook, C., & Light, P. (1999). Information technology and the culture of learning. In J. Bliss, R. Saljo & P. Light (Eds.), *Learning sites: Social and technological resources for learning* (pp. 183-193). Oxford, UK: Elsevier Science Ltd.

Daniels, H. (2001). *Vygotsky and pedagogy*. New York: RoutledgeFalmer.

Engeström, Y. (1987). *Learning by expanding*. Helsinki: Orienta-Konsultit Oy.

Engeström, Y. (1999). Situated learning at the threshold of the new millennium. In J. Bliss, R. Saljo & P. Light (Eds.), *Learning sites: Social and technological resources for learning* (pp. 249-257). Oxford, UK: Elsevier Science Ltd.

Holland, D., Lachicotte, W., Skinner, D., & Cain, C. (1998). *Identity and agency in cultural worlds*. Cambridge, MA: Harvard University Press.

Lagrange, J-b (2005). Transposing computer tools from the mathematical sciences into teaching. In D. Guin, K. Ruthven & L. Trouche (Eds.), *The didactical challenge of symbolic calculators: Turning a computational device into a mathematical instrument* (pp. 67-82). New York: Springer.

Lincoln, Y. S., & Guba, E. G. (1985). *Naturalistic inquiry*. Newbury Park, CA: Sage.

Maull, W., & Berry, J. (2000). A questionnaire to elicit the mathematical concept images of engineering students. *International Journal of Mathematical Education in Science and Technology, 31*(6), 899-917.

Praslon, F. (1999). Discontinuities regarding the secondary/university transition: The notion of derivative as a specific case. In O. Zaslavsky (Ed.), *Proceedings of the 23rd conference of Psychology of Mathematics Education* (Vol. 4, pp. 73-80). Haifa, Israel:PME.

Sazhin, S. S. (1998). Teaching mathematics to engineering students. *International Journal of Engineering Education, 14*(2), 145-152.

Wenger, E. (1998). *Communities of practice: Learning, meaning, and identity*. Cambridge: CUP.

Wertsch, J. V. (1998). *Mind as action*. New York: Oxford University Press.

Chapter 12

School Practices With The Mathematical Notion Of Tangent Line

Márcia Pinto and Valéria Moreira
Universidade Federal de Minas Gerais

Abstract: We seek to understand the process of learning mathematical notions as forms of practice in the classrooms of two different technical school courses. More specifically, we investigate students' and teachers' experiences and use of the mathematical concept of tangent line in these different contexts. We use empirical data collected through non-participant observation, analysis of students′ written responses to a questionnaire, and semi-structured interviews with groups of six students from each course observed. We take a situated perspective of learning which enables us to see these classroom activities as genuinely mathematical, though distinct. Through our analysis, we describe aspects of what we see as the common direction of learning mathematics in the two vocational course lessons; this is found to be closer to 'being mathematical' in work settings than in school mathematics classrooms.

Key words: situated learning, local communities of practices, vocational school classrooms

1. INTRODUCTION

The main focus of this chapter is an inquiry developed from a situated learning perspective (Lave and Wenger, 1991), designed to investigate learning related to the mathematical notion of tangent line. It is a reinterpretation of results from the analysis of data collected by the second author (Moreira, 2004; Moreira and Pinto, 2003). We seek to answer questions arising from a study conducted with first year university students[33] in which we focused on the notions of concept image, as the whole cognitive structure associated to a mathematical idea, and concept definition, as the form of words used to designate a mathematical concept (Tall and Vinner, 1981). In that study we partially reproduced research already conducted by those authors, applying it to students in our own country.

Amongst our results, we became interested in those related to the notion of tangent line, as students surprised us when we interviewed them to elucidate their answers to a questionnaire. Reproducing Vinner's research instrument in part (Vinner, 1991), we had invited students to identify, if possible, the tangent line(s) through the point P on specific curves we had selected in advance.

Students had already attended a calculus lesson where the lecturer introduced the mathematical notion of the tangent line as the limit position of secants to a curve at a point. As often happens when an individual experiences new ideas, it appears that interviewees had not taken up the secant method. They responded giving explanations that involved movement, which had possibly occurred in their earlier physics lectures, or describing a procedure of "adjusting a circular arc at a point". In the latter, instead of perceiving the curve as being locally straight, as we generally suggest in our calculus course, some students spoke freely about 'the centre of a curve at a point' when referring to the centre of the adjusted circular arc, identifying it before drawing the tangent as the 'perpendicular to the radius'.

Such ideas reminded us of mathematical notions we experience only later in advanced mathematics, which we never take into account when teaching the first year calculus course. We believed that students would naturally generalize the definitions and the procedures of determining tangents to circles from their elementary schools. By this we mean the idea of considering curves as if built of small arcs of circles, osculate circles, which is as consistent within the practices of mathematicians as the idea of imagining them locally straight, made up of small line segments. Procedures they would adopt to carry out the specific task of drawing the tangents would, we imagined, be straightforwardly related to students′ experiences with circles at the elementary school.

In our country, the educational system includes kindergarten schools (mostly private) and provides basic education consisting of elementary and secondary schools, corresponding to 6-14 and 15-17 age groups respectively. Professional education at a basic level is provided by technical schools, which are largely public and offer vocational courses at elementary and secondary levels. Vocational courses are mostly attended by students from

[33] This research project was supported by CNPq, from 1999 to 2000.

working class backgrounds. After a recent curriculum reform, technical secondary schools have to offer their vocational courses to students who had already finished secondary school successfully, or who are simultaneously attending the secondary school in the same or another institution. In secondary school, students attend classes on biology, chemistry, physics, mathematics, geography, history, Portuguese and a foreign language. Some technical schools were, and are still, amongst the institutions providing the best secondary school education in our country.

Being aware of the technical school backgrounds of the participants in our research, examining some of their technical school design papers and exercise books and drawing on our own teaching experiences, we conjectured that the images revealed by students relating to curves and tangent lines may not have been made explicit by their teachers or by any other instructional material, and were more likely to be due to experiences with *technical* practices in their technical secondary school rather than to experiences in a school mathematics lessons. For us, those unexpectedly evoked images shared by various students reveal a social character of learning while suggesting an apparent diversity of practices with school mathematics between the various subject classrooms at the school institution. In the research literature, such diversities have not been investigated.

Encouraged by those results, we elaborated a research project (Moreira, 2004) aiming at exploring technical school contexts where (we conjectured) those students had experienced such ideas, while searching for an understanding of the process of learning mathematical notions through investigating practices in the classrooms of diverse technical school vocational courses.

Reflecting on instances that, we believed, revealed a learned curriculum through experiencing and sharing ideas other than the one planned by the teacher, we sought a theoretical shift to provide a deeper understanding of learning in such terms. Jean Lave´s theoretical perspective would contribute to answering the questions raised, as it reconceptualises teaching in schools taking into account "learners, learning, as the fundamental phenomenon of which teaching may (or may not) be a part" (Lave, 1996, p. 157).

2. RESEARCH FRAMEWORK

Building on previous research, Lave and Wenger (1991) present a theory where individuals become knowledgeable through participating in culturally bounded social practices. The unit of analysis in Lave and Wenger's theory of learning is neither the individual, nor the social institutions, but the community of practice itself.

Lave (1996, p. 150) suggests that no matter what the context is "learning is an aspect of changing participation in changing communities of practice" in which individuals take part. What 'learning at school as a state of change' means should account for all the relations and values between one's peers and teachers, and actions, thoughts and feelings, all of which are being learned, though rarely focused on by explicit teaching.

Nevertheless, what actually constitutes communities of practice and practices in the school classroom is still not obvious. In spite of believing that many features from Lave and Wenger's theory could be re-signified in schools, Winbourne and Watson (1998) observe that in the case of the mathematics lessons, "teachers are not engaged in learning mathematics" and also "pupils' participation is often passive", in contrast with features from communities of practice, such as "all participants in the practices being engaged in the same activity" (p. 94). To theorise forms of shared practice of mathematics in school institutions, Winbourne and Watson developed the notion of *local* communities of practice, bounding Lave and Wenger's notion by referring to the time and space of its constitution. The theoretical perspective is built on research in school mathematics classrooms taking into account contexts in which there is intentional teaching. Certain incidents and configurations which can occur in school mathematics classrooms are identified and described as indicating the existence of *local* communities of practice, where *local* refers to time and space boundaries which could be "said to be at least an indicator of effective teaching" (p. 101) (or effective learning). Features defining *local* communities of practice (p. 103) refer to pupils' developing an identity, in this case as being mathematical, supported by a social structure of the practice, with a "public recognition of developing competence within the lesson", individuals working with a common purpose "towards an achievement of a common understanding" (p. 103), and shared ways of behaving, language, habits, values, and tool-use. The authors' description requires students' and teachers' active participation and engagement in the same activity in the classroom.

In this chapter, we use the above framework to problematise the learning of mathematics at school, especially the learning of the mathematical notion of tangent line, by taking the classroom practices as the unit of analysis of our research. Our aim is to investigate which experiences with the mathematical concept of tangent line are shared by students and teachers in the different technical school classrooms. It shares similarities with the aim addressed by Bingolbali and Monaghan in this volume: what it means to learn about a mathematical concept in different settings. We also share an interest in the development of related mathematical conceptions, tangent lines and derivatives, though in different academic levels. Nevertheless, our research focus differs: rather than broadening the perspective to account for the institutional setting and the impact of the diverse course affiliation upon the mathematics classrooms, our study widens its initial exploratory focus by taking into account other school subjects than mathematics, in school classrooms in the same institution, where students still experience mathematics. We propose to investigate the learning of mathematics in each context, i.e., we are supposing that the learning of mathematics should not be analysed solely by observing students' experiences in mathematics classrooms. Instead, we are proposing to observe it as a process of participation in a diversity of school classroom experiences, which would include other subject classroom lessons, all of which change who students are (see Winbourne and Watson, 1998).

To make this argument, we first asked ourselves whether it is appropriate to take into account the students' experiences with school mathematics in the various subject lessons as participation in *local* communities of practice of mathematics. Though Winbourne and Watson (1998) suggest a positive answer in some circumstances, we were challenged by the complexity of designing the research. Its design is discussed in the next sections. Supported by Winbourne and Watson (1998), we then investigated the practices inside three technical school vocational course classrooms. Restrictions on the length of the chapter lead us to consider the data collected in two of them. Finally, we analyse the experiences shared by students and teachers in classrooms, describing each them as distinct mathematical practices.

3. MATHEMATICS, SCHOOL MATHEMATICS AND *LOCAL* COMMUNITIES OF PRACTICE

We set out in this study to investigate the experiences with the mathematical concept of tangent line which are shared by students and teachers at technical schools. The first complex challenge we face in its design is related to our understanding of what knowledge is, in general, and what mathematical knowledge is, in particular. From a situated perspective, the teaching and learning experience is to be understood as participation in an ongoing social practice, within which an aspect is "becoming more knowledgeably skilled" (Lave, 1996, p. 157). We can describe "small-scale 'becomings'" if we accept the notion of *local* communities of practice (Winbourne and Watson, 1998) and draw upon its features (ibid., 103), especially the one which refers to "shared ways of behaving, language, habits, values and tool-use". To inform the constitution, or not, of *local* communities of practices, we planned to observe classroom routines taking into account the role played by the teacher and the students, and didactic material from which learners may have experiences even when this was not made explicit during the lessons.

The second challenge in our study, related to the first, is to go beyond the mathematics classroom and to ask ourselves whether it makes sense to consider various other subject classrooms. In fact, attuned with Lave's perspective which considers "crafting identities in practice"(Lave, 1996, p. 157), an important dimension of teaching and learning in practice, one defining feature of Winbourne and Watson's (1998) notion of *local* communities of practice is the development of a mathematical identity as its common direction of learning. Those researchers saw instances in a mathematics classroom practice when learners and teacher sense they are aligned; in other words, there may be instances when they are all "functioning mathematically" (ibid., p. 103). We conjecture that 'being mathematical' in the various school subject classrooms would have a diverse, albeit interwoven, meaning from the development of the common direction of learning in mathematics classrooms. We mean that each of these school classrooms, both of mathematics and of another school subject, have distinct objects of study, roles, functions, practices; therefore, they appear as practices with suggest different perspectives on 'being mathematical'. To inform this dimension, we planned to focus on lecture course goals when approaching mathematics, drawing upon what is being shared in classroom. Our initial plan was to observe mathematics lessons from the Regular Secondary School Course, in the same technical school, where tangent lines and derivatives would be approached. The expectation was to compare, in the same institution, the practices of a mathematics classroom and the mathematical practices in another subject. A strike in our educational system hampered the research design. To overcome such constraints, we need to bring our own experience as researchers in mathematics education, mathematicians and mathematics teachers to the task of identifying what might be noteworthy or different about the lessons observed, deliberately looking for potential differences in order to show how the learners and teachers observed might differ from those in a mathematics classroom. Adler (1998) expresses her view that the learning of mathematics at school is to be understood as a specific practice where mathematics is learned through the language in use in the classroom. It would also include recognizing and developing specific ways of using language, meaning that discursive analysis would be appropriated to investigate learning in school settings. Adopting her perspective to frame our examination of the mathematical opportunities in the classrooms, we propose to explore course goals by drawing a special attention to the mathematical language in use and the meanings shared by students and the teachers within the practices observed.

Reflecting on the research path we followed as mathematicians, mathematics teachers and mathematics educators in order to better explain the phenomena we were seeing in the new situation, we shifted to these new frameworks to speak about meanings and identities, values and goals we had developed. We refer to the practices we have each experienced in communities of practices of mathematicians and of mathematics teachers as 'mathematics', or 'mathematical knowledge', and, distinctly, 'school mathematics'.

Procedures of data collection and data analysis systematically explore the combination of the four components mentioned above: goals in approaching mathematics, shared meanings and the mathematical language in use, classroom routines, and didactic materials, as a framework for investigating practices as *local* communities of practice and to address our questions on the learning of mathematics across subjects within schools.

4. SCHOOL PRACTICES AND SCHOOL MATHEMATICS

The whole study is a two-year qualitative research project, with data collected at a technical (secondary) school during the first academic semester of 2003. The technical school is attended by students aged 15 to 17 years, predominantly from working-class backgrounds. Methods of data collection are non-participant observation of eleven regular course lessons, analysis of students' written responses to a questionnaire handed out by the researcher at the end of each period of course classroom observation, and semi-structured interviews with groups of six students from each regular course observed, selected on the basis of their responses to the questionnaire. The analysis of the regular course lessons refers to data collected in five lessons of the Technical Design classroom from the Regular Secondary School Course, four lessons of Project classroom from the Highway System course, two lessons of Mechanical Design classroom from the Mechanics course. The three regular courses were previously selected and field notes were taken when the activities referred to the notion of tangent line. Most of those lessons are video recorded. As in Moreira, Sampaio, Cardoso, Almeida, Prado, Zumpano, and Pinto, (2000), our questionnaire partially reproduced an existing research instrument (Vinner, 1991). The first question in the questionnaire asks students what a tangent line is. The second and last question invites students to draw the tangent line to each of the six curves in a picture: "Consider the six curves below and the point P in each of them. Draw the tangent line through P to each curve, if it exists. If in your opinion there is more than one single tangent line through P, draw all of them. If you believe there is an infinite number, make a note on this fact, and draw some of them."

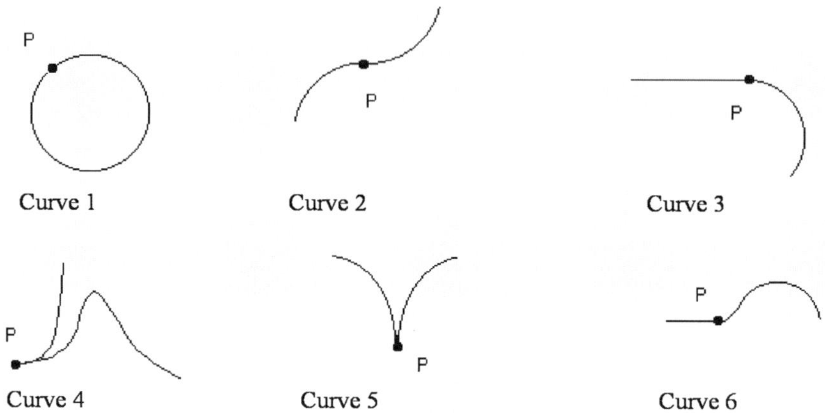

Figure 11-1. Could you identify the tangent lines through P?

We interviewed groups of students to investigate the experiences shared among them and to get in touch with students' perspectives on the mathematical notion in focus. During the interviews, students were handed the questionnaire, and invited to share and discuss their responses with the other interviewees.

Data analysis is presented by introducing an overall description of the classroom practices in each course, contextualising the episodes selected from data collected during non-participant classroom observation and from semi-structured interviews with groups of students. Pictures presented in this chapter were taken from the blackboard during classroom observation. Excerpts from data collected in the contexts of the Regular Secondary School course and of the Highway System course are selected for presentation, due to their power in illustrating mathematics being restructured. The whole study is reported in Moreira (2004). Participants in the research received a special invitation to attend the public session where the whole study was presented and submitted for external examination.

4.1 The Technical Design Classroom

Technical Design is a basic subject attended by all students in their first year at the Technical School. Teachers from the mathematics department share the responsibility for the classes. Classes have, approximately, thirty-five students grouped according to their vocational course option.

The course content syllabus mainly covers elementary geometric constructions, which are actually not included in the current mathematics syllabus of the vast majority of primary and secondary schools in our country. Practical activities and exercises brought by the teacher often refer to technical course projects. Mathematical notions are occasionally recalled during the lessons.

Lessons take place in ordinary school classroom rooms, where students attend all the other regular secondary school lessons: students' individual desks face the blackboard and are organized in rows. The teacher is not rigid with that classroom organization, allowing students to reorganize the space as they like, and to work in pairs, groups, or individually.

The lesson routine starts with explanations given by the teacher, who uses the blackboard to make notes previously prepared and considered by her as necessary to develop the subject. At the end of the lesson she proposes some tasks, during which the students apparently follow the steps defined by the teacher in her previous explanations.

The tasks proposed in the classroom are mainly taken from the course exercise book, which is similar to a geometric construction book, though emphasising technical aspects and presenting a structure to support projects in a hypothetical workplace setting. Some tasks are specially conceived by the teacher, brought in on a separate sheet and handed out. The teacher calls the latter 'short technical design projects', and they are part of the course evaluation.

All the students have the coursebook, and their own notebook to work out the tasks during the lessons. Some of them have a proper sketch book. All of them own a ruler, a compass, a triangle and a protractor, which are often used.

The episode which follows is extracted from data collected in the last lesson recorded, when the topic being introduced was *arc tangential to a line*.

In the previous lesson, the teacher had recalled what 'to be tangent' means while drawing and representing two tangents to a circle through one external point: "To be tangent is to touch at one point, and one point only". She commented on the importance of the notion for a wide number of technical projects: "… pay attention to a road border … this is called tangential … when perfectly constructed, no one knows where the straight line ends and the camber starts, where the camber ends and the straight line starts ...” At the end of the lesson, the teacher comments on “the lines on a circle” – meaning its radius, diameter, tangent, chord, arc and secant. She then defines orally “tangent touches the circumference at one point, called the tangential point, the T, and it is perpendicular to the radius”, representing it graphically while speaking.

The lesson routine starts as on the day before. The teacher addresses the whole class orally to explain the lesson content, facing the students, and using the blackboard. She writes down the lesson subject title: *Arc Tangential to a Line*, and introduces the content informally, explaining the notion as follows:

“In order to get an arc tangential to a line, with no mistakes, to get it right, look, an arc tangential to a line is this, here, isn’t it, a line tangential to a circumference in this case ... (she draws figure 2 on the blackboard during her speech)

Figure 11-2. An arc tangential to a line through the tangential point T

... this single point, where the line touches the circle, will be our tangential point.” (she uses her finger to indicate the point named T, drawn in the picture)

She then erases the name T in the picture on the blackboard, while saying: “I will not even write T because there are people thinking that writing the T is sufficient, isn’t it? To get an arc tangential to a line correctly, with no mistakes, I must follow a rule which says the following ...”

The teacher suddenly interrupts her own speech, and before discussing the announced rule, she decides to show what she meant by “correct tangential, with no mistakes”. She erases part of the picture drawn on the blackboard, leaving the picture below (figure 3). She then completes: “... if we erase all the construction lines we don´t know where the tangential point is.”

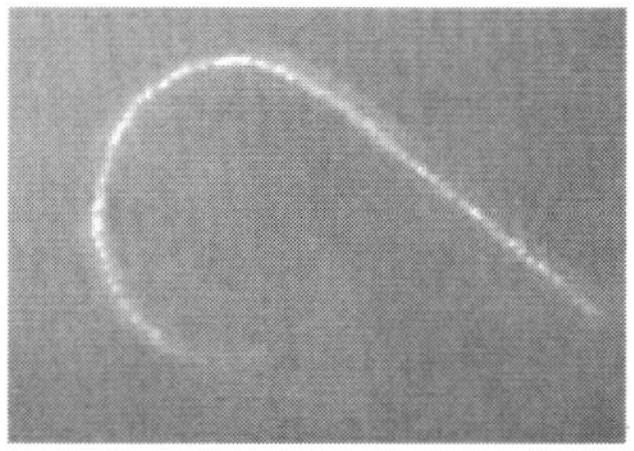

Figure 11-3. The result of a precise construction

To conclude, she enunciates the rule: "Therefore to get it right, with no mistakes, the rule is rather simple. Look, I will speak first, and then you write it down, okay? Centre of the arc (she marks it in the picture as below, while explaining), drawing freehand here for a while, and the tangential point must be in the same perpendicular. That's all. Okay? This is the rule of an arc tangential to a line."

She then dictates the rule she enunciates, which in fact teaches the steps for ruler-and-compass construction, as expressed above. Students are expected to take notes in their books, which apparently they do: "In order to get an arc tangential to a line it is necessary that the centre of the arc and the tangent point lie in the same perpendicular. The set arc-line must constitute one single line."

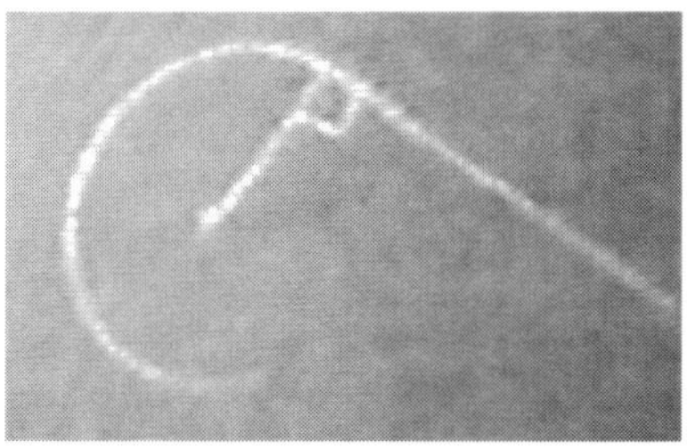

Figure 11-4. The ruler-and-compass construction of an arc tangential to a line

The teacher's explanations in this episode have characteristics which are common in other lessons observed. Firstly, we mention the evoked notion of rigour and precision, meaning rigour and precision in (geometric?) constructions. The teacher constantly emphasises that the drawings must be "right, with no mistakes". Such features seem related to procedures of ruler-and-compass geometric constructions, where rigour and precision in plotting are required, making a distinction between the context of the Technical Design lesson and that of an ordinary mathematics lesson. In fact, the resultant pictures are distinct from the freehand drawings which are commonplace in mathematics classrooms. Thus the overall procedures followed by the Technical Design teacher are supported by a notion of rigour and precision (in picture representations) which differs from that supporting the subject content in mathematics lessons. Although we are not presenting any evidence from mathematics classrooms to support our observation, our experience in school classrooms in our country is that, in the latter situation, naming an object is sufficient to state it exists, and naming points in the pictures is a common practice among mathematics teachers when highlighting parts of their freehand drawings on the blackboard. In this episode, the teacher erases a point T in her drawing on the blackboard "because there are people thinking that writing the T is sufficient", definitely expressing a distinction between the current practice and a mathematical one.

Second, the lesson subject name: *Arc Tangential to a Line*, an expression borrowed from the geometrical construction content, suggests a dynamical movement, *arc moving towards a line*, instead of the version *line moving towards an arc*, and also an inversion of the procedure of drawing a tangent line to an arc, namely *given the line, draw the arc*, instead of *given the arc, draw the line*. The latter versions in each case which predominantly circulate in mathematics lessons are explicit during the teacher's comments "an arc tangential to a line is this, here, isn't it, a line tangential to a circumference in this case …" suggesting a slip to a mathematics lesson context, or a possible interchangeable use of both expressions.

Curiously, the teacher accepts, and suggests, "using freehand here for a while" and "exercising your sight". Such flexibility also distinguishes her classroom practice in the Technical Design lessons from those ordinary ones in geometrical construction lessons, where all the objects must be geometrically constructed using ruler and compass.

A focus on actions, rules and procedures is, interestingly, implicit in the first teacher's definition of what "to be tangent is", which is the format of a verb, in place of defining the noun tangent, thus referring to a mathematical object. When she does it, she is restricted to the case of a circle, which is the curve they are actually working with: "this single point, where the line touches the circle, will be our tangential point."

Six out of thirty seven students who responded to the questionnaire were invited to participate in an interview with the second author as researcher. During the discussion of the first question in the questionnaire: "Write down what a tangent line is", the researcher observed that all the students were attempting to enunciate a procedure to draw the tangent lines to a circle. They were recalling simultaneously both conditions for being a *line which touches the circle just at the tangential point* and for *being perpendicular to the radius through it*. There was an apparent agreement in the group until the researcher's intervention: "And what happens if it is a curve?"

The interviewees Matheus (M), Robson (R), João (J) and Breno (B) accept the researcher's invitation to account for a broad notion of curve. They started talking at their same time, bringing their own ideas into the discussion.

M: ...in the case of a circle, but if it's an irregular curve, I never thought about it till now.

R: No, in the case of an irregular curve, you will, you will have....

Researcher: What do you mean by a regular curve and what do you mean by an irregular curve?

M: Ah, I don't know, it's one that isn't exactly a circle.

J: It's one that doesn't follow a circle, that isn't a circle, it's just a part of it, just an arc of a circle ... and if you take, for example, an irregular curve could be a parabola, then it would not have, as a parabola would not have a centre, we could not draw a centre for a parabola, we could not draw a circle in a parabola, it would not have a centre, only an imaginary centre, because each point would be a centre in itself, and would have...

B: ...then we would have to build a circle for each point in the parabola...

J: ....as if there was a circle in each point of the parabola...

We found that such classification of regular and irregular curves supported a strategy used by students among those in the group who responded to all the items in the second question in the questionnaire.

In that respect, the researcher observed that all students considered the curves 2, 3 and 5 as regular curves. The procedure to respond to the question is, firstly, filling in an imaginary circle for each of them, or for part of them. Second, drawing the tangent line as the one that touches the curve at a single point, forming a right angle with the radius through the centre of the imaginary circle, which could be determined. Curves 4 and 6 were classified as irregular curves. Some students declared themselves unable to draw a tangent line in such cases, while others did not entirely abandon the earlier strategy, and reformulated it; they developed an idea that accounts for an irregular case, determining a circle touching the curve at P (or, in other words, an arc tangential to the curve at P) in the first place. Then they would be able to determine its centre (named by the students as the centre of the curve), and once they had done it, one could easily recall and apply the procedure already discussed for the regular case. Notice that the procedure in this case is local, meaning that it is focused on that specific curve.

Students' strategies for the 'irregular curve case' are apparently tied to the procedures for drawing an arc tangential to a line, developed in the classroom. In fact, the evoked strategy suggests an action similar to the procedure: *in order to get an arc tangential to a line it is necessary that the centre of the arc and the tangent point lie in the same perpendicular. The set arc-line must constitute one single line.*

Students appear to be creating a circle (an arc tangential) touching the irregular curve (like touching a line) at the point P, and using its centre to draw the line perpendicular to the arc radius through P.

The strategy described above was not common among the group of interviewees. Some students in the group accepted it during the interview, while others declared they were still in doubt as to drawing the tangent lines in the case of irregular curves.

4.2 The Highway System Project classroom

The Highway System Project classroom is attended by students aged 16 in their second year of the Highway System course. There were nineteen students in the class where this research was conducted.

The Project course syllabus consists of the development of an activity of plotting roads on topographic maps. Plotting procedures were learned in previous courses. In this same academic term, students also implement a project in a placement, locating a proper road.

The course activities run in a special classroom setup, with individual drawing boards for students facing the blackboard and the teacher's desk.

Lesson routines follow a practical-theoretical frame, articulating explanations and practical activities, with a predominance of the latter format. The space organization in the classroom seems to induce a collaborative dynamic; the side by side position of the drawing boards supports students′ communication and collaboration.

The teacher conducts the activity giving initial instructions and discussing procedures he believes are necessary to implement the activity. He also leaves procedures to be developed and discussed by students on their own, opening up the possibility of other strategies being developed, or students′ use of strategies already developed in their previous classrooms.

The entire activity consists of plotting roads on a topographic map handed to each student, individually. Their active participation, sometimes under supervision of the teacher and sometimes without a teacher's presence, is a characteristic of the classroom lesson. When in doubt, students consult each other, constantly discussing the task and working purposely together. They often meet in small groups around a student's drawing board, looking for agreement or for explanations from those who had already concluded the activity, or part of it. Sometimes discussions involve the entire classroom when students get stuck or when they perceive a diversity of results in their work.

No textbooks are required, nor is any literature suggested. Students have in hand their lesson notes taken during the teacher's explanation, and tables with data and specific measurements for plotting cambers, constructed and systematized as course work in a previous course. They have the topographic map, pencil, a calculator, and tools for geometric constructions in paper, i.e. a ruler, a compass, a triangle and protractor.

In the third lesson observed, the teacher explains the procedures for plotting the road on the blackboard, paying special attention to the cambers of the roads. In previous lessons, students had already plotted the tangential points between the straight lines and the cambers, both of them being parts of the roads projected. The picture below is taken from a student topographic map.

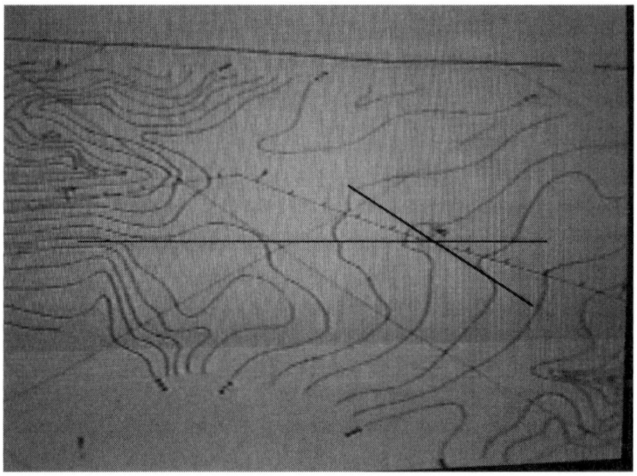

Figure 11-5. Projecting roads on topographic maps

When the lesson starts, students are handed such a map, although it only contains the level curves and the tangential points between the straight lines and cambers, which will be plotted. In order to draw the cambers, students at first must plot the straight lines, which would be the tangent lines to the cambers at the tangential points. Such straight lines are included in the picture.

Teacher: (…) and these straight lines will be tangent to these curves at an initial point which we will call PC [PC means "curve points", and refers to the centre of the circle which will be tangential to both straight lines already plotted in the topographic map] and a final point which we will call PT. (…) [PT refers to the two points where the two lines are tangent to the circle, and are in fact the initial point and the final point of the camber.]

Teacher: [While drawing on the blackboard, he addresses the whole classroom] Are you already working it out in your Plotting course?
Students: Yes.
Teacher: So, I don't need to talk a lot about it. You know the formulas, and everything, isn't it?
Students: Yes.

At the beginning of his speech, the teacher takes for granted students' awareness of the notion of tangent line. Interrupting his own explanation, he addresses the whole class as follows:
Teacher: You all know what a tangent line is, don't you? We have a circle [he draws a circle on the blackboard while speaking], or, it doesn't really matter if it's a circle or another circular picture, okay? Any curved picture. It could be a circle, it could a spiral, it could be any curved picture, she would know better how to say what a curved picture is [he refers to the researcher, facing her], it could be an ellipse. If you have a line which touches this picture [he draws a tangent line to the circle he had drawn on the blackboard] at a single point, only a single point, we say this line is tangential to the curved picture, isn't it? And in case of roads, what we are going to do is to take these lines and adjust them with a curved picture, in this case a circle, or an arc of a circle …

Notice that the teacher relates the notion of tangent line to the notion of tangent line to a circle. He also emphasizes the idea of a tangent line as a line which "touches the picture at a single point". He makes an attempt to discuss the mathematical notion generically when observing that "it doesn't really matter if it's a circle", though turning it back to the specific when referring to a curve as any "circular picture". We also suggest his unawareness of the mathematical discourse when referring to "circular pictures or curved pictures", or his conscious reconstruction of the mathematical language and definitions. In the end, he brings his discourse and his definition into his practice when naming the procedure of plotting the tangents by using a technical jargon such as "adjusting them [the tangent line] with" the curve. Such an expression is often used by the teacher during the lessons.

Notice that he refers to the presence of the researcher in his speech, recognising her as a mathematician who "would know better how to say what a curved picture is". It indicates that he considers the researcher as someone in a better position to talk about the mathematical concept. He does not feel diminished when recognising he is not an expert in mathematics. Other than that, he seems to be distinguishing between two kinds of mathematics: a conceptual one, performed by mathematicians, and an instrumental one performed in those technical work practices he takes part in. Overall the teacher seems to always relate the tasks and procedures during the lessons with those from other lessons attended by students, and with those from the workplace.

During the interview, Victor (V), Carla (C), Laura (L), Marta (M), Daniel (D) and Hilton (H) were handed the questionnaire they had previously responded to. The researcher invites them to comment on its first question, which asks "Write down what a tangent line is".

(…)

V: Mine [tangent line or definition of tangent line] are straight lines that stretched cross at a single point.

Researcher: Sorry?

V: They are stretched straight lines, umm, that cross at a single point, making an angle which will be used to define the kinds of cambers which will be plotted in a project.

Researcher: Could you please explain a little … if someone doesn't know how to plot roads and cambers in Projects …

(laughter and comments from students)

V: It will cross at a single point, umm … at, at the straight line, from umm … the circle. Therefore, umm, when you're plotting you'll have another straight line which will cross, umm … also with this other camber that I .. oh, please, help me!

C: Oh, like this. I believe that, in the context of, like, geometry, in actual mathematics, it is just the one [the straight line] which touches a single point at a curve, at a circle and everything. In the case of plotting roads they would be, like, it is straight lines, it is, it is like it is not in the field, it is like imaginary. They would be the procedures you develop, they cross each other and they are also called tangents these procedures that you carry out, in topography, when plotting roads, we use them as tangents.

L: Yes, then you go there and plot the points, carry out the procedures and then they cross each other and they are considered tangent, for us, you know, in the case of our course.

M: What I'd learnt about this term in my seventh grade is that it [the tangent] touches at a unique point of the circle, isn't it? This is what I'd learnt there, since mine …

V: Yes, tangent is that.

Interviewer: Is that the same tangent, when you are plotting highways?

M: In the case of, of…

V: … plotting …

C: …cambers, in the case of whom would be the tangents …

V, C, L, M, D and H: …the external tangents …

D: …tangent in such different way would be, as I say, a straight line, isn't it?

M: Yes, they would be straight lines, isn't it? Straight lines which cross each other.

Initiating the discussion, Victor is not intimidated by the interviewer's implicit comment: "Sorry?" on his answer nor by the interviewer's mathematical background. He continues, supporting his definition with procedures he worked out in his technical experience. In this respect, his idea of devising tangents before defining the curve is an interesting aspect of his experience, reminding us of other mathematical notions such as integral curves.

Victor continues his explanation responding to the interviewer, who insists that he clarify his thoughts. Asking for other students' help, Victor is assisted by Carla and Laura.

Both Carla and Laura suggest a distinction between the "actual mathematics" learned in mathematics classroom and the mathematics reconstructed with their current practices, referred to by Laura as (the mathematics) "in the case of our course". Carla also suggests a distinction between the "actual mathematics" and her technical experience observing that, in practice, tangents are "imaginary". We conjecture that, for her, this is not the case in geometry or in "actual mathematics". This could suggest a distinction between practising mathematics and using mathematics in another practice that poses a mathematical instrument as imaginary (or abstract?) when dealing with applications.

Marta joins the conversation, being assertive about her mathematical meaning for the word tangent: "What I'd learnt about this term in my seventh grade is that it [the tangent] touches at a unique point of the circle", though restricted to the case of a circle. Victor had no choice other than agreeing with Linda's mathematical notion.

The interviewer's intervention recalling the experience of plotting highways provokes a collective construction of a concept definition of tangent, situated in the students' experience, where, for them, tangents are the straight lines drawn as an aid to plot roads, in a curious inversion of the statement that the straight lines aiding such project must be tangents (to the cambers). Victor's definition is directly linked to the practice of plotting cambers. The lines, which are drawn to guide plotting the cambers, are designed to cross each other in the topographic map where the highway is plotted. The angle between the two lines is used to calculate the arc size of the camber, as an arc of a circle which will be designed between the two straight lines. These straight lines are tangent to the cambers and both are parts of the highways which are being plotted.

Although students declared they were all talking about the same tangent line, the many experiences evoked in this episode suggested that distinctions are made by them when the notion is worked out in each of those contexts – that of mathematics lessons and that of the Highway System lessons.

5. THE MATHEMATICAL EXPERIENCES SHARED IN THE CLASSROOMS OBSERVED

This research was developed to investigate the learning of a specific mathematical concept as forms of practices in the classrooms of diverse technical school vocational courses.

We started our study from a personal perspective from which we had doubts about the mathematical nature of experiences evoked by students. A situated perspective of learning (Winbourne and Watson, 1998) had enabled us to represent the activity in fruitful ways that suggested that we might actually see the activities as genuinely mathematical, though distinct. Knowing and learning are accounted for in terms of participation in communities of practice, which means that we consider the learning of a concept as tied to, and made explicit by, school practices.

David and Watson (in this volume) see local communities of practices as "just a way to describe certain incidents in which personal mathematical development, identity, is in tune with the mathematical classroom practices" (p. 5). Here, we get a lively picture of the classroom practices through following the six defining features of local communities of practice, as developed by Winbourne and Watson (1998, p. 103). From such a perspective, we expect to describe some experiences shared by students through an analysis of a "small scale becoming" (ibid., p. 102), meaning that learners' participation in the vocational course classroom practices, seen as local communities of practice, changes who they are.

Following the six defining features of local communities of practice, we are finding active participation of the great majority of the learners during the lessons, bounded by space and time. Although there are qualitative differences among school classroom routines, we focus on the striking commonalities we perceived in the mathematical experiences shared in the classrooms. We describe the experiences which are being shared, getting a sense of a common direction of learning in the practices we observed, through building on the analysis of the shared ways of behaving, language, habits, values and tool use, and of the developing mathematical competences, as recognized within the lessons observed.

5.1 The shared ways of behaving, language, habits, values and tool-use

In participants' behaviour and other shared values which we observed mathematics is appreciated, though regarded as external to the actual classroom practices. Teachers' remarks about the researcher and the author of the text book, respectively, acknowledge both of them as expert mathematicians, though identifying them as external to the current practice. This behaviour perhaps explains the comfort with which interviewees re-signified mathematical meanings in the presence of the researcher who was acknowledged as a mathematician. Yet the Technical Design teacher developed a step-by-step geometric construction on the blackboard, while accepting students 'exercising their sights' and using freehand drawings at their individual desks. In essence, what matters is the final product which is being developed, within admissible errors. Rigour refers to the precision of the final product, meaning its concrete construction with 'no mistakes'. Such expressions have singular meanings. In contrast with mathematical practices, naming an object does not rigorously guarantee its existence in the Technical Design lesson. Thus, the notions of rigour and precision when using mathematics in both observed lessons differ from the notion of rigour both in formal mathematics and in the practice of geometric construction.

Language used in the observed classrooms includes non-mathematical technical terms and gives an indication that both students and teacher modify the mathematical discourse while re-signifying mathematical meaning within their practice. In fact, as we could see when interviewing students from the Highway System classroom, learners seem to be 'borrowing' concept names from mathematics practice and collectively transforming meaning to indicate their use in a technical procedure.

5.2 The developing mathematical competences, as recognized within the lessons observed

In each classroom, during the lessons observed, questions mainly refer to knowledge of geometric construction relating to work project contexts. Reasons for learning mathematical notions are straightforwardly presented, since students are always given a flavour of the context where such mathematics will be applied in their immediate future. In general, problems mirror authentic technical work context tasks rather than the practice of mathematicians.

The definitions of tangent line which are made explicit by the teachers restrict the notion to tangent lines to circles. In one attempt to account for a generic definition, the teacher steps back and recalls arcs of circle in his examples, drawing ellipses and a spiral. In fact, practices in all lessons just require the notion of tangent line to arcs of circles, and teachers' directions in these lessons seem to indicate an instrumental intention when approaching mathematics. We also notice a curious definition presented in the Technical Design first lesson for the verbal form 'to be tangent', instead of the noun 'tangent'. It recalls action rather than a mathematical object, implicitly suggesting procedures and methods as the main mathematical knowledge component of such a context. Definitions presented by students during the interviews show the meanings for tangent lines produced by them are tied to the technical design and plotting procedures. Lines with curved ends, or lines crossing at a single point, are given in statements tied to the various practices these students experience and which, in this study, learners distinguish from their earlier mathematical practices at school.

Other mathematical statements presented in the classrooms by the teacher are mainly prescriptions of procedures to be carried out to develop geometric constructions required by specific projects. In fact, they are presented with no formal justification and they are accepted by students, who follow them as steps to develop the constructions. Mathematical notions and procedures are informally introduced and discussed.

Although suggesting mathematical tool-use as the main mathematical meaning shared within these practices, we noticed interesting inversions of common mathematical procedures and actions indicating that the practice in the observed contexts may not always be restricted to the mathematical procedure applications usually performed in mathematics lessons.

In fact, the 'given a curve find the tangent at a point P' common practice in school mathematics classrooms contrasts with procedures and actions performed within the observed lessons. In the Highway System classroom project, tangent lines represent parts of the highways and the procedure to develop the highways project consists of drawing the lines first, and then designing the curves, as cambers, to which the given lines must be tangents. So the procedure to be carried out is aligned with 'draw the tangents, determine the curve'. The same applies to the 'arc tangential to a line' construction developed during the Technical Design lessons, which also suggests moving an arc towards a line, while school mathematics usually works the other way around. In such practices, tangent lines are being taken at first as a support to develop the task, as tools for drawing curves, which as cambers are part of the highways.

5.3 The common direction of learning: functioning mathematically, across school classrooms

At this point, we must say a word on the teaching and learning of mathematics described in this study. Our understanding is that the participants are engaged in a mathematical activity which is distinct from the mathematical activity in school mathematics classrooms. For this reason we warn the reader that, from our view, crude comparisons between those teachers and mathematics teachers are not appropriate. On the other hand, there are instances in the interviews which could be understood as indicating interviewees' nice grasp of the structure of the practices of mathematicians. We do not experience similar responses from students in our everyday mathematics classrooms!

In fact, during the interview, learners explain their idea of imagining a circle closer to the curve at the tangency point in order to overcome a mathematical problem by reducing it to cases where old strategies would apply. For those students, the classification of the curve as regular or irregular corresponds to the solution of the problem of drawing its tangents. The method of completing the curve to design a circle if possible resembles procedures developed in classroom, while the strategy of sketching the osculate circle and the curvature centre freehand to determine the tangent goes beyond the strategies we had observed in the Technical Design lessons and in didactic materials.

The analysis above suggests distinctions amongst goals, language and also values which are implicit in the mathematical practices observed. Those from our ideal school mathematics classrooms mirror the practices of mathematicians, or reflect the national educational curriculum orientations for mathematics development in mathematics classrooms at school. We presented instances where mathematics is practised as an instrument to solve problems suggested by other subject matter questions, building a different meaning for 'functioning mathematically'.

On the other hand, we perceive similarities amongst the values and goals in the mathematical practices we observed and those described by researchers who investigate the use of mathematics in work settings. For instance, Magajna (1998) describes three mathematical methods as part of the practices he observed which he claims differ from procedures commonly learnt at school. Even so, inherent in the practices in the vocational classrooms described in this chapter are step-by-step mathematical procedures which are followed by students when developing their tasks and projects where "they may not understand the mathematical tasks s/he carries out and often, if everything can be foreseen, no mathematisation is desired…" (p. 66). Furthermore there are instances where "the occurrence of a mathematics-related task is not expected" (ibid., p. 66) since freehand drawings are accepted in students' practices, even when the rigorous methods of school mathematics are available.

According to our observations, the focus seems to be on knowing how to construct and use the mathematical instruments. To some extent we saw students and teachers involved in the same activity in both cases. Other roles played by teachers are different in each classroom observed. In the Technical Design classroom we notice a traditional school classroom routine, apart from for the fact that collaborative work is encouraged by the teacher with no strict rules for space organization in classroom. All students apparently rigorously follow the teachers' plan for the lesson. For the Highway System classroom the teacher plays the role of a supervisor. After fixing a plan and giving instructions, students are left to develop the task on their own, which they do. A correct procedure is bounded by general agreements among students, reminding us of descriptions of technicians using mathematics in a workplace (Magajna, 1998).

To summarise, the common direction of learning mathematics in the two vocational course lessons, even though they differed, was closer to 'being mathematical' in work settings than in an ordinary school mathematics classroom.

6. SOME IMPLICATIONS FOR TEACHING

Mathematical participation in the two observed lessons did not always correspond to what we expect in school mathematics. For instance mathematical meanings and inquiries which could enrich school mathematics lessons were brought into the discussion, such as a notion of centre of curvature, sketching tangents before determining curves, considering a tangent as 'a line which touches the curve at a unique point', all these notions seem naturally exhibited by procedures of technical design practices or plotting projects which require identification of curves and arcs of circles. In contrast, in calculus lessons, curves are at first worked out as if intuitively locally straight.

Students in our study and some workers seem to believe that there is an actual school mathematics in school mathematics classrooms which is different from the knowledge in other subject classrooms or working settings (Magajna and Monaghan, 2003). This can be taken to mean that they are not really 'functioning mathematically' within their practices. If instead we think about their practice as experiences within a distinct community of practice of mathematics, as we did, assisted by Winbourne and Watson (1998), we can rethink vocational and services courses through perceiving them as much more complex than a simple context for modelling or applying mathematics. In fact, and within the limits of this study, it seems the language used in vocational course classrooms where mathematics is a tool for other subjects indicates that both students and teacher re-signify mathematical meaning through modifying discourse and goals when approaching the content. In this sense, we may think of a diversity of school mathematics practices in school, in particular, in a technical school. 'School mathematics', therefore, cannot be assumed to be homogeneous since the practices differ across classrooms.

ACKNOWLEDGEMENTS

We are grateful to all those who made suggestions and supported us in writing this chapter, particularly to Anne Watson and Peter Winbourne.

REFERENCES

Adler, J. (1998). Lights and limits: Recontextualising Lave and Wenger to theorise knowledge of teaching and of learning school mathematics. In A. Watson (Ed.), *Situated cognition and the learning of mathematics* (pp. 161-177). Oxford: University of Oxford, Centre for Mathematics Education Research.

Lave, J. (1996). Teaching, as learning, in practice. *Mind, Culture and Activity, 3*(3), 149-164.

Lave, J., & Wenger, E. (1991). *Situated learning. Legitimate peripheral participation.* Cambridge: Cambridge University Press.

Magajna, Z. (1998). Formal and informal mathematical methods in working settings. In A. Watson (Ed.), *Situated cognition and the learning of mathematics* (pp. 59-70). Oxford: Centre for Mathematics Education.

Magajna, Z., & Monaghan, J. (2003). Advanced mathematical thinking in a technological workplace. *Educational Studies in Mathematics, 52*(2), 101-122.

Moreira, V. G. (2004). *Comunidades de prática da matemática no ensino médio técnico.* Unpublished Master Dissertation. Faculty of Education. Universidade Federal de Minas Gerais, Brazil.

Moreira, V. G., & Pinto, M. M. F. (2004). Technical school students' conceptions of tangent lines. In *Proceedings of 28 conference of the International Group for the Psychology of Mathematics Education* (Vol. 4, pp. 33-40). Haifa, Israel: PME.

Moreira, V. G., Sampaio, V., Cardoso, R., Almeida, A., Prado, H., Zumpano, A., & Pinto, M. M. F. (2001). *Investigando a Transição da Matemática da Escola Elementar para a da Universidade.* Poster Presentation. Scientific Week, Universidade Federal de Minas Gerais, Brazil.

Tall, D., & Vinner, S. (1981). Concept image and concept definition in mathematics with particular reference to limits and continuity. *Educational Studies in Mathematics, 12*(1), 151-169.

Vinner, S. (1991). The role of definition in teaching and learning mathematics. In D. Tall (Ed.), *Advanced mathematical thinking* (pp. 65-81). Dordrecht: Kluwer.

Winbourne, P., & Watson, A. (1998). Participating in learning mathematics through shared local practices in classrooms. In A. Watson (Ed.), *Situated cognition and the learning of mathematics* (pp. 93-104). Oxford: Centre for Mathematics Education Research, University of Oxford.

Chapter 13

Learning Mathematically As Social Practice In A Workplace Setting

Brian Hudson
Umeå Research Centre for Mathematics Education, Umeå University

Abstract: This chapter reports on a small-scale case study involving 15-16 year old secondary school students participating in a vocational module under the General National Vocational Qualification (GNVQ) scheme that operated in England during the late 1990s. The development was a pilot study involving experience in the workplace in a small-scale light engineering context. An initial aim of the study was to explore the potential of the setting for the development of numeracy practices. The theoretical framework adopted is based on a social perspective on learning and a view of learning mathematically as social practice. Of particular interest were the differences between everyday and school mathematical practices. The analysis focuses on differences in the practices between the settings of workplace and school in particular. Finally issues to emerge from this study are discussed in relation to the wider context of policy and practice. These include issues of relevance, questions of purpose, learner confidence and approaches to assessment in mathematics.

Key words: social practice, learning mathematically, workplace learning, vocational education, relevance of mathematics, purpose in mathematics, learner confidence, assessment

1. INTRODUCTION

This chapter focuses on a module that included experience in the workplace in a small-scale light engineering context for Year 10 (15-16 year old) students in the UK in the late 1990s and is based substantially on the work reported in Hudson (1998). The factory forms one division of a multinational company that specialises in the manufacture of products for the electronics industry. The General National Vocational Qualification (GNVQ) which operated at that time included a focus on the 'Application of Number' as one of the Core Skills elements. Accordingly a particular focus of interest in the study was the potential of the setting for the development of numeracy practices.

2. THEORETICAL FRAMEWORK

In developing a social perspective on learning mathematically, the work of Vygotsky (1962) has provided a fundamental starting point. A key assumption underpinning such a perspective is that socio-cultural factors are essential in human development. Intellectual development is seen in terms of meaning making, memory, attention, thinking, perception and consciousness which evolves from the interpersonal (social) to the intrapersonal (individual). In discussing the influence of such a perspective, Lerman (1996) describes language as providing the tools for thought. He argues that language is not seen as giving structure to the already conscious cognising mind; rather the mind is constituted in discursive practices.

Central to such a perspective is the notion of 'activity' which Crawford (1996) highlights as meaning personal (or group) involvement, intent and commitment that is not reflected in the usual meanings of the word in English. She draws attention to the fact that Vygotsky wrote about activity in general terms to describe the personal and voluntary engagement of people in context – the ways in which they subjectively perceive their needs and the possibilities of a situation and choose actions to reach personally meaningful goals. The work of Vygotsky was built on by thinkers such as Leontiev (1978) and Davydov and Markova (1983) who made clear distinctions between conscious 'actions' and 'operations' which are relatively unconscious and automated. Operations are seen as habits and automated procedures that are carried without conscious intellectual effort. So that *activity* corresponds to a motive, *action* corresponds to a goal and *operation* depends upon the conditions.

The social practice theory offered by Lave and Wenger (1991) and further illustrated in Lave (1988, 1996) is also consistent with this view and offers additional insights. This perspective offers a view of learning as a process of participation in communities of practice, which is at first 'legitimately peripheral' in relation to any new practice but that increases gradually in engagement and complexity. Learning is seen to be located in the processes of co-participation, as opposed to within the heads of individuals. The learner acquires the skill to perform by actually engaging in the process, under the conditions of legitimate peripheral participation (LPP), to a limited degree and with limited responsibility. Those participating in the community are seen as learners and learning, as such, is distributed among co-participants and is not seen as a one-person act. With regard to understanding, this is not seen to arise out of the mental operations of a subject on objective structures, but rather it is located in the increased access of learners to participating roles in expert performances. Learning can be a feature of various practices and is not seen to be limited to examples of training and apprenticeship. For example, the production of language can be seen as a social and cultural practice. Lave and Wenger's notion of LPP can be seen both as a way of engaging and also as an interactive process in which the apprentice engages by simultaneously performing in several roles. Learning is seen as a way of being in the social world rather than as simply a way of coming to know about it. Learners are actively engaged not only in the learning contexts but also in the broader social world and learning presupposes engagement in activity, in the strong sense of the word, without which no learning will occur.

3. RESEARCH METHODOLOGY

The approach to this study was that of participant observer involving attendance at as many of the factory visits as possible. This involved working alongside the students as far as possible although in most situations this was as an onlooker and active participant in discussion. Data was collected by the use of field notes together with a video camera to capture the detail of the activity. In addition, a focus group interview was conducted with the students and semi-structured interviews with four of the staff involved in the process were also carried out at the end of the programme. Three of the four adults were staff mentors; the fourth was the Operations Manager who was the driving force behind the initiative with the College.

In deciding on the interview schedule, Jean Lave's contribution to the Oxford Seminar *Situated Cognition and the Learning of Mathematics* in 1996 was of particular influence; here she proposed that the study of learning elsewhere than school offers clearer windows on what learning is all about. Other work of significant influence was that of Lerman (1996) who suggests that much greater attention might be given to an awareness of 'the differences between everyday and school mathematical practices and meanings, and between different, mostly workplace out-of-school practices and meanings'. Accordingly a particular line of questioning focused on 'differences' and 'similarities' between school and workplace mathematics.

Furthermore the adult mentors were asked to reflect on and to say how 'good' they judged themselves to be at mathematics as 'retellings of performance events' (Lave and Wenger, 1991).

4. FACTORY PROCESSES

The main work of the factory is centred on the production of components for electronic devices such as satellite dishes. These are machined parts that, in the main, are produced as an automated production process using high technology computer controlled lathes. In order to monitor the various processes within the factory, samples of output typically are taken on the start-up of a new process and then at regular intervals after that. The results of this process are designed to allow for the resetting and adjustment of the machines, if necessary, ensuring consistent quality of output and avoiding the chance of producing scrap material as a result of a defective process. Statistical process control (SPC) is a central feature of the working practice of the operation. This approach is based on that of Ford (1984) which emphasises a strategy of prevention 'before the event' to avoid waste rather than simply one of detection 'after the event' which is seen to tolerate waste and to be uneconomical, wasteful, expensive and unreliable.

5. STUDENT ACTIVITIES

The student learning environment was designed around a series of eight activities, each of which was scheduled to take place in an afternoon session over the Autumn Term of 1996. The group was made up of twelve students who were identified as being in need of additional academic support. The decision to offer the opportunity to these students in the first place was taken on the grounds that the traditional academic curriculum was not best serving their needs and interests, although there was an intention to expand such opportunities in the future to a wide group of students. Each visit involved a short introductory meeting involving all the students and adult mentors in the office of the Operations Manager who gave the group his full attention. All participants were fully briefed about the nature of the activities and the students were given an opportunity to ask questions at the outset prior to accompanying the adult mentors to the work place.

Several of the student tasks related to the statistical process control (SPC) operation which underpinned many of the functions within the factory. In particular one of the major methods for carrying out this monitoring process is through the use of a Process Control Chart as illustrated in Figure 1. This chart can be seen as an example of a structuring resource (Lave 1988; Pozzi, Noss and Hoyles, 1998).

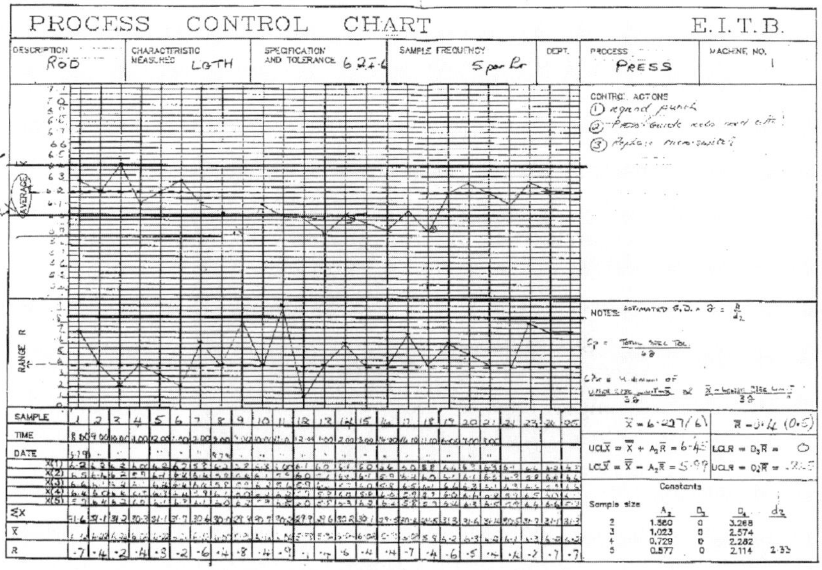

Figure 13-1. Process Control Chart

The completed chart illustrates how 25 sets of 5 samples were taken over a period of a few days during one week. In each case the sum, mean and range is calculated, and the mean and range are plotted on the corresponding charts above.

This chart displays the results of a monitoring process based on the use of a variety of handheld devices for measuring, pocket calculators for calculating and more traditional tools in the form of pencil and paper for recording purposes. These tasks were well within the capabilities of the students taking part in the project.

More technologically automated processes in the factory involved the use of devices such as digital verniers and digital micrometers linked to mini-processors. These devices needed to be calibrated initially, based on information from the technical specification of the part, after which a series of measurements were taken and the mean, range and standard deviation are automatically computed and printed out. Operation of these processes required more advanced technical skills that were generally beyond the current capabilities of the students.

The students were grouped in pairs and each group worked with an adult mentor who was an employee of the factory. They worked alongside factory workers in their day-to-day activities involving the assembly, storage, despatch and quality control procedures of the factory. For example, the students would select a set of 5 samples at random from the production line, measure each one with a vernier, record the results together with the timing, process these to find the sum, mean and range and plot the results onto the chart. In these processes they were supervised and supported by adult mentors.

With regard to the more technically sophisticated processes they observed the work of the adults as they operated the machines. In turn, these adults explained the processes that they were engaged in and offered the students appropriate opportunities to operate the machines to a limited extent.

At the end of each visit the group re-assembled in the office of the Production Manager. Each student was asked to describe his or her general response to the experience and to give an evaluation of his or her own performance. After initial reluctance to participate, all the students eventually relaxed and took an active part in these plenary activities. They seemed unused to working in this way but all involved commented on the value of this particular aspect of the experience.

6. THE MATHEMATICAL DIMENSIONS OF THE WORK PRACTICES ON THE FACTORY FLOOR

In order to convey some of the mathematical practices of the work in the factory setting I will refer to the words of Jane, who is a 'Leading Hand' on the factory floor. Her role involves organising the workforce in a particular section of the factory. She had been employed at the factory for about ten years:

> There are bits of maths in quite a bit of my job – my job actually is inspection and you have to tally works orders up to make sure they are right. You have to make sure all the amounts are correct, so you've got different forms of arithmetic like on the back of a mix order you've got the parts made, parts ... and then you have to carry on parts to the next operator, you have to make all your parts tally all the way through. And it can be very complicated sometimes, because sometimes you make parts but we don't send them all through if we haven't got a full tube or anything ... but it's tallying everything up so at the end we can have a proper tally.
>
> Especially there's maths especially the SPC in capability studies but at the moment one of the girls downstairs is doing some tests on some parts ... but we are not using the mini processor, we are using the hand-written capability studies and that involves you've got your writing down, measuring ..., reading that off ... rounding up and rounding down to ... 3 or 4 figures – then you've got the use of tally charts, you have to read the tally charts, we've got to put it into frequency, percentage, etc. and then it's got to be copied on a graph and then we work out all from this. I can work out all the percentages of possibilities of things going out of control etc. ...
>
> As I say we've got the use of the vernier, which is, round up or round down, which ever which way you want to use it but we mainly use that. Now we've gone off the sheets and gone onto computers a lot. There's a lot less mentally, you've got to put in your numbers correctly, otherwise you end up with something totally out of control which happens sometimes.
>
> There's also reading of your graphs, reading from information, using your scales, that also involves maths mostly to a smaller degree, cause you could be measuring the outer parts, then there's a way you can use your scales to measure them more accurately. So you have keep on

there's a way of doing it where you are just putting a few parts, putting a few more and press a few buttons and it'll come out.

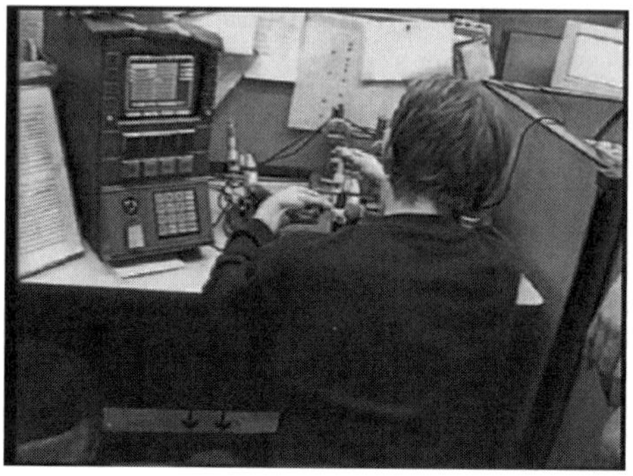

This account confirmed the observations in the factory of a wide range of mathematical practices that could be related to mathematics in the National Curriculum of the time, which included a component called 'Using and Applying Mathematics'. There was much evidence of using and applying mathematics in practical tasks and real life problems. This component had subheadings about: 'Making and monitoring decisions to solve problems' which were manifested here in reviewing progress and checking and evaluating solutions, and 'Communicating mathematically', which was enacted through a need to understand mathematical language and notation, to use mathematical forms of communication (including diagrams, tables, graphs and computer print-outs), to interpret mathematical presentation in a variety of forms and to examine and evaluate these critically. Another curriculum strand was 'Number', which was enacted through the need to understand and use relationships between numbers, to understand and calculate averages and to develop methods of computation including calculators and calculating devices. In relation to the 'Algebra' strand, there was a need to understand and use formulae and expressions, and to interpret and evaluate these in real life situations using computers and calculators as necessary. Finally with regard to 'Handling Data' there was the need to process and interpret data, to interpret a wide range of graphs and diagrams and to evaluate results critically.

7. DATA ANALYSIS

A number of key issues emerged through the process of data analysis. In particular issues of the relevance and purpose of mathematics came to the foreground, as did issues of abstraction and connection with the real world and also issues related to personal confidence in the subject. The students also highlighted issues related to assessment and testing in particular. These issues are highlighted through the analysis of the mentor interviews and focus group interview with the students in particular.

7.1 Adult mentors' retellings of (mathematical) performance

The adult mentors were Linda, Janet and Jane. Each was interviewed at the end of the project using a semi-structured interview approach. The first issue was that despite being very capable in their workplace roles, all three adults exhibited a low level of confidence with abstract mathematics. The second issue to emerge related to the relevance and purpose of studying mathematics in school. In particular Jane, who was the most able mathematically, had struggled to pass her General Certificate in Secondary Education (GCSE) in mathematics at the third attempt and yet recalled the ease and enjoyment with which she had worked with statistical ideas in her Advanced Level geography course where it was related to a real life context and to people in particular:

> I think it's like people relate to ...like ...people where it's put into relation to people or things but where it's figures it tends to overload me sometimes I think ...But A level geography it was that side I enjoyed that far more than the physical side of geography, where it was related to people, cities etc. Why people do this and why they do that ...

Janet was less confident in her mathematical ability:

> But yes you do have to be pretty good at maths, it isn't my strong point, I've got a calculator.

In reflecting upon her experience of school mathematics, she emphasised the idea of doing 'exercises':

> I think in school you just like getting your exercises right, it weren't like finding things like in a drawing like we do. I've never come across that till I came to work here. You didn't actually measure anything ...I prefer it as it is now.

Linda also saw herself as not being very good at mathematics in school, although she felt that her mathematical ability had improved since working at the factory. Also, she saw the use of a calculator as a basic mathematical skill and not as a sign of her inadequacy, as it seemed in the case of Janet:

I'm not numerical, I've never done maths, I wasn't very good at it at school, I've got better since I've come here. It's got a lot easier since I've been in stores, than whatever I did at school so I think it's good.

When asked how she coped with arithmetic, she emphasised the need to use the calculator:

As long as I've got a calculator there, which you have to have because your customer demands that he has that quantity and because every single thing is logged on to a computer, if you miss one piece you know about it, do you know what I mean? It's so very spot on, immaculate and everything that you've got to spot on ... I use a calculator but you never did when we were at school so you've got to learn how to use a calculator. I mean some kids haven't got a clue how to use a calculator so I think you should be taught how to use one properly.

7.2 Focus group interview with students

The issues arising from the interviews with the adult mentors were reflected in the feedback from the students. However a further issue to emerge from this stage of data collection was the students' emphasis on testing and assessment. This feedback was obtained from a semi-structured focus group interview with all the students at the end of the project. They were asked firstly to think about what was different in the factory setting from the mathematics done in school:

BH: How was it different from the maths that you do in school? What's different about it?

Student A: It's rubbish at school.

BH: OK. So why do you say that it's rubbish at school?

Student A: It's boring.

Student B: It's harder.

BH: So it's harder at school. OK. Why is it harder?

Student C: You don't get any homework.

However when pressed to say more the responses emphasised the idea of 'testing':

BH: OK. Let's go back to the idea that it's harder in school. Why is it harder in school?

Student A: Because they're testing you in school.

This comment about 'testing you in school' was echoed strongly by others in the group. When asked about being tested in the factory, the response revealed a strongly perceived difference between the two settings:

BH: So they're testing you in school. Were they testing you in the factory?

Student A: Not really ... they weren't testing you were they? They were showing you how to do things.

When pressed to say more about why it is boring in school, the responses emphasised routine, lack of engagement, passivity and inactivity:

BH: Someone said it was boring in school. Why is it boring in school?

Student D: You get it all the time.

Student A: It's just boring.

Student E: It's not practical ...

Student F: You just sit down don't you?

When asked about things that they had learned in school being applied in the workplace, there was some sense of the relevance of their school mathematics in terms of its content:

BH: Are there things that you have learned in school that you were using in the factory?

Student B: Yeh. Statistics ... how to do graphs ... the average weight of these like sweet things

Student F: Making sure it was the right weight ... making sure it was not outside what it should be.

8. DISCUSSION

In the initial analysis of the responses from the adult mentors (Hudson, 1998), a particularly relevant aspect of social practice theory was Lave and Wenger's (1991) thinking about the notion of 'engagement' and in particular the description in the Forward by William Hanks that

> Learners … are engaged both in the contexts of their learning and in the broader social world within which these contexts are produced. Without this engagement, there is no learning. (p. 24 my underlining)

In reflecting upon her experience, Jane distinguishes between her enjoyment of the mathematics in her A Level geography when it was about 'people' in contrast to being just about 'figures', which might be seen as abstract mathematics which 'tends to overload me'. However she proceeds to emphasise purpose also i.e. 'Why people do this and why they do that'. This sense of purpose reflects Lave and Wenger's notion of 'engagement' and is consistent with activity in the strong sense of the term as highlighted by Crawford (1996). In her recollections, Janet seems to emphasise the lack of purpose in school mathematics i.e. it is 'just about getting answers right' (in school) and not 'like finding things out like in a drawing like we do'. She emphasises that she had 'never come across that' (sense of purpose) until she 'came to work here'. A further relevant aspect of Janet's view is the way in which she sees the calculator as a tool that is taken for granted. Linda also emphasises the use of a calculator. However she stresses the need to use a calculator for a purpose i.e. 'because your customer demands that he has that quantity' and 'it's so very spot on, immaculate.'

In recalling what the students were saying there was a strong sense of conviction and general agreement about how they found mathematics to be 'boring', 'not practical' and just about 'testing'. The suggestion that school mathematics is not practical is consistent with the responses from the adults i.e. mathematics without a purpose. The expression of boredom conveys that overwhelming sense of waste when one is not engaged with something and yet unable to escape from it. However the view that 'it's harder ...because they're testing you in school' also conveys some of the impact of the assessment system upon this particular group of students. The 'testing and examination' culture associated with accountability and external control as described by Gipps (1994) was very apparent through these comments. She contrasts this culture with that of an assessment culture associated with teaching, learning and formative assessment which seems to have been far more evident in the factory setting than in that of the school for these students.

One of the reasons for developing the link with the school by the factory at that time was the poor take up of opportunities to work in manufacturing within the local area. A deep resistance was perceived, especially on the part of local parents, to a vocationally orientated curriculum. However at the end of the programme two of the students in the group were very interested in the possibility of taking up apprenticeships at the factory and were thought to be very suitable candidates, despite the fact that they were not seen to be succeeding in school. This is indicative of a wider problem not addressed within an education system that is preoccupied with performance indicators, testing, targets and school league tables.

The accounts from the adults of their school mathematics conveyed a generally low level of confidence and yet in the workplace they were using mathematical skills appropriately, effectively and with confidence. This raises serious questions about what the school system is achieving in terms of a mathematics curriculum 'for all'. A number of echoes could be found in the comments from the adults with what the students had to say about their current experience of school. For example, the relationship with the 'real world' seemed to be powerfully engaging, as did the idea of doing mathematics with a purpose in a practical setting. Given the ongoing debate about the role of the calculator, it was especially interesting to note Linda's comments on her use of the calculator as a tool and also on her view of the need to teach students how to use such tools effectively.

In reflecting on this study ten years later, it seems that the issues raised are just as relevant today as they were at the time. In more recent work (Ongstad, Hudson, Pepin, Imsen, and Kansanen, 2005) we have emphasised a 'communicational' perspective to approach thinking about the design of teaching and learning situations in mathematics which focuses on mathematics *as* communication. This is based on a view that whenever we engage with mathematics through textbooks, or follow exposition concerning mathematics, or listen to explanations, we meet mathematics as specific and concrete utterances, that is, we see or experience mathematics *as* communication. This approach is combined with the concept of 'positioning', which acknowledges that mathematics can be 'seen' in different ways which place emphasis on different aspects e.g. the rational, the practical and/or the affective. It recognises that students and teachers may approach the subject predominantly with either the 'head', the 'hand' or the 'heart', or a mixture of these. The key question then becomes where we position ourselves as teachers and learners *in relation* to these aspects. In our view (Ongstad et al. 2005) it is the aesthetic/emotional aspects which have been neglected in the process of reform over recent decades. This is despite what we have known since the Cockcroft Report (1982) which highlighted the way in which a teacher in every lesson conveys "even unconsciously, a message about mathematics which will influence [the pupil's] attitude" (para 345). Thom (1973) contends that "all mathematical pedagogy … rests on a philosophy of mathematics", however poorly defined or articulated it might be. Other research shows that differing conceptions on the nature of mathematics have an influence on the ways in which teachers and mathematicians approach the teaching and development of mathematics (Cooney, 1985; Thompson, 1984). With regard to the implication for ways of knowing, the work of Carlgren (2005) is seen to be very relevant. She describes the need for a shift from *knowledge* as substance to *knowing* as a contextualised relation involving dispositions to act and qualities of knowing as embedded in the habits of social practices.

9. CONCLUSIONS

This study highlighted issues of engagement in and questions of relevance and purpose of mathematics on the part of participants which has echoes more widely. For example the findings of this study resonate with the issues raised in the Smith Report (Smith, 2004) i.e. the long term decline in the number of young people continuing to study mathematics post-16 in the UK, apart from Scotland. A lack of confidence in mathematics emerged as a common issue on the part of both mentors and students. This, combined with a reaction against the idea of being constantly 'tested' may help to give some clues as to the reasons for such a lack of engagement in mathematics and to a lack of interest to study it further post-16. The Smith inquiry also reports on a shortage of specialist mathematics teachers; the failure of the curriculum, assessment and qualifications framework in England, Wales and Northern Ireland to meet the needs of many learners and to satisfy the requirements and expectations of employers and higher education institutions; and the lack of resources, infrastructure and a sustained continuing professional development culture to support and nurture all teachers of mathematics. These issues can also be seen to be reflected internationally and in particular across the western industrialised world at the time of writing. The needs for fresh perspectives on practice and new approaches to policy making seem even greater and more urgent at the time of writing than ten years earlier.

REFERENCES

Carlgren, I. (2005). The content of schooling – from knowledge and subject matter to knowledge formation and subject specific ways of knowing, *ECER 2005 – European Conference on Educational Research*. Dublin: University College Dublin.

Cockcroft, W. (1982). *Mathematics counts: Report of the committee of inquiry into the teaching of mathematics in schools*. London: HMSO.

Cooney, T. (1985). A beginning teacher's view of problem solving. *Journal for Research in Mathematics Education, 16(5)*, 324-336.

Crawford, K. (1996). Vygotskian approaches in human development in the information era. *Educational Studies in Mathematics, 31*(1-2), 43-62.

Davydov, V. V. & Markova, A. K. (1983). A concept of educational activity for children. *Soviet Psychology, 21*(2), 50-76.

Ford Corporation. (1984). *Statistical process control: A guide to the use of statistical process control to improve quality and productivity*: Ford Corporation. Product Quality July 1984 EU880.

Gipps, C. (1994). *Beyond testing*. London: Palmer.

Hudson, B. (1998). Learning mathematically as social practice in a workplace setting. In A. Watson (Ed.), *Situated cognition and the learning of mathematics* (pp. 71-80). Oxford: University of Oxford, Centre for Mathematics Education Research.

Lave, J. (1988). *Cognition in practice*. Cambridge: Cambridge University Press.

Lave, J. (1996). Teaching, as learning, in practice. *Mind, Culture and Activity, 3*(3), 149-164.

Lave, J., & Wenger, E. (1991). *Situated learning: Legitimate peripheral participation*. Cambridge: Cambridge University Press.

Leontiev, A. N. (1978). *Activity, consciousness and personality*. Englewood Cliffs, NJ: Prentice Hall.

Lerman, S. (1996). Culturally situated knowledge and the problem of transfer in the learning of mathematics. In *Research group for social perspectives on mathematics education*. London: University of London, Institute of Education.

Ongstad, S., Hudson, B., Pepin, B., Imsen, G., & Kansanen, P. (2005). *Didaktik and/in mathematics education. Studying a discipline in international, comparative and communicational perspectives*. Oslo, Norway: Oslo University College.

Pozzi, S., Noss, R., & Hoyles, C. (1998). Tools in practice, mathematics in use. *Educational Studies in Mathematics, 36*, 105-122.

Smith, A. (2004). *Making mathematics count: The report of inquiry into post-14 mathematics education*: London: DfES.

Thom, R. (1973). Modern mathematics: Does it exist? In A. G. Howson (Ed.), *Developments in mathematics education: Proceedings of the second international congress on mathematics education* (pp. 194-209). Cambridge: Cambridge University Press.

Thompson, A. G. (1984). The relationship of teachers' conceptions of mathematics teaching to instructional practice. *Educational Studies in Mathematics, 15*(2), 105-127.

Vygotsky, L. S. (1962). *Thought and language*. Cambridge, Mass: MU Press.

Chapter 14

Analysing Concepts Of Community Of Practice

Clive Kanes and Stephen Lerman
King's College London, London South Bank University

Abstract: This chapter is based on the notion that Lave and Wenger (1991) and Wenger (1998) work with similar, although characteristically different concepts of 'community of practice' and our goal is to compare and contrast these. We point out the relative strengths and weaknesses of each, illustrating our arguments with research examples drawn from the literature. We conclude by indicating ways in which each perspective informs research in the mathematics education community and to directions in which they might be developed to support our understanding of teaching and learning across a range of contexts.

Key words: community of practice, activity, legitimate peripheral participation, mathematics education, identity, tensions, contradictions

1. INTRODUCTION

In Lerman (1998) the revolutionary notion of the situatedness of knowing, meaning and acting and the centrality of identity in learning of Jean Lave and Etienne Wenger (Lave and Wenger, 1991) were acknowledged and discussed. In this, Lave's early anthropological studies led her and Wenger to argue that knowledge is located in particular forms of situated experience, not simply in mental contents, and that knowledge has to be understood relationally, between people and setting. Learning is about participating in a social practice, she argued. And thus, for instance, the problematic notion of transfer, predicated as it is on the move from abstract decontextualised knowledge that can be applied across a range of situations, is transformed into a problem of when and by what means full participation in a target practice is achieved. As Boaler's (1997) book indicated, unless teachers teach explicitly for transfer it is not likely to be a resource available to students. Furthermore, in the social practices studied by Lave and Wenger learning skills in context becomes a process of identity formation; people take on the identity and social relations of that occupation or activity. This raises important issues for the mathematics classroom and especially for researchers as we try to make sense of who succeeds in mathematics classrooms and who does not, and why, and of how learning can be interpreted in a situation where the focus of the activity appears to be purely that of knowledge acquisition, as in what is often referred to as a 'traditional' classroom.

However, for all its appeal, the notion of 'participation' in a social practice, also known as a 'community of practice' (Lave and Wenger, 1991; Wenger, 1998), has been criticised for its definitional haziness, the absence of an account of learning that traces both individual and social indications of learning, the restricted focus of participation around stable rather than rapidly changing forms of social practice, and so on. Lave and Wenger's (1991) theory, in particular, also does not address adequately the complex and multi-layered nature of the development of pupils' identities in the school mathematics classroom. Pupils in the mathematics classroom may have a goal of learning mathematics, which may best be understood as developing a school-mathematical identity. But they are also engaged in identity work in relation to friendship groups, developing sexuality, conforming to or resisting school practices, indeed what we might call performing the self. Moreover, as Lerman (1998) pointed out, the model of a social practice in which there is a master or participants, and into which newcomers wish to become apprenticed, does not at all reflect the reality of the mathematics classroom. Few pupils are interested in becoming mathematicians or mathematics teachers. New models are required, as reflected to an extent in Lave's writings in the mid-1990s (1996, 1997). Moreover, Lave emphasises explicitly the need to separate teaching from learning and to focus on the latter. However, the schoolteacher's identity is precisely about the intention to teach, and indeed a growing body of literature on mathematics teacher education draws on the notion of a community of practice to study the specificity of local teaching practices and of the processes of teacher education (Graven, 2004; Lerman, 2001).

These brief observations suggest that the current development of the concept of community of practice does need clarification and analysis in a number of respects. What does seem clear, however, is that Lave and Wenger (1991) and Wenger (1998) work with similar, although characteristically different, concepts of 'community of practice'. In this chapter, we call these *Community of Practice Type 1* (CPT1) and *Community of Practice Type 2* (CPT2). The purpose of this chapter therefore is to come to a sense of comparison and contrast between these two. It is divided into a number of sections. The next section attempts an offering of what is characteristic in each of CPT1 and CPT2 and how these have been used in analysing and framing empirical studies, principally in mathematics education. This then affords a starting point for discussion of the differences among these types of community notions and the relative strengths and weaknesses of each against theoretical and empirical criteria. The last section, building on these, suggests where these theories can and need to be developed, and speculates on what future theoretical moves may be of help.

We wish to make a final observation here in this introduction. We take a community to be something like a group of people connected by circumstance or purpose, but on a trajectory to share meanings and values and to collectively create new forms of life. As such, however, we would want to distance ourselves from the notion that 'communities of practice' totally saturate our lives. In particular, everyday life, even life that is structured, like play, or going on holiday, is not necessarily a community of practice or even part of one. This is not to say the scenes of lived life may not be set partly within 'communities of practice', but these on reflection may be rapidly changing, growing or contracting or may be indeterminate. At the very least we want to suggest that these questions are open but are important as they give more or less support to deterministic models of human existence. In short, we do not subscribe to the view that social life can be reduced to a study of communities of practice – at most, for us, the concept is helpful to come to grips with certain coordinated, collective, purposeful activities and their interaction.

2. DISTINGUISHING AMONG THE CONCEPTS OF COMMUNITIES OF PRACTICE

In this section we distinguish between CPT1 and CPT2. We emphasise, however, that we regard CPT1 and CPT2 as being of equal though different value. Despite apparent similarities, their approach to learning, as we will argue, is radically different, and this means they give rise to very different kinds of research questions, expectations and problems. In what follows we offer windows into some characteristic features of each. Later in the chapter we discuss implications of these for research and offer suggestions for further research into 'community of practice' concepts.

2.1 A characterisation of CPT1

Wenger offers the most coherent account of this form of 'community of practice'. In this he assumes that the space in which practice works is pre-eminently meaningful to a participant and this includes not only a particular sense of things (physical, conceptual, procedural, behavioural), but also encompasses a distinction between affective, intellectual and values orientation for things. However in a community of practice not all meanings are compatible and they do not all address problems equally well. It thus arises that meanings of participants have been subjected to the processes of negotiation in the community. In addition, because the criteria of legitimacy themselves are precisely the ones the community has developed a meaning for, these criteria are located within the community rather than elsewhere. Of course the community may need to grapple with an artefact produced elsewhere, and be judged by external criteria, but even then, the community develops its own characteristic meanings around these. These meanings may provide the basis for sustained resistance or compliance. However, this analysis begs the question: What does 'making meaning' mean? For Wenger the answer appears to be 'participation' in the community of practice, for this describes the "social experience of living in the world in terms of membership in social communities and active involvement in social enterprises" (p. 55). However, participation in this sense would seem not to account for the individual way in which members of a community build their own meaning for their participation. This latter, he calls 'reification' by which he means "the process of giving form to our (sic) experience by producing objects that congeal this experience into "thingness"" (p. 58). For Wenger, reification and participation are viewed as complementary concepts naming the process whereby individual and community experiences are shared and lead to the production of shared ideas and concepts. Together they explain, for him, what negotiation of meanings is about.

As an instance of CPT1, Wenger's book sets out a very elaborate description of the functioning of a workplace social practice, defining through that example three characteristics of practice: mutual engagement of participants; negotiation of a joint enterprise; and development of a shared repertoire of resources for creating meaning. We suggest that the example reinforces the idea that persons begin on the periphery and move towards the centre through taking on the practices and to some extent the identity of the community of practice. The practice changes through external pressures or internal modifications – the latter often in mitigation of the former. However, to identify the range of resources and influences on the practice, the sources of the tensions that lead to change and the development that takes place, calls for analytical tools not currently available in CPT1 formulations.

Nevertheless, CPT1 offers important tools for mapping out key aspects of a community of practice. In the case of mathematics classes, for instance, the study of Winbourne and Watson (1998) has proved illuminating.

2.2 A characterisation of CPT2

In contrast to CPT1, Lave and Wenger's concept, CPT2, is built around tension, conflict and discontinuity in practice and production. This latter term, production, relates to mathematics teaching in normal classrooms, for instance, by including outcomes such as learning, passing exams, producing student reports, and achieving high scores on league tables – as well as students who now understand concepts and process they were once not able to. For these authors, the community of practice operates within the theoretical frame of what they call 'legitimate peripheral participation' (LPP) in which all participants are construed as peripheral to the community of practice. Whilst they talk about full participation for established participants, and partial participation for beginners, for Lave and Wenger, LPP is about all participants (whether full or partial) traversing the periphery of practice and some moving from partial to full participation. This latter is one sense of how participants' trajectories 'advance'. Another is the work, often invisible, that needs to be done in the practice in order to keep it maintained. Effort is always required to maintain a given position – the practice continually needing to be productive in its ostensive sense. However, in addition, very often there is also the need to manage conflict and tension around new technologies, new expectations, goals and aims, new kinds of knowledge (Wedege, 1999), changes driven by advancing age and the resulting shift of skills profiles across the community as time passes, and so on. Though kinds of participation are also associated with the potential to advance the community of practice, some that are usually not identified in these situations are the concepts of sequestration, transparency, continuity-displacement, and so on.

To take an example that illustrates powerfully our sense of production, consider the process of a new teacher or teachers entering into an established school mathematics department. The individuals and the department, as a community, will, in order to 'produce' mathematics education in its various forms, typically experience the tensions and conflicts to which we have referred, and moving to full participation, or merely maintaining a given position, will all require great effort and will bring about changes for the individuals and the community. We note here, however, that CPT2 can be sometimes confounded with CPT1. In consequence of this we note the possibility of over emphasising states of harmony within communities of practice – as if real life, even that which is apparently stable, were not in reality turbulent and heavily inflected by competing ends. For instance communities of practice relating to mathematics education might include: particular mathematics classrooms, e.g. class 8B or 10C; the set of staff teaching mathematics in High School X; professional associations of mathematics teachers; school teaching and administrative staff; producers of mathematics textbooks and other teaching resources; the *National Curriculum* and all those maintaining it, certifying compliance, etc. Within these, studies drawing on the CPT2 framework identify and trace out the trajectories of participants as they engage in producing mathematics education outcomes (as above), and encounter a wide range of tensional issues such as sequestration (e.g. the masking to beginners of the examples and illustrations by which full participants in a community understand general statements), transparency (the situation where, for the sake of clarity, not all is revealed about a concepts of practice at one time), continuity-displacement (the need for newcomers to maintain the practice, yet the implications of introducing new full practitioners into the practice can threaten existing members).

We note that there has been some very interesting research developing the idea of communities of practice in order to study teacher education practices and classrooms (e.g. Graven, 2003). Researchers have studied the conditions under which communities of practice emerge or do not, how they might be maintained, and what are the effects on pupils and teachers, often incorporating the sense in which pupils are developing identities in a range of overlapping practices, not just the one that the teacher might be interested in.

2.3 Examples in mathematics education

At this point it may be helpful to offer more mathematics education illustrations of CPT1 and CPT2 frameworks at work. We choose instances relating to mathematics classrooms.

A CPT1 approach: As an example Pallas (2001) cites the contrasting cases of mathematics teachers in inner city Kennedy High School and suburban Truman Academy. In this he makes the point, though all are mathematics teachers, given their different contexts, one might expect a significantly different set of shared experiences in the different sites. Pallas suggests that teaching topics such as probability and statistics would be rather different in each of the institutions: Kennedy may develop the tradition of teaching these topics to the non-college bound, whereas at Truman students traditionally head for post-high school tertiary education. Thus in one of these schools topics may grow to become part of an applied maths skills course; in the other, they are a part of advanced high school mathematics instruction. Moreover mathematics teachers in these institutions either will or will not agree or disagree, or express interest or otherwise, in the way these topics are differently rendered. Thus the "meaning of probability and statistics in the context of the applied math course, a local event, partly defines the Kennedy math teachers' experiences of being a math teacher" (p. 8), and this means, for CPT1, that the identities of teachers in the different communities of practice are characteristically different.

In CPT1 successful classroom experiences for students might be described in terms of success in moving to full participation, co-operating with the teacher and others similarly apprenticed in harmonious mathematical and pedagogic activity. This description goes beyond purely cognitive explanations of 'successful' learning in that the importance of the norms of the classroom community and those of mathematics are in the foreground of notions of 'successful'. As researchers we look for students' alignment in their language and doing, and in their apparent mathematical thinking, with those of the teacher. As a result, some students develop an identity as a 'good student', 'mathematically able', 'a brain', etc., whereas others develop less advantaged identities around more negative attributes e.g. "I can't do mathematics", etc. What this kind of account lacks, however, is a sense of the institutions around both the teacher and student in contexts necessarily broader, but containing the classroom experience itself, and a sense of how these maintain and regulate the life of classroom practice. Themes of collaboration, cooperation and conflict are played out in the classroom as a consequence of how students and teachers respond to the constraints and opportunities they are afforded – yet, because CPT1 analyses tend to treat incoherence or disjoint behaviours in a practice merely as aberrations or departures from otherwise normal performance, we argue a more rich theoretical environment, in the form of CPT2, may sometimes be needed. Nevertheless Solomon (1998), in addressing what she refers to as the ritual *versus* principle distinction as it relates to mathematical knowledge, makes use of the notion of 'induction' into a community of practice and the role of the mathematics teacher's 'epistemological authority' to move beyond such reductionist notions of mathematical knowledge. It is notable, we think, that Solomon's engagement with community of practice requires also a notion of 'authority'. Below we draw attention to the absence in CPT1 of a theoretical treatment of power relations.

A CPT2 approach: Here the tensions and conflicts for the individuals and the community, the work required to remain on the periphery, the changes to the practice itself, and the mediating artefacts are all elements of the analysis of the mathematics classroom. Indeed, Lave has shown (1996) how students may well develop alternative practices for doing mathematics of which the teacher may not be aware. Students may exhibit a range of trajectories as they remain on the periphery or play other parts in the classroom. The teacher is not 'the master' seen in the many studies carried out by Lave and others, and this calls for elaborate research strategies to characterise these various trajectories. Teachers constitute a community possessing stronger or weaker borders within a school or institution, and these in particular cases, and in general, are under more or less constant revision, reappraisal, negotiation and mandate. Within such a regime, more experienced teachers and less experienced teachers respond in different ways to the curriculum challenges set before them. More experienced teachers can exploit the 'tricks of the trade' in order to simplify processes around assessment; they know what 'really counts', and can draw on expertise in a large repertoire of pedagogic methods and approaches. All this may mean that the potential to respond to new technologies and media may either advantage or delay them. However, not all of this knowledge, and certainly not all of it at once, is made available to newer teachers. Indeed, for these teachers the range of 'survival skills' and strategies for everyday work may be somewhat restricted. Now all of these differences have different effects on what is 'produced': learning, league table performances, etc. Moreover, as the boundary disputes (who teaches what classes, with what tools, in what spaces, etc) fluctuate in levels of intensity, teachers move from being younger to older people, from less to more experienced teachers, and from experienced teachers to other positions in the school, to other careers or to retirement.

CPT2 offers a view of such dynamics. In the process of holding back knowledge, Lave and Wenger call this *sequestering*, introduced above; a 'folk epistemology' they suggest around the familiar distinctions between abstraction and concretisation is created. The driving force here is the disconnectedness and fragmentation that sequestering creates in a social practice; and thus abstracting, sequestering and participating are aspects of the same social process. Thus withholding important information, supplementary ideas and approaches may fragment a beginning teacher's knowledge of mathematics teaching and may lead to it being offered to students in a fragmented way.

Another related concept, also introduced above, is that of *transparency*. Here the clarity with which the beginning teacher sees the mathematics pedagogic point of a particular approach or the mathematical need for a given design is dependent on the kind and scale of sequestration within the social practice as well as the dialectical relationship between 'productive activity' and 'understanding'. For instance, Adler (1999) makes interesting use of transparency in how the dual functions of 'talk' establish relations of visibility and invisibility that create dilemmas for teachers working in multilingual classes in South Africa. In another interesting study, Hemmi (2006) shows how Swedish students studying university level mathematics encounter concepts of proof. In referring to findings around proof, Hemmi writes of the "intricate dilemma about how much to focus on different aspects of proof at a meta-level and how much to work with proof without focusing on it, both from teacher and student perspectives". In this study, Hemmi observes how proof can present itself to students as a 'mysterious artefact'.

In addition to the issues of sequestration and transparency, Lave and Wenger identify a contradiction between the social relations of production and the social reproduction of labour, the so-called contradiction of continuity-displacement (see above). In the first they are referring to relationships that provide mathematics teaching and in the second to relations that provide mathematics teachers. They see these as inherently contradictory. As established teachers grow older they leave teaching, and this creates a demand for new teachers. In addition, society is moving and has new notions of mathematics teaching, mathematics teaching technologies, new kinds of assessment requirements and so on. At first, newcomers are disadvantaged, mainly because of their relatively limited knowledge and skills and relative marginalisation against notions of expertise and objective educational products. In time, however, the professional knowledge of experienced teachers, especially in environments of great change in pedagogic technologies, curriculums and learning theories, may all become increasingly obsolete, (unless, like newcomers, they apply themselves to building new knowledge and skills profiles). In addition, established teachers may aspire to other positions or leave the service entirely. Thus newcomers are required for the sake of continuity, but their introduction brings the threat of displacement.

For CPT2, these dynamics are captured in the concept of LPP. Here participation in a social practice such as mathematics teaching in a particular setting, say, involves traversing a trajectory peripheral to the practice. Later we explore further this notion of peripherality; here, however we note that LPP gives rise to inherently troubled, conflicted and problematic social, conceptual and practical relations. Driving these conflicts are contradictions such as those of continuity-displacement, and dilemmas of transparency and sequestration.

3. COMPARING AND CONTRASTING CTP1 AND CPT2

In this section we would like to analyse the relationships between CPT1 and CPT2 and in the process come to some view about the contrasting positions they occupy around key questions in mathematics education research such as the nature and place of specifically mathematical and curriculum knowledge, the way these kinds of knowledge are enacted and transformed, and the processes involved. In our analysis we have chosen to examine the role structural suppositions play in each of CPT1 and CPT2, the alternative view they have around the dynamics of learning, and a final comment about to what extent communities of practice saturate social life.

3.1 Community of practice: concepts of learning, identity and boundary

In looking at CPT1 and CPT2 perhaps the most striking initial difference is that they appear to draw on different kinds of empirical material. In the first, these are contemporary Western, first-world type scenarios; in the second, we are asked to consider alternative narratives relating to apprenticeship in traditional practices within indigenous cultures, notions of apprenticeship in both traditional first world settings and in alternative settings. Looking at these scenarios as a set, we note that the first appear intended to open up the issues of how work is done, whereas the second are more holistic; they explicitly refer to multiple transitions describing the trajectory of a beginner to the ultimate destiny of a master practitioner. In each case, the analyses offered identify what they regard as salient processes, and introduce concepts associated with these. In the sense of ethnographic tradition from which they grow, these conceptualisations are intended as either provisional or more or less solid theoretical categories around which their data can be articulated, and the more general phenomena they are concerned with can be framed. However, here a difference arises. In the first, the stated purpose is to generate a social theory of learning around the poles of social structure/situated experience and practice/identity, whereas in the second the goal was to elaborate a theory of situated learning around 'legitimate peripheral participation'.

Now a useful way to further analyse these alternative approaches to learning is to distinguish the extent to which each are influenced by structural ideas. By definition the key criterion for a structural approach for us will be the use of categories (tacit or explicit) belonging to a person or people around which personal and shared meanings are claimed to be formed. Though we feel this idea may be of help in tracing the critical difference between CPT1 and CPT2 we need to state explicitly, however, that we do not regard either CPT1 and CPT2 as purely 'structural' analyses (see Wenger, 1998, pp. 281-282). Nevertheless, we *do* regard CPT1 as more nearly structuralist than CPT2. And this claim is made at least plausible by an examination of the (published) empirical base relating to each and the general plan of analyses of each, as above. In CPT1, the data heavily emphasise the development of personal and shared meanings and the theoretical framework is constituted by systems of dualities dissecting social space. In contrast CPT2, focuses principally and especially on developmental processes (apprenticeship) across dissimilar settings and the concept of LPP, itself somewhat provisionally defined (as we point out below), is offered as an elaboration of the trajectory of participants rather than as a (grand) theorisation of learning.

In the space available we will follow selected aspects of these different theoretical moves; alternative concepts of 'community of practice', 'learning' and 'identity' as they are encountered in each of CPT1 and CPT2 are discussed.

The concept of 'community of practice': Within these projects the concept of 'community of practice' evolves in quite different ways, playing quite different roles. In the case of the first, it is introduced as the primary overarching concept around which learning is taken up, segmented and repositioned. For Wenger,

> Over time, ...collective learning results in practices that reflect both the pursuit of our enterprises and the attendant social relations. These practices are thus the property of a kind of community created over time by the sustained pursuit of a shared enterprise. It makes sense, then, to call these kinds of communities of practice (Wenger, 1998, p. 45).

Thus the community of practice concept is a way used by Wenger to embrace the social, individual, structural and situated conditions of learning.

In the second case, membership of 'community of practice' entails 'participation at multiple levels' (Lave and Wenger, 1991). Indeed,

> Nor does the term community imply necessarily co-presence, a well-defined, identifiable group, or socially visible boundaries. It does imply participation in an activity system about which participants share understandings concerning what they are doing and what that means in their lives and for their communities (ibid., p. 98).

So for Lave and Wenger, community of practice names objective activity driven in the direction of particular and palpable human interests within which the analyses of a theory of LPP are developed as a particular modality of learning. Wedege (1999) illustrates the power of such an approach in her talking with a 75 year-old women around the course of her life, and her knowing and not-knowing of mathematics in different situations and contexts.

The concept of learning: In CPT1 the concepts of participation and reification offer a complementary structural account of learning. Participation is intended to "describe the social experience of living in the world in terms of membership in social communities and active involvement in social enterprises" (Wenger, 1998, p. 55), and reification, as indicated previously, refers to "giving form to our experience by producing objects that congeal this experience into 'thingness'" (p. 58). For Wenger, the processes form a duality in which a "fundamental aspect of the constitution of communities of practice, of their evolution over time, of the relations among practices, of the identities of participants, and of the broader organisations in which communities of practice exist" (p. 65) are expressed. However, positing of this duality privileges, we think, the structural, rather than functional aspects of learning. Moreover, this view of learning construes it as a process that may or may not occur as a result of 'membership' of community of practice.

In contrast, however, CPT2 offers a view of learning which is a completely integral aspect of practice. Lave and Wenger write:

> There is a significant contrast between a theory of learning in which practice (in a narrow, replicative sense) is subsumed within processes of learning and one in which learning is taken to be an integral aspect of practice (in a historical, generative sense). In our view, learning is not merely situated in practice – as if it were some independently reifiable process that just happened to be located somewhere; learning is an integral part of generative social practice in the lived-in world (1991 pp. 34-35).

They go on to add that the notion of LPP is a "descriptor of engagement in social practice that entails learning as an integral constituent" (ibid., p. 35). Thus we see that in contrast to CPT1, in CPT2 the focus is not directly on learning, but on social processes in which learning takes place. Now this, in contrast to CPT1, is not a structural approach to learning; instead the approach is more ethnographic and anthropological in flavour, seeking to collect and make meaning around the artefacts of a community in order to tell and understand, we think, something like a narrative of learning events and processes.

Thus we argue that CPT1 is a gesture towards a theory of learning *per se*, whereas in CPT2 the purpose is more ethnographic, it is to trace the actual trajectories around learning within communities of practice and drawing on this to speculate on hypothetical trajectories in more general settings. Put simply, if the orientation of the first is taken as abstract/theoretical, then that of the second is concrete/empirical. (This is not to say features of both do not occur in the other, of course.)

The concept of identity: So far we have restricted our discussion to Wenger's social structure/situated experience axis in his analysis of the community of practice. We now turn our attention to his account of the practice/identity axis. In this, we pay particular attention to Wenger's theory of identity development. Here his approach is explicitly structural; it consists of the two interacting 'components of identity': identification and negotiability. For him the first refers to the process through which "modes (sic) of belonging become constitutive of our identities by creating bonds or distinctions in which we become invested" (Wenger, 1998, p. 191); the second refers to the "ability, facility, and legitimacy to contribute to, take responsibility for, and shape the meanings that matter within a social configuration" (ibid., p. 197).

His rationale for invoking the concept of identity is that it "serves as a pivot between the social and the individual, so that each can be talked about in terms of the other" (p. 145), and thus the process of identification, for him, is both a social and individual process binding individuals to the social, and the social to individuals. Negotiability concerns his concept of 'economies of meaning' and is "among individuals and among communities ... shaped by structural relations of ownership of meaning" (ibid., p. 197).

In CPT2 however, discussion of identity arises out of the acknowledgement that the "cognitive focus characteristic of most theories of learning ... only seem to focus on one person" (Lave and Wenger, 1991, p. 52). And these accounts are driven, Lave and Wenger note, by "reference to reified 'knowledge domains'" (ibid., p. 53), yet these as abstractions (and therefore removed from the social practice from which they arise) cannot serve as a basis for understanding learning as integral to rather than 'happening to be located somewhere within' social practice. Thus they say it is imperative to start with the domain of social practice. Consequently, if by learning one means to acquire knowledge, then learning must be equated with participating in a specific social practice and potentially "becoming a full participant, a member, a kind of person" (ibid., p. 54). Therefore, in Lave and Wenger's view, "learning and a sense of identity are inseparable: They are aspects of the same phenomenon" (ibid., p. 115).

Thus it would seem that primary distinction here between CPT1 and CPT2 is that in the first identity appears as a site for 'creating bonds and/or distinctions in which we (sic) become invested' (who are 'we' here?) and in which identity is formed passively and inductively; whereas in the second, it is seen as a process depicting activity of the 'decentered' person (i.e. the idea of identity as inductively derived, coming to regard oneself as a person of a particular kind) and the active person as a focus point for a 'rich centre of agency in terms of [the] 'whole'' person. In contrast to CPT1, in CPT2, these active and passive processes are regarded as implying one another. Lave and Wenger write:

> It is by the theoretical process of decentering in relational terms that one can construct robust notion of "whole person" which does justice to the multiple relations through which persons define themselves in practice (ibid., p. 53).

Lave and Wenger conclude: "This implies that changing membership in communities of practice, like participation, can be neither fully internalized nor fully externalized" (ibid., p. 54). This last is a significant point, because it goes to the issue of the status of boundary between communities of practice. What it is implicitly suggesting is that CPT2 do not have boundaries in the sense CPT1 may. We now look more closely at this important feature of CPT2.

The concept of boundary: CPT1 attempts a theory of the interrelationship among communities of practice. To this end Leigh Starr's concept of 'boundary object' is co-opted (being a reified object in different communities of practice) and brokered (being a form of participation in different communities of practice). Culpepper (2004) makes use of this kind of thinking in her study of how a deliberately formed community of beginning mathematics teachers supports the development of a teacherly 'pedagogic voice', often so much in conflict with the established mathematics teaching practices in placements. This uses CPT1 to tackle a real and significant problem in mathematics education – in doing so emphasis is shared among social and individual variables (such as personal disposition, for instance) and this very duality, perhaps, lends analytical significance to the notion of boundary objects and brokers.

In contrast, in CPT2 the concept of boundary is treated as problematic. For as Lave and Wenger explicitly say, "Nor does the term community imply necessarily co-presence, a well-defined, identifiable group, *or socially visible boundaries*." (italics added, 1991, p. 103). They explain that in their view the key notion for LPP is not boundaries, but shared 'understandings' among participants "concerning what they are doing and what that means in their lives and for their communities" (ibid., p. 103). Thus, in CPT2 what is salient are the activities that have life-meaning and community-meaning. Meaning here cannot be taken as mere significance, as in the 'meaning' of a word, but significance in the larger sense of 'implications for that which we love and care for in our hearts and minds'. In this strong sense of 'meaning' a community of practice is experienced as unbounded, and this creates the sense in which it has no boundaries. In some measure, the point here again arises from a methodological difference – whereas for CPT1 the theory arises as a process of abstraction, in CPT2 LPP arises as a tool or 'descriptor' to access and depict the concrete relations of the people at work in specific communities and social practices.

3.2 Problems arising from these comparisons

Arising from our analysis above and our broader work, we would now like to set out what we see as some of the key tasks needing attention. We propose to look at each of CPT1 and CPT2 in turn, and then consider matters that both approaches to community of practice may need to address.

3.2.1 Developing CPT1

Firstly, and this is a general comment, the published empirical grounding for key concepts and their relationships would appear to need a good deal more empirical and theoretical analysis. Consider, for instance, the notion of 'community of practice'. As noted above, Wenger explains this seminal notion in quite approximate terms. For instance 'collective', 'enterprise', 'property', 'community' – each of these rich words carry connotations which appear to anticipate each other, but it is not said whether the concept is to apply to all or only part of our lives (for instance is standing in a queue a community of practice, or sitting for 2 hours in a surgery?), and whom the 'our' refers to is left open. Should CPT1 be understood as a specifically Western phenomenon, or rather as an attempt to describe human behaviours beyond cultural boundaries?

Likewise for identity Wenger appears explicitly to refer to the 'individual'. But what is an 'individual'? Is he referring to any of: person, personality, consciousness, name, subjectivity, body, ego, id, 'I', etc? Whereas each of these has their own connotations and theoretical frameworks, Wenger's examples (drawn from Vignettes 1 and 2, in his 1998 book) offer individuals as third person representations within a narrative genre. What 'authorial' investments are made in this narrative? Such an analysis would seem crucial to us in order to best come to conclusions about the learning processes observed.

Similarly, many other key terms, for instance identity, meaning, negotiation, participation, and so on, are neither explained empirically, nor have their definitions motivated empirically and theoretically.

Secondly, it is not clear to us, either in the data or from the informing theory, how detailed theoretical structures and their interrelationship have been obtained. For instance, on what empirical evidence is his theory of learning as engaging the dual process of participation and reification based?

Why (apart from definition problems) are these the only salient variables at this point of the CPT1 approach? Why and how do these processes interrelate in practice? And importantly, to what extent do they re-run the ground of other theorists; say Vygotsky's theory of the zone of proximal development? (For Vygotsky the learner's trajectory is shaped both by co-participation with a more competent interlocutor and the process of conceptual development.)

Thirdly, in CPT1 relatively scant coherent attention is given to the development of communities of practice. This is so despite the view adopted around brokers, and the notions of discontinuity and generational change and despite work such as Wenger, McDermott and Snyder's (2002) that addresses developmental processes. In this work, 'seven principles' conducive of cultivating communities of practice are set out. In summary, these are: design for evolution; open dialogues among public and private spaces; inclusion of different levels of participation; a focus on value; combining familiarity and excitement; creating a rhythm for the community (Holm, 2003, p. 4). In these instances, however, the work is heuristic, and formulaic, rather than coherent with theoretical notions made use of elsewhere in the approach. We consider what may be at issue here may be the adequacy of the theoretical approach itself, which is, as we have previously noted, essentially structural. This means that analyses are biased towards the maintenance of a reproduction of communities of practice, not their production and recontextualisation. This appears to us a serious issue, as the field of mathematics education is under great pressure for change, and it is undergoing great changes of technology, assessment and curriculum reform, new kinds of institutional demarcations and interpenetrations and interfaces (mathematics school work driven digitally into the home, and conversely, for instance), new participants (both in the roles of student and teacher), and so forth. All of these are, or can create new communities of practice and transform and develop existing ones. For instance, Holm (2003), working in Switzerland, works with the community of practice notion, in the sense of CPT1, around the development of eLearning methods in teaching mathematics in universities. He finds that the features of a community of practice conducive to addressing the learning and pedagogic problems encountered in his research site are: bringing peers from different institutions together at a given education level; sharing of knowledge; meaningful interactions; collaboration and cooperation among participants; mutual trust and respect; and sharing in knowledge and expertise around use of the internet. Whilst we agree these elements may be encouraging, for us the question is whether it is necessarily the case that encouraging the formation and development of a community of practice necessarily leads to such indicators of educational wellbeing.

Fourthly, the extent to which CPT1 theorises change appears to be limited to endorsing the view that participants amend their perspectives in rational and apparently useful ways. Such a liberal view of change abstracts from the objective presence of power and interest, and optimistically concludes that these will not ultimately distort the outcomes in regressive ways. This view seems to us quite unrealistic. Below we will suggest that both CPT1 and CPT2 need to be made more critical in the sense of being able to recognise, comment on, and intervene in and around issues of power and interest.

3.2.2 Developing CPT2

We suggest that CPT2 requires clarification of the LPP concept. The notion of the peripherality of participation seems to capture the notion of motivation or intention around a driving purpose that is somehow permanently deferred, or delayed, or out of reach. We would suggest the following metaphor. Take hunger – one is always hungry; there is always the need for the next meal, and at the meal, for the next mouthful, until one stops. But it soon returns. In a sense one is always peripheral to eating, the best one can do is eat, take-up a trajectory – and whether one eats plainly or lavishly, whether one eats simply or is a gourmand, there is no 'central position' in the practice of eating – every act of eating gives rise to another for which the contents of the meal again must be decided upon, procured, prepared and consumed, and so on, and at each step more or less risks encountered and contingencies dealt with appropriately.

3.2.3 General remarks

CPT1 attempts a theory of the interaction of communities of practice whereas CPT2, as noted above, adopting a more ethnographic approach, does not posit such a structure. There would seem to us to be a need to understand more clearly the implications of these alternative views. One way to construe this problem can be visualised by a metaphor. In CPT1 the critical metaphor is one of enclosure defined by boundaries. These may change over time, but the notion of enclosure remains central. In contrast, CPT2 works with the idea of *lines*, taken as lines of *trajectory*. Thus we see in the very notion of the LPP that all participants, whether beginning or full, move on lines or trajectories on the periphery of social practice. In this sense, in CPT2, all legitimate participants are, to use CPT1 language, brokers, and all objects are boundary objects. In CPT2 all of these take up trajectories on the periphery of social practice. To use a mathematical metaphor, in CPT1, participants and reifications exist as *scalar* quantities, albeit, with dimensions and processes of interchange. In contrast, in CPT2, participants exist as vector quantities, having both dimensions *and* direction. Thus, Lave and Wenger write,

> Giddens (1979) argues for a view of decentring that avoids the pitfalls of "structural determination" by considering intentionality as an ongoing flow of reflective moments of monitoring in the context of engagement in a tacit practice. We argue further that this flow of reflective moments is organized around trajectories of participation. This implies that changing membership in communities of practice, like participation, can be neither fully internalized nor fully externalize (1991 p. 54).

Although we clearly see here Lave and Wenger's primary 'descriptor' LPP at work, we suggest, however, that this suggestive language needs to be more fully worked out. In particular, what does not emerge clearly is what impels change within a social practice. It is true LPP works out the elements of a theory of generational change, and, drawing on further anthropological studies, change as what it means to be a legitimate participant in a mathematics-related social practice (Lampert and Blunk, 1998). However, CPT2 does not talk about other sources of change, for instance, change as driven by contradictions within the social practice, and between components of a social practice at different stages of historical development, and between components of alternative social practices (Engeström, 1987).

This, among other ideas, leads us to think that a theory of mediation is needed in both CPT1 and CPT2. The nature and role of artefacts and tools is hazy in both. Indeed it is not clearly recognised that media both facilitate a given task as well as generate new ones. This recognition has been deeply influential in the sociology, psychology and philosophy in the twentieth century, and plays a key role in such diverse work as that of Vygotsky, Habermas, Foucault, Rorty and Lacan. In the works of these and other theorists, mediation is played out by language, speech, symbols, interests, power, physical and mental tools, desire, etc. Without a theory of mediation, CPT1 and CPT2 will find it difficult to contribute a theory of change processes that captures the transit among social, cultural, and individual forms of human subjectivity. Of the two, CPT1 may find this hardest, as this theory recognises change only from the objective point of view (participants doing X to Y) and thereby tends to render change merely as process outcome. In CPT2 change, like the participants and their objects, is construed subjectively (participants, X, Y moving in trajectories over time) and thereby CPT2 construes change as the integral state of the social practice.

As we have noted in CPT1, the notion of 'negotiated meaning' is critical. This for us merely underscores a problem both CPT1 and CPT2 share, and this is that both seem to assume that social interactions are not distorted by power and interest. True, Lave and Wenger specifically cite the process of displacement-continuity, whereby full participants purposely do not pass on the whole story to newcomers for fear of peremptorily losing their position of precedence. This is a form of power they exercise. However in neither CPT1 nor CPT2 is there the recognition that communications can be systematically distorted. Recent studies, for instance Fuller, Hodkinson, Hodkinson and Unwin (2005), whilst acknowledging the CPT2 as continuing to provide an "important source of theoretical insight and inspiration for research in learning at work", have strongly emphasised the limitation of its current application in "contemporary workplaces in advanced industrial societies and to the institutional environments in which people work" (p. 49). What they find is at stake is the lack of a theoretical purchase on power relations and how they mediate social spaces and practices. For Habermas, the task of critical social science, by implementing 'theorems of enlightenment' within the context of studies of the pragmatics of discourse, is to unmask distortions of communications and embark on transformative practices around new conditions of communicative performance. Another helpful frame in this regard is offered by Foucauldian methods of power/knowledge analysis. In this style of analysis empirical work takes place directly on discourses of communication in order to make visible the techniques of power operating through them. It would seem to us that notions of communities of practice might in various instances benefit from such analyses of power relations or, as Foucault calls them, 'regimes of power'.

In the next section of the chapter, we turn to a brief exploration of further alternatives of how the project outlined in the previous paragraph might be taken up.

4. DEVELOPING THE CONCEPTS OF COMMUNITY OF PRACTICE

As indicated in the previous section, one outstanding problem with both CPT1 and CPT2 is the under-theorisation of the social processes, in particular those around power and interest, they each presuppose. Issues of concern arise around engagement with tension, change, development, multiple forms of mediation, and the multiplicity of identities of the actors in the mathematics classroom. Thus, for instance, there is a need to identify and analyse the processes of regulation, the 'glue' that holds a community of practice together and affects different social groups differently, to look and understand how technologies of social practices work, and the means of communication by which purposes are shared and actions coordinated. In the following we propose two possible theoretical approaches to these tasks.

4.1 Sociology of structures

One approach we feel might be promising is that of the British sociologist of education Basil Bernstein. For Bernstein the dominant communicative principle in the classroom is the interactional which regulates "the selection, organisation, sequencing, criteria and pacing of communication (oral, written, visual) together with the position, posture and dress of communicants" (Bernstein, 1990, p. 34). The communicative principle offers recognition and realisation rules which need to be acquired by communicants in order to achieve 'competence'. Different pedagogic modes have different effects on social groups and Bernstein's theory explains how.

The framing of the pedagogic interactions can range from strong to weak. In the latter case the pedagogy is what Bernstein calls invisible, that is, means of gaining the approved discourse and being able to demonstrate the acquisition of that knowledge are hidden from the students. Middle-class children, however, have generally acquired these rules from their home life and are therefore not disadvantaged by the weak framing, whereas working class children have not and consequently find themselves in a position where they cannot demonstrate their knowledge. Research (e.g. Cooper and Dunne, 2000) shows that the setting of mathematics questions in everyday contexts is a form of invisible pedagogy in that pupils who have not acquired the appropriate way to read such questions may find themselves responding in everyday mode and not the 'esoteric' school mathematics mode that is required. As teachers we tend to assume that pupils have picked up the correct reading in informal ways, and we are rarely explicit about those recognition and realisation rules. Publication of this phenomenon is sometimes interpreted as calling for a return to traditional teaching, since here the framing is strong and the rules visible. We know, however, that such classrooms fail most students for a range of reasons. In particular, if children cannot meet the requirements of reading, coping with the pacing of school discourse, and so on, at the early stage of their entry into schooling they are likely to find themselves in an unending spiral of remedial situations, through which they are publicly identified and because of which they fall further and further behind (Bernstein, 2004; pp. 204-5).

Research shows that working within a progressive paradigm, that is, where the pedagogy is invisible, but mitigating the weak framing through strengthening some of the features of the pedagogy can make a substantial difference to the success of disadvantaged students (e.g. Fontinhas, Morais and Neves, 1992). These are the kinds of analyses and research strategies that are enabled by the richly descriptive framework of Bernstein (and other sociologists of education). Lave's perspective, from the early days when working on the anthropological project in Africa with Michael Cole, where she formulated her ideas, is rooted in Vygotskian cultural psychology. Bernstein's theories are a suitable fit because he too works with Marx's notion of consciousness being a product of social relations, and power, in the case of education symbolic power rather than economic power, being differentially distributed. However, Bernstein's goal is to describe practices and their regulatory mechanisms, providing indications that they can change, but not to identify how they might change and what tensions are at play, except at the macro level, that might bring about change.

4.2 Discursive practice

Another alternative, one that contrasts with Bernstein's structuralist account, is that of Walkerdine. Walkerdine (1997) suggests that what is missing in Lave's analysis of the subject in practices is subjectivity, the regulation of individuals within discursive practices. The account that Lave gives of the operation of practices focuses on those practices. Although Lave and Wenger present a more nuanced account in terms of LPP, in which a range of ways in which people might engage in an activity or in a community is suggested, there is no discussion about why and how individuals might engage in diverse ways. They might appear to choose freely how they participate but this is to hide the social regulation of discursive practices. Certainly in later publications by Lave (1996), Wenger and others there appears to be a goal for the learning that is characteristic of the practice and apprenticeship into that practice is to a large extent universal in its application to individuals. However, Walkerdine shows how the notion of 'child' is produced in the practices of educational psychology (1988; see also Burman, 1994), differentially positioning those who conform – white boisterous males, and those who do not – non-white people, girls, quiet boys and so on. Significations matter, they are not neutral meanings; situating meanings in practices must also take into account how those significations matter differently to different people. Practices should be seen, therefore, as discursive formations within which what counts as valid knowledge is produced and within which what constitutes successful participation is also produced. Non-conformity is consequently not just a feature of the way that an individual might react as a consequence of her or his goals in a practice or previous network of experiences. The practice itself produces the insiders and outsiders. Analysis of apprenticeship in particular workplace settings appears not to reveal differing subjectivities produced in the practice. Power, in particular, is seen as a property of old-timers, not of the discourse. A discourse analysis using discursive theory incorporates affect into knowledge and power, as Evans' (2000) study demonstrates, and reveals tensions between discourses expressed as subjectivities. In this way it also offers a language for the shifting identity of an individual as shifting of discourse through signification, which is always located in particular discourses and carries affect in the form of emotional charge. Evans offers as an example a student who is generally confident and indeed successful in solving mathematics tasks in that she is positioned in what he calls a school mathematics discourse; when the context of a mathematics question calls up a different positioning, even when using what appear to be the same mathematical terms, the student feels much less confident and is in fact less successful. The 'same' terms actually signify differently because of the social practice called up for the individual.

4.3 Final remarks

It is clear from the preceding analyses that CPT1 and CPT2 have both much in common, yet are different in salient ways and this suggests they have different, though related uses in research around learning, and learning of mathematics in particular.

Commonalities are a shared sympathy for focusing principally on the 'ties that bind' rather than characteristics of the individual. In this regard they stand together as an alternative to cognitive theories from early forms of constructivism in Piaget, the early Bruner, Dienes and onwards to versions of social constructivism put forward in the early 1990s by Cobb and others. Regarding the latter, for instance, we call to mind the early critiques of social constructivism that pointed out the all but absent account of the 'social' structures and collective concepts notable in these works.

However we have also pointed out differences. Among these is a more significant preoccupation with eliciting expositive theoretical structures of learning (CPT1) rather than eliciting more discursive narratives and trajectories of learning (CPT2). In terms of the literature drawing on these resources, these differences become quite marked. Those drawing on CPT1 tend to venture toward the formulaic applications of CPT1 in pursuit of practical solutions to curriculum problems (see, for instance, Holm, 2003; around mathematics e-Learning or Reeves, Herrington and Oliver, 2004; around instructional design for multimedia). Alternatively, and more rarely, they enhance and support the development of CPT1 by contributing substantially to its empirical grounding, whilst at the same time illuminating their own studies through CPT1 theory-referenced analyses. We cite Burton (2004), who studies mathematicians' 'own knowing of their discipline' via their work as researchers, as an example. In interviewing 70 research mathematicians, Burton constructs a rich database; analysis of this is rendered in a variety of ways, however the CPT1 notion finds an important position both in terms of description and analysis. In contrast, studies engaging CPT2 tend to make use of this resource in order to develop 'customised' theory, or crystallise problems or dilemmas of instruction – see Adler (1999) and Hemmi (2006) cited above.

Associated with this is a very different methodological stance. In the first, the work is concept-building exhibiting elements of both inductive and deductive reasoning, although, as we have noted, the published empirical work of CPT1 may need significantly further development. In the second, the work is primarily ethnographical in the sense of being the record of encounters with diverse, though thematically-given sites. Here we find the contrasts to be quite stark: indication of the empirical work supporting theory building in the first is scant, whereas in the second the choices possess a methodological rigour and anthropological focus. Thus, the first posits abstract categories (e.g. the dualism of participation/reification) and builds a theoretical account of learning out of these, whereas the second starts with the social practice and seeks discursively to elucidate the threads – trajectories – of learning found within. However, this is not to say that the first does not also have use for the notion of learning as a process, nor that the second has no use for abstract categories (like 'identity', for instance). We argue the salient differences are in how these alternatives are differently valued. Another important difference between CPT1 and CPT2, clearly also linked with the foregoing, is the different representations of learning they offer. In the first, as we have discussed, this is by the metaphor of boundary. To know is to belong *within* and know that one so belongs. However in the case of CPT2, learning is represented as a linear trajectory, one that transits the periphery of social practice. These differences suggest different epistemologies of practice. In the first, practice exists as an abstraction from the social world, whereas in the second, the practice is concrete and is the world in its socially organised form. Thus, in further explaining their approach, Lave and Wenger write

> The theorist is trying to recapture those relations in an analytic way that turns the apparently 'natural' categories and forms of social life into challenges to our understanding of how they are (historically and culturally) produced and reproduced. The goal, in Marx's memorable phrase, is to 'ascend (from both the particular and the abstract) to the concrete' (1991, p. 38).

Thus, the LPP as a 'descriptor' of learning activity is taken as an 'analytic way' or strategy to turn apparently 'natural' categories of learning (including for instance knowledge, concepts, meaning, understanding, negotiation, etc) into socially constructed and maintained (and guarded) tools. Whereas in CPT1, the direction of analysis is towards the abstract, in CPT2 it works the other way; it is to the concrete. In one, the direction it is towards representation, in the other, it is towards the concrete.

In particular, studies which wish or need to ensure that the social dimension of mathematics education is foregrounded or more clearly noticed find either CPT1 or CPT2 compatible alternatives. Studies which need or wish to be engaged in theory building around learning or aspects of pedagogy or teacher education and development or other curriculum work may find CPT1 of particular interest. Studies of this kind, for instance, trace relations in student learning among the variables of participation and reification or delineate the processes of identity development arising by means of identification and negotiability, for instance. In addition, structural studies of boundaries and boundary relations among alternative but co-existing communities of practice may have interesting applications in mathematics classroom-based analyses as well as in other salient sites (e.g. mathematics teacher staff-rooms). In contrast, studies that take an ethnographic line may be better supported by CPT2. In this alternative the research questions would be about how trajectories of learning are progressed within definite times and places. For instance, it may be a longitudinal study of a particular learner or learners as they engage with mathematics pedagogies in a school setting, or of a teacher or teachers transiting a term or terms of teaching with a class or classes and in the process transforming or reshaping the practice of their mathematics teaching in the environment of the social practice of their school. In such instances the descriptor of learning LPP would be used as a rubric or theoretical device or spur designed to fillip the ethnographic discourse and dispel any sense that one can develop as a mathematics teacher by simply getting 'inside' teaching mathematics – as if it were capable of such a reduction, or can develop as a researcher of mathematics education by inventing new kinds of representations for it. Roth and McGinn (1998) consider the implications of the methodological and conceptual approach of CPT2 in a study around research training in mathematics and science education. In this they adopt the inherently critical reflective stance illustrated in CPT2 as it might apply to research in mathematics education and apply it to *education* in research in mathematics education. Here Roth and McGinn see the beginning researcher traversing a trajectory peripheral to the practice of mathematics education research, and they see the primary contradiction facing this pedagogy as being that,

> it must enculturate students in ways of using both tested instruments for constructing reality (problematics, concepts, techniques, methods) and at

> the same time, a formidable critical disposition to question ruthlessly those instruments (p. 229).

It is this 'critical disposition' which we see as an important legacy of Lave and Wenger's work in CPT2.

ACKNOWLEDGEMENTS

We wish to express gratitude to the reviewers and editors and acknowledge the helpful additions suggested by Dr Jeff Evans in the section on discourse above.

REFERENCES

Adler, J. (1999). He dilemma of transparency: Seeing and seeing through talk in the mathematics classroom. *Journal for Research in Mathematics Education, 30*(1), 47-64.

Bernstein, B. (1990). *Class, codes and control* (Vol. IV: The Structuring of Pedagogic Discourse). London: Routledge.

Bernstein, B. (2004). Social class and pedagogic practice. In S. J. Ball (Ed.), *The RoutledgeFalmer reader in sociology of education* (pp. 196-217). London: RoutledgeFalmer.

Boaler, J. (1997). *Experiencing school mathematics: Teaching styles, sex and setting.* Buckingham, UK: Open University.

Burman, E. (1994). *Deconstructing developmental psychology*. London: Routledge.

Burton, L. (2004). *Mathematicians as enquirers*. Dordrecht: Kluwer Academic Publishers.

Cooper, B., & Dunne, M. (2000). *Assessing children's mathematical knowledge.* Buckingham, UK: Open University Press.

Culpepper, S. (2004). *The role of communities of practice in supporting first-year teachers learning to teach mathematics in urban schools.* Unpublished dissertation submitted to the University of Pennsylvania.

Engeström, Y. (1987). *Learning by expanding*. Helsinki: Orienta-Konsultit Oy.

Evans, J. T. (2000). *Mathematical thinking and emotions: A study of adults' numerate practices*. London: Falmer.

Fontinhas, F., Morais, A., & Neves, I. (1992). Students' coding orientation and school socializing context in their relation with students' scientific achievement. *Journal of Research in Science Teaching, 32*(5), 445-462.

Fuller, A., Hodkinson, H., Hodkinson, P., & Unwin, L. (2005). Learning as peripheral participation in communities of practice: A reassessment of key concepts in workplace learning. *British Educational Journal, 31*(1), 44-68.

Giddens, A. (1979). *Central problems in social theory: Action, structure and contradiction in social analysis*. New York: SUNY Press.

Graven, M. (2003). Investigating mathematics teacher learning within an in-service community of practice: The centrality of confidence. *Educational Studies in Mathematics, 57*(2), 177-211.

Hemmi, K. (2006). *Approaching proof in a community of mathematical practice.* Unpublished PhD thesis, Department of Mathematics, Faculty of Science, University of Stockholm (http://www.diva-portal.org/su/theses/abstract.xsql?dbid=1217).

Holm, C. (2003). eLearning as a catalyst for a new teaching and learning culture? *SCIL - Congress: Shaping Innovations*. University of St Gallen.

Lampert, M., & Blunk, L. (Eds.). (1998). *Talking mathematics in schools: Studies of teaching and learning.* New York: Cambridge University Press.

Lave, J. (1996). Teaching, as learning, in practice. *Mind, Culture & Activity, 3*(3), 149-164.

Lave, J. (1997). The culture of acquisition and the practice of understanding. In D. Kirshner & J. A. Whitson (Eds.), *Situated cognition: Social, semiotic and psychological perspectives* (pp. 17-35). Mahwah, NJ: Laurence Erlbaum.

Lave, J., & Wenger, E. (1991). *Situated learning: Learning as situated peripheral participation.* Cambridge: Cambridge University Press.

Lerman, S. (1998). Learning as social practice: An appreciative critique. In A. Watson (Ed.), *Situated cognition and the learning of mathematics* (pp. 33-42). Oxford: Centre for Mathematics Education Research, University of Oxford.

Lerman, S. (2001). A review of research perspectives on mathematics teacher education. In F. L. Lin & T. Cooney (Eds.), *Making sense of mathematics teacher education* (pp. 33-52). Dordrecht: Kluwer.

Pallas, A. (2001). Preparing education doctoral students for epistemological diversity. *Educational Researcher, 30*(5), 6-11.

Reeves, T. C., Herrington, J., & Oliver, R. (2004). A development research agenda for online collaborative learning. *Educational Technology Research and Development, 52*(4), 53-65.

Roth, W. M., & McGinn, M. (1998). Legitimate peripheral participation in the training of researchers in mathematics and science education. In B. Atweh, B. John, A. Malone & J. Northfield (Eds.), *Research and supervision in mathematics and science education* (pp. 215-230). Mahwah, NJ: Lawrence Erlbaum Associates.

Solomon, Y. (1998). Teaching mathematics: Ritual, principle and practice. *Journal of Philosophy of Education, 32*(3), 377-390.

Walkerdine, V. (1988). *Counting girls out* (2nd ed.). London: Falmer.

Walkerdine, V. (1997). Redefining the subject in situated cognition theory. In D. Kirshner & A. Whitson (Eds.), *Situated cognition: Social, semiotic and psychological perspectives* (pp. 57-70). Mahwah, NJ: Lawrence Erlbaum.

Wedege, T. (1999). To know or not to know – mathematics, that is a question of context. *Educational Studies in Mathematics, 39*(1-3), 205-227.

Wenger, E. (1998). *Communities of practice: Learning, meaning, and identity.* Cambridge: Cambridge University Press.

Wenger, E., McDermott, R., & Snyder, W. (2002). *Cultivating communities of practice.* Harvard: Harvard Business School Press.

Winbourne, P., & Watson, A. (1998). Learning mathematics in local communities of practice. In A. Olivier & K. Newstead (Eds.), *Proceedings of the 22nd annual meeting of the International Group for the Psychology of Mathematics Education* (Vol. 4, pp. 177-184). Stellenbosch, South Africa: University of Stellenbosch.

Chapter 15

'No Way Is Can't't': A Situated Account Of One Woman's Uses And Experiences Of Mathematics

Sandra Wilson*, Peter Winbourne**, and Alison Tomlin***

Abstract: Three authors bring a range of different perspectives to bear on the experiences of one of them. The chapter explores the complexity of hearing voice and tackles the methodological difficulties of making and understanding personal accounts: of using mathematics in non-formal settings, of learning mathematics in formal educational settings, of planning for and understanding the learning of others. It presents multiple voices and a shared multi-voice analysis of Sandra's changing mathematical identity over time and place. We found that – no matter where we looked - none of us could properly account for Sandra's persistence with mathematics education, despite a history peppered with stories of the sort that might reasonably make people despair of mathematics education.

Key words: life history, voice, representation, adult education, mathematics, discourse, narrative methodology

1. INTRODUCTION

1.1 One by One

I am Sandra Wilson. As a child I had many problems with maths. I am dyslexic. The teachers only seemed to help the clever children: I did not get any help as a child. So it was very hard to work out anything to with sums.

---

* former astudent and co-researcher in literacy and numeracy courses for adults, **London South Bank University, ***former researcher in the Department of Education and Professional Studies, King's College London

As an adult I started going to two workshops to discover there was a way I could do the sums. One teacher was curious about how I did a certain sum and asked me some questions. I said I go to the bank and pay back the money that I borrowed. It was a very powerful thing to learn that I could do some of the sums my way rather than walk out of the classroom.

1.2 Placing

I am very scared of zeros. This is how I work my zeros: I make them into people, like the first digit in units 0-9 I think one person, next two digits husband and wife, next digits husband and wife with child, next digits husband and wife with child and baby. As it gets bigger my family grows.

1.3 Measurement

My daughter was moving her bedroom around. I said I have no tape to measure for the placing, I said have you got a ball of string, we will measure that way. She laughed. We took the string measurement of the wardrobe, and took it to the new place to string measure. She fell about laughing. The whole room was done that way. She does it all the time now. Another time we had to measure with a piece of string and iron cord, to get the right measurement.

Sandra

2. WRITING THIS CHAPTER

Sandra's account of her mathematics is the only bit of the chapter written by one person alone. The rest of it started with the three authors talking together, but at different times and places: Sandra and Alison, and Peter and Alison, most often in someone's kitchen. We taped or made notes of the conversations, then whichever of us was not there asked questions: about the meaning of particular words ("What's hermeneutics?") or stories ("You didn't enjoy that? I thought you did"), and about the overall picture.

Here, in an extract from the first tape, Alison reads aloud some notes she had already sent to Sandra with proposals for how to write the chapter. Italics show written text that Alison was reading aloud; plain text shows Alison's and Sandra's discussion, and the start of the re-writing process.

Alison: So, *either you read the draft of the article or I read it aloud and you tell me what you heard, so you put it in your own words. This is to check whether we're all getting the same meaning...*

Sandra: Yes.

Alison: *And when we do that, you may be changing the meaning. I check if I agree with what you're saying and maybe change the writing.* Do you see what I mean? Just to make sure that we're all getting there. *You may say that we should miss some out altogether.* And it might be because you think "well, that bit just doesn't make sense and you don't need it and it's boring" or it might because you think "well I don't want that saying in public". Because it looks different, doesn't it, when you see it in writing instead of said. It looks, things look different to me. *We keep going the same way until we are sure it's okay. We tape these conversations too and then I've put: I will want Peter to check that the explanations I've given to you are right.* Because he knows more about the theory. So if I'm trying to check something about this situated cognition business, because I'm unsure, I'm going to need Peter to check that…

Sandra: Yes.

Alison: *Meanwhile, Peter and I have conversations working through the ideas, including your ideas. And then we start all over again until we all think it's okay.*

Sandra: That's good.

Alison: It's going to last forever.

Sandra: Did you get that bit that I sent to you?

Alison: Yeah?

Sandra: I don't like the way it is.

This chapter is unusual, for academic writing, for starting with stories and putting the theory later. Before we get further, you should know four things. First, mathematics is central here because that is what the book is about, but it's no more central to Sandra's life than to most people's. (You may think it's central to everyone's, but that's a different story.) Secondly, we talked about mathematics in the context of theories of situated cognition. We travelled in this order: Peter, who likes the ideas of situated cognition and is an editor of this book; Alison, who had a basic understanding but little reading; and Sandra, who had never heard of it. This was Alison's explanation to Sandra: 'cognition' means thinking, learning or understanding, and saying it's 'situated' means you do maths in different ways in different settings.

Third, we have different contexts for our own ideas of mathematics and personal history. Peter and Alison have a ten-year history, starting when Alison was a student and he was her teacher. Alison and Sandra have a seven-year history, starting when Sandra was a student and Alison was her teacher. That relationship shifted when Sandra became involved in research projects. She was a key organiser of a conference for mathematics students, which she discussed at a conference on research and practice (Gray, Kattah, Lesley, Sandra, Tomlin, and Tracy, 1999); she has worked on a study of how people do calculations (Wilson and Tomlin, 1999); and she has spoken at a university seminar on how to involve adult basic education students in research. When Alison and Sandra talk, they move between what happened yesterday or in a class five years ago; between reminiscing about the class where they first met and discussing what Sandra is doing now to help her grandchildren with mathematics.

We have re-ordered our conversations for this chapter, and cut out a lot of what has little to do with mathematics. But in making those cuts, we have also cut a lot of the context – the situation – for the conversations behind the chapter: the cups of tea, and discussions about health, mothers, children, work, holidays, politics and bicycles.

Fourth, we have been clear from the beginning that we should all agree to and understand everything that was written.

We cannot give the whole of the picture. It's not just that we haven't enough space here; we (including Sandra) cannot see the whole picture ourselves. We start with some stories which illustrate Sandra's maths.

3. STARTING FROM ZERO

I am dyslexic ... I am very scared of zeros.

Sandra has a particular form of dyslexia: we say 'particular' because although we do not want here to discuss Sandra's medical history, it is important for this story because it crops up at every point of her education and work (she's a former office worker in a local authority), and in the ways she helps her grandchildren. She and Alison met in a basic mathematics class set up for Alison's research project, in a community centre in South London. The first week, Sandra was the only student. Alison remembers the moment when Sandra just stood up and left the room. (Why? What's wrong? Has she just gone for a cigarette, or gone home? Should I find her or will I make it, whatever it is, worse?)

They had been working on place value, and Alison had suggested using a calculator to repeatedly multiply or divide numbers by 10, to find out what happens. What happened was a lot of zeros, and that's why Sandra left the room. Sandra told Peter about her fear of zeros, so that later when he wrote her a question about it, he added "Sorry" beside it. Here are Sandra and Alison talking about doing 'take away' calculations when there is a zero in it, using father (hundreds), mother (tens) and child (units). Sandra said until she learned that way of thinking about it, "I would never have been getting sums right":

> Because I could not pick up what anybody was saying to me, I couldn't see it. But to get that after, I had to learn the way and somebody as I was growing up must have said to me make it a family, and it stuck with me.
>
> Because in school days we had a maths teacher, I remember that class as well as it was yesterday. They were very strict and that. If you could do your sums, it was ok. The teacher taught you. If you weren't sure teacher still taught you. But if you didn't know at all, you were sent to that corner and that was me every time in that corner, with a dunce's hat. [A dunce's hat is made of paper, with a D written on it, so that the child is deliberately humiliated.]

Sandra helps her grandson Lewis, who has difficulties with zero like her own:

> Every Friday I'd do some work with him – English and sums and that – and as I discovered with Lewis, he's "I can't do it, I can't do it". I said "Look, no way is can't, you have a good try". And I could see that he couldn't get it. I said "Nanny has another way, would you like to do it?" He says "yeah", so I showed him but with great difficulty, he could not grasp what I was telling him.

Sandra showed Alison her method with this calculation:

$$\begin{array}{r} 357 \\ \underline{-\ 142} \end{array}$$

Sandra: So the 2 is you (Lewis), and that's Mummy and Daddy. 2 away from the 7, and how much does that leave you? Are you going to tell me? and he was using his fingers. I said "Lewis, you're 5 and there's Mummy and Daddy. So the answer is 215".

But there would be no need for a mummy and daddy if there wasn't a zero. We tried 350 take away 142, with Alison checking, "You're not turning into a jelly at the prospect of zeros?"; Sandra had moved away from the table. Sandra uses the equal addition method of subtraction: "I've to go to the bank here, so I make that [the zero] into a 10. You're taking the 2 from 10".

'Going to the bank' is a variation of 'borrowing' in order to 'make it [the zero] into 10'. Taking 2 from 10 gives 8 in the units place. In the equal addition method, 1 is then 'paid back' to the 4 in the bottom line.

Sandra: I tell Lewis, "Right, you're 8, but you've got to pay back. 1 and 4 is 5 and 5 from 5 is zero", and then he's worrying about the zero. I said, "Remember, Lewis, just like the other sum, there's the child, the mum and the dad. The mum is zero."

Why should this work? Sandra and Lewis are doing the same calculation as they would without making the columns into a family. Does Lewis find it easier just because his grandmother is spending a lot of time with him? "Not at all. I know the child. It's just that he had the same problem as me. He could not do the sums if there's a zero". And when Alison was doubtful: "I'm sure, if you are not! You relate, it's related to each person. We all like Mummy and Daddy and child".

At other times Sandra judges that you can ignore the zeros in a calculation, and then put them in later. When Lewis had to add two numbers, both ending in zero, Sandra said he should pretend the zeros were not there and just add it at the end. An example is adding 450 and 320: if you add the numbers without the zeros, you get 77. Then you replace the zero, to get 770. He said "But Nanny it won't work out right!" She said, "Just do it!" Lewis did it, and it worked for him, though Sandra comments that she didn't know whether it was 'real' or not. But that time it didn't work for her. When Lewis added the zero at the end:

> Lo and behold, Sandra's thing came again, down the shutters; couldn't do it. But Lewis was coping with it this time; I don't know was it my support that I was giving to him that he was able to do it, I don't know, but he got the sum written down.

As we talked, Alison started writing down the sum, then realised what she was doing, and checked: "So, as we speak, zeros are getting to you?" They were, so we stopped writing the sums down. When Sandra was helping Lewis, she had to walk about:

Sandra: I'm walking about. I just didn't want to know, but I wanted to help him as well.

Alison: Did he know that you'd got up because you couldn't stand zeros, or did he think you were making a cup of tea or something?

Sandra: No, he knew I was walking about. I don't know what was going through his little brain. "Nanny", he says, "I've done it!" and I said "Yes, that's right. Now, write it again and this time put a zero in it." Which he did. If there's going to be a zero underneath, it just blocks. My mind is zooming in, and there are the two zeros on that page, it's all I can think of, just not to go near them …

Zeros block her; her mind zooms in to them; all she can think of is not to go near them. 'Cognition' isn't the right word for a nightmare. The 'situation' has layers, shadows and echoes, overlaid with determination. It includes the calculation itself – the column layout, the fact of having a zero; Lewis's anxiety; Sandra's relationship with her daughter (whom she helps in every way possible) as well as her grandson; there's an echo of her friend David, who first suggested 'going to the bank'; Lewis' belief he can't do it, and Sandra's rescue of him while losing herself. Running through everything is Sandra's own fear of zeros, and her memory, fifty-odd years old but still raw, of appalling viciousness in her own schooling. She changes a calculation to a story about a family, so that zero is 'related to each person'. When all else fails, she walks about, or leaves the room if she must. Peter heard (literally, on the tape recording) modest heroism and self-sacrifice in this story: Sandra is putting her grandson's understanding before her own safety; she values his progress in something she cannot bear to think about.

Sandra's difficulties with zeros do not come just from being in the classroom of a vicious teacher. Nevertheless, one thing that puzzles Peter and Alison is why Sandra has persisted with mathematics education despite a history peppered with stories of the sort that might reasonably make people despair of mathematics education.

4. THE BLOCK OF FLATS

Next we tell two stories, Alison's and Sandra's, about two mathematics lessons starting from a photograph of a block of flats.

4.1 Alison's version

I got the idea from a conference workshop (Haacke, 1999). The presenters showed a photograph of a high-rise block of flats, which they had used as a source for mathematical questions. The block, which I suppose was Dutch, looked much like those in British housing estates, and when Sandra and another student, Sue, asked me what I had learned at the conference, I showed them the photograph and suggested we try out writing questions about it. Although neither of them had tried such work before, they immediately asked these questions, which I wrote down:

> How many flats are there? What is the height of the block? What is the width of the block? What is the area of the block? Is it built on springs? How many windows? How many balconies? How many people live there? Family block or single people? Where is it? – city or out of town?

The following week the group (Sue, Sandra and two others) worked on answering the questions; Sue had done them at home. Most of the questions are clearly 'mathematical' in that the solutions would be expressed in numbers. One, *Is it built on springs?*, comes from civil engineering; we discussed why Sandra asked it, and learned that the high-rise block outside the classroom window was in fact sprung. The question *Where is it? – city or out of town?* also led to discussion; the students associated high-rise blocks with inner-city areas, but the block in the picture had trees around it. The other questions all revolve around measurement and size, and demand some knowledge of such buildings. For instance, Sandra and Sue argued that each flat would have one balcony, not two; and there was some discussion about whether the wider balconies served two flats.

The group also looked at Sue's answers. She had decided how many flats there were by counting the visible balconies, but the others said there would be more flats on the far, invisible side. Sue had measured the photograph to get the height (in centimetres); we initially thought she had made a mistake, but she said she wanted to practise using a ruler. Drawing on their knowledge of public housing policies, the others disagreed with Sue's view that the block would be for single people. As well as looking at the photograph, the group looked out of the classroom window at a similar block, and estimated, for example, the height of each storey, using closer objects (trees, people, the classroom) for comparison.

This wordless, numberless image led to creative, energetic mathematical work including calculation, measurement and estimation (of the image, people and classroom), ratio, and 3D spatial thinking. It built on students' knowledge of the real world while retaining ambiguity so that solutions could be challenged but not dismissed. The questions were not 'real world' problems, in that no-one needed the answers; they were written (by the students) solely for the mathematics class, but they became 'real' through the students' engagement with them; and the image itself was perhaps less abstract than the words and numbers of traditional word problems, both because it is an object that itself can be touched, measured, and so on, and because it became a representation of the real block outside the classroom.

4.2 Sandra's version

I remember when Alison first handed that picture round. Sue got the sums; she told Alison something about that building, something about the windows. I didn't know what it was all about. I remember you could see the front and one side of the block, but I didn't want to know. It was just a picture to me. There were no figures around the building that I could use, and I couldn't think of how to do it. I do remember it was a nice picture, and I remember Sue doing it. My friend lived in a block just like the one in the photograph, across the road from the education centre, so I know the building slightly moves with the winds and it's built on a spring.

In maths it's much better to work with figures, just plain figures. In other words, I'm no good on 'problems', with words.

4.3 What's the situation?

We hope you were convinced, as you started reading this section, by Alison's account of work on the block of flats. Alison believed it true when she included it in her PhD (Tomlin, 2001), and only found out when we were discussing this chapter that her version was wrong. She did remember Sandra saying she had had enough of Alison's 'political worksheets', as Sandra put it. Alison had thought that comment was about the more obviously 'political' work – for example, discussing statistics used in newspaper articles. But Sandra's example was the block of flats. As Alison said, discussing this chapter: "This is news to me – I've always thought that bit of work was blistering good and that you enjoyed it as well."

Who knows what is happening in a mathematics class? Alison thought that Sandra, like Sue and the other students, was contributing to writing questions, gaining some mathematics skills, using her knowledge of public housing and working on questions she contributed to writing. She wrote that the problems became 'real' through students' engagement with them. Jean Lave argues that situated learning theory supports mathematising everyday experience:

> [It] generates a different model of appropriate 'traffic' across the bridges between school and the other sides of life. It conceives the process of learning as one in which math culture is collectively generated in the classroom in such a way that it changes relations for school [students] between everyday experience and their mathematical practice. Perhaps [students] are the right bridge builders to construct relations (for themselves) between school math culture and the rest of their lives. (Lave, 1992, p. 87)

We think Jean Lave is right, but so is Sandra. Alison thought the students were working on Lave's 'collectively generated' questions, and making links between mathematics education and the rest of their lives. It is only now, as we write this chapter, that she has discovered she was wrong, or wrong at least as far as Sandra is concerned.

There are questions here of trust, good manners and power. Sandra trusted Alison enough to stay in the class despite finding some work unhelpful (or, in the case of the zeros, distressing); she was polite, and teachers are in charge. Lave may be right, but we puzzle over how to establish the meanings, in particular classrooms, of the terms she uses. How do teachers test the accuracy of a feeling for whether a class is acting collectively? How do we know what to take as the setting for the everyday experience of students, or their mathematics practice? How do we know whether we are constructing bridges or jumping through hoops?

Alison: Well I'm really sorry Sandra, because I didn't know. I'm sure I'm a sensitive teacher …
Sandra: Well you are.
Alison: Yeah, but I didn't realise at all.

5. PARTICIPATION AND PERSISTENCE

Sandra's stories represent, for Peter, evidence of her changing participation in culturally designed settings, where such changing participation IS, as Lave (1993) and McDermott (1993) suggest, what learning is. Sandra tells us about settings that include formal and non-formal educational settings, and domestic and other apparently non-educational settings. We have stories that provide glimpses of the most violent exclusion from and, Peter supposes, 'non-participation' in the formal educational settings of her youth. We have stories of Sandra's participation in the formal educational settings of her more recent years; these show a person who supports and teaches both her fellow students and some of her tutors (though more of her tutors could learn much from listening to her). Seeing mathematics in a particular setting or practice does not necessarily make that setting or practice mathematical; nor does it mean that this mathematics is central to the lives of those participating in the practice. What makes a setting or practice mathematical?

5.1 Peter's view

I think what makes a setting mathematical is related to ideas of community of practice: participating in mathematical activity is what makes a setting mathematical. I have written about this elsewhere in this book. I think it fits with the kind of distinctions Boaler (2000) makes, pointing out as she does that the practices of school mathematics include few that are essentially mathematical, and many where 'interpreting cues, seeking structured help and memorizing school procedures' are far more central. If this way of thinking about mathematical settings is accepted, then some of the formal educational settings in which Sandra has participated have been mathematical; a lot, including some called mathematics or numeracy, have not. I think that some of the more domestic, family settings, the ones where Sandra tells of her activity with her grandson, Lewis, for example, have been much more mathematical than all of the formal educational settings of her childhood (from which she was, in any case, excluded).

I will give an example of the way in which I have made judgements about the nature of the settings of Sandra's activity. In the table, the left-hand column is a transcript of some of a conversation between Sandra and Alison. The right-hand column is a record of what I have made of this conversation after listening to it and reading the transcript a number of times and, of course, talking about it with Sandra and with Alison. I have used italics to help to distinguish the commentary from the conversation; arrows represent continuity within the commentary.

Here Sandra and Alison were comparing mathematics classes in different adult education centres that Sandra had been to.	*Here is what I heard and thought as I listened to the tape of Sandra and Alison talking*
Alison: I don't know whether I did this, but I intended to get people talking to each other in the class quite a bit. The woman in Centre A, was she expecting you to do things individually? Kind of you and the worksheet.	Alison and Sandra were talking about a class that Sandra had been to at another centre some years before. It was a class of which Sandra had said, 'With that [teacher], if I had stayed there I wouldn't be doing anything.'
Sandra: She'd just hand you out the worksheet and you had to do it and not talk to each other; but we did. It was the only way we could get anything done.	*I wouldn't call this teacher's classroom a mathematical setting....*
Alison: Is talking to each other in the class important?	↓
Sandra: Yes. You can lean on each other or you just can't do it. You ask somebody else, they will tell you a way to do it.	*....it could be...*
Alison: I'm trying to get myself inside that teacher's head. It's hard, because I don't know why you would want to torture people … I suppose she might think ..if you're not doing it for yourself, by yourself you're not really learning..	
Sandra: We all had a crack at doing it ourselves. A page was never blank. The sums were there, but not really knowing how to do it, we just had a crack at doing it.	↓
Alison: So if you got the right answer, you wouldn't be sure you could do it the next time?	

Here Sandra and Alison were comparing mathematics classes in different adult education centres that Sandra had been to.	*Here is what I heard and thought as I listened to the tape of Sandra and Alison talking*
Sandra: No. This […] straight down the page and flinging it back at you; she liked doing that, I think. She knocked you down a bit.	
Alison: So, when students are helping each other..	*……but it does sound like a sad continuation of the practices Sandra had experienced as a child at school*
Sandra: That's very important that is.	
Alison: If you were helping somebody else… would you show them the answer, or tell them how to set about it?	*There is activity going on that sounds like the kind of thing I would want to see in a mathematical setting….*
Sandra: How to set about it; how to do it. Never the answer. They could understand how they were getting round to doing the answer.	
Alison: If you're showing someone else how to do it, do you try to show them by the teacher's way or by your way?	
Sandra: By my way.	
Alison: … And you trust other students? If somebody else is helping you, you trust them?	*and Sandra sounds like a teacher…*
Sandra: Yes. With a big capital T.	*….but I recall some earlier conversation between Sandra and Alison when they were talking about this same setting:* Alison: What did the students do? Did you all just struggle? Did you help each other out of the class or inside the class? Sandra: We helped each other inside the class. On our tea breaks we didn't want to go back in. There was terror raging. Alison: Terror? So lots of other students were fearful as well as you? Sandra: Yes, yes. I wasn't the only one. *and I think I see a mathematical practice in that classroom that was strangely subversive; it was powered by Sandra's extraordinary persistence and the persistence*

Here Sandra and Alison were comparing mathematics classes in different adult education centres that Sandra had been to.	*Here is what I heard and thought as I listened to the tape of Sandra and Alison talking*
	of her friends and drew on relationships established in all sorts of other settings. I doubt this teacher saw that persistence. To me it is quite amazing, *but could this teacher have seen this persistence as some kind of threat?*
Alison: When you were in Centre A, were you remembering school while you were in the lesson? Or are you just remembering it now because I'm asking you about it? Sandra: No. I did remember it. Alison: So, had all that business about the dunce hat come back to you? Sandra: Oh, yes, very much, yes. Alison: What does that do to you when you're trying to do sums? Sandra: You lose your confidence, your support is gone. Every Monday, or whenever it was, I'd go home and have a good cry to myself. Alison: Oh, I am sorry.	*This part of Sandra and Alison's conversation is striking ....* ↓ *[Sandra sounds tearful to me.]* ↓ *.....because it makes Sandra's persistence all the more amazing. Where does it come from?*
Alison: Everybody in Adult Education ought to know that if you go into a class it's because you've had a bad time in school. …I find it just extraordinary that you had that terrible experience at school, then I know you're saying the English was OK, but a terrible experience of maths at Centre A, and then, despite all of that, you keep trying; you go to the Orchard Centre and then you go to Bede. Many a normal, sensible person would have thought, 'I'm not going to maths again.' So, why do you keep going when you have such awful things done to you? Sandra: Because I want to improve my life as an adult. Like this counsellor inside me… "Sandra, you've had a lot of knock backs,	*Could it be anything to do with the more positive experiences she's had as an adult?* ↓

Here Sandra and Alison were comparing mathematics classes in different adult education centres that Sandra had been to.	*Here is what I heard and thought as I listened to the tape of Sandra and Alison talking*
your emotions have been turned round, your feelings are all over the place". Now I want to put the English in the right place … saying "I CAN DO." Alison: But, before you had the counsellor, you were doing that for yourself. You were taking yourself off to improve yourself, without the counsellor, without any help from anybody. Sandra: Yes. But that's what it was all about: I CAN DO Alison: But where has that come from? I mean you have had a life, where, not to put too fine a point on it, you ought to be a wreck by now, and you're not. That's what I find astonishing about you: you should have given up forty years ago, or more. You should have just thought "stuff it" and you haven't. And I don't understand how you keep going. You're amazing. I really do think you're amazing, you know. Sandra: Talking about amazing, I've gone [back] to Bede, and one of the teachers took me for an interview. I just thought she we was asking me questions, but, no, it was an English test you get, spelling test you get. …But I did what I thought was the best, that's for the computer class. And I want to go for creative writing to keep my brain ticking over. I want to apply for that one.	
	Maybe these more positive experiences don't account for that persistence? Maybe they have been positive because they have allowed Sandra's persistence to be visible, recognised, seen? …

6. LEARNING, FORGETTING AND MATHEMATICS

What does 'learn' mean? In everyday English (rather than the language of research) it usually suggests getting some new knowledge or skills that will last for some time, and perhaps forever. When you learn to drive, for example, or to sew, or speak a new language, you expect to keep that skill, unless you do so little of them that your skills fall into disuse. Alison studied mathematics to university level, did well in the examinations, but has never used higher-level mathematics since; it's not surprising that she now cannot recognise the work she did for the university course. Unlike Alison, Sandra has kept on studying, even when she had good reason to abandon it. Both in mathematics classes and in ordinary uses of mathematics in her own life, her confidence in her skills in mathematics comes and goes. How does that relate to the setting?

Sandra said that, as a teacher, Alison took the boredom out of the paper copy. One thing she proposed was students writing their own 'word problems' (Gerofsky, 1996); Alison's aim was that the students should get inside the problems, rather than feeling attacked by them (Tomlin, 2002, 2005), that students might build the kind of bridges to which Lave (1992) refers. Sandra wrote this as one of a set of problems for fractions practice:

> I have 2/3 bag of ready-mix cement left and I am given ½ bag by my neighbour. How much do I have altogether?

She first thought of cutting an apple into pieces, then changed it: "I was cutting the apple up in different pieces, and there was something on TV about cement, so I just put the cement in". Writing her own problems did not make them easy:

> Because of the dyslexia I knew the block was there, but I keep pushing myself so that I can not give in. "Sandra, because you've got that problem, you can't do these things". I just keep pushing against it. So one day, like tomorrow, I'll be able to do it, sort of thing.

She had to push herself, but her own problems were better than those in textbooks: "I see problems, I sweat, I panic".

As a teacher, Alison knew Sandra was good at working out money in her head. Sometimes as Sandra struggled over a calculation in the class, Alison would suggest looking away from the page and thinking of the numbers as money. Sandra could often get the answer, and then go back to the problem on paper and change the context back to the original.

Following a mini-stroke, Sandra's memory "has been cut up a bit", so that she is not always confident of being able to check change when she is shopping. It is still, however, easier than on paper:

> You have a list [of what things cost] in your head and you're turning it into a sum. Not on paper. In the paper, it doesn't connect. If you are physically doing it, you can connect to it. On paper, your brain cannot make a connection.

Sandra's daughter Mandy sometimes asks her mother to help her check her weekly budget; Mandy comments "You get it before I've all the numbers written down". Some of the figures are the same from week to week, and Sandra builds on her knowledge of the usual budget to help her mentally tot up the figures. Similarly, Sandra described having given Mandy money to buy something, working out what change to expect and getting it exactly right:

> I worked it out to keep my brain ticking over all the time – you know your memory's going and you want to hang on to it.

But doing mental arithmetic relating to budgets and shopping is often much easier than handling money in a shop. Since the stroke, she is not always confident of handing over the right money, so for a bill of, say, £2.50 she may offer a £10 or £20 note rather than struggle over finding the right change.

Sandra calls this "having the confidence". When David was ill, she successfully paid his rent, and was "*really* pleased" with herself. But sometimes she cannot keep pushing against the difficulties:

> It's not to do with energy. It's just I know that I absolutely can't do it. Other times, I wake up and I say I know. I'll be very very pleased with myself. I've learned to do a sum while I was out *and* I've got the right change. But other times just nothing happens.

"I've learned to do a sum while I was out": Sandra observes herself learning; she (sometimes) works on a calculation even when it is unnecessary, in order to hang on to her memory, for mathematics as well perhaps as for other things.

These stories raise questions about the role of teachers: it's easy enough to recognise that students may not criticise teachers' work (for example, the block of flats), but harder to think that teachers may be more or less irrelevant. At times when Sandra knows she cannot work on mathematics, it is, she says, nothing to do with the teacher. It is her own self-confidence: "I could have some good days and some bad times".

For teachers, this is humbling. "It's nothing to do with you": teachers cannot change Sandra's confidence with mathematics. This changes our perception of the context of Alison's and Peter's reaction to Sandra's persistence with mathematics education despite all the difficulties she has found there. Sandra's early education has "marked her for life"; now she has to "find a technique to work it in"; it's Sandra doing that work, not her teachers. Teachers may make less difference to people's learning than we like to think.

Teachers can sometimes make space for students to settle into working even when the world is against them, and they can also make it possible for students to support each other:

> You may come and feel you can't do it, but the teachers support you when it comes against you; they help you to do it, or see it differently. We would explain it together, and then you say "Oh yes, I know", and then you would be able to do it. … I would ask the other students, and I hope they would ask me as well.

So students can build up trust in their relationships with some teachers. Sandra would give a new teacher four or five sessions, then if the class wasn't working, she would back out; she and other students recognise, too, if teachers themselves are preoccupied or ill.

What do teaching and learning mean here? Sandra wants to work things out for herself, and understands that's what other students want too. We have seen that students would never tell each other answers, but rather show each other methods. Faced with unsympathetic, arrogant or violent teachers, students support each other both inside and outside the classroom; and when they feel unable to study effectively, they value teachers who help them see problems differently. Sandra treats her teachers much in the way she wants them to treat her: she recognises when they have days when they cannot work effectively. Outside the classroom she talks about "learning" when she goes shopping, yet there are days when she "absolutely can't" do calculations which at other times would be relatively straightforward.

This is not tied to particular situations in some tidy way. A situation is not made up only of its outward signs: a shop, a worksheet, calculations, a TV programme, particular numbers, teachers' actions. It includes all those, but each situation also has its own history. We have mentioned many elements that may go to make up a mathematical setting. We have seen Sandra creating a mathematical setting against the odds, subverting a teacher's non-mathematical classroom. We have also seen her create a mathematical situation where there was no need, when she worked out change while Mandy was at the shop. The former was in collaboration with her fellow students, the latter was solitary. Her creation of opportunities for mathematics is to do with Sandra's sense of self, and we come back to that later.

There are other sorts of situations too, within the world of writing and reading mathematics for the sake of learning it: the presence or absence of zeros; the patter for an algorithm; concrete mix or an apple to 'carry' a fractions problem. The importance of zeros comes from Sandra's dyslexia; the patter for an algorithm is learned from a friend and passed on to a child; a word problem that starts its life in the apple of children's textbooks shifts to the concrete mix of a TV programme. There is, too, all the mathematics Sandra does in the public world (including shopping), and it is this that has been most knocked back by recent illness.

These situations are not pocketed off from each other. We cannot understand the concrete mix question without a sense of Sandra's active collaboration in mathematics classes. Her determination to help the friend who first suggested 'going to the bank' overcame her fear of managing financial transactions in the public world.

We come back to the question of what 'learning' means. Sandra said "I've learned to do a sum while I was out": perhaps most of us would say we had "managed". The learning isn't mathematics itself; it's the confidence to make use of mathematics in the outside world, and that confidence comes and goes, tied in to the whole of Sandra's life, in ways that are beyond what we (including Sandra) can properly describe.

We focus here for a moment on knowledge and thinking, rather than learning. Sandra is conscious of her secure mathematical knowledge and ability (working out Mandy's accounts, for example), but that may be lost in particular settings: "I see problems, I sweat, I panic" is a description of the impossibility of thinking. We have seen that she both relies on and helps others to get to solutions to mathematical problems, whether in the classroom or outside. She also talks to herself: "Sandra, you've had a lot of knock backs", "Sandra, because you've got that problem …" Without any rejection of teachers, she has become her own, pushing herself to get beyond her difficulties and recover from a history which has included shocking abuse from the world of education.

7. THE STORIES WE TELL AND DON'T TELL

We have been telling stories that we think have something to do with the way that Sandra has experienced mathematics and help us to look more closely at the settings of some of that experience. We have tried to be careful to point out how mathematics has been only a small part of Sandra's life, but we can't help but make it seem to have been more important than that. In the same way, none of us has any reason to suppose that Sandra is so very different from other people, but our focus on her and the settings of her activity may make it look as if we do.

For Lave (1996) and for us, the adoption of a perspective of situated cognition represents a choice informed by political and social values. The decision to see people as learning within a social and political context springs not only from a recognition of its explanatory power, but also from a desire to do so: it feels to us like the right thing to do. It springs from a need to address questions like:

- why is it that some people appear disposed to go along with what formal learning has to offer whilst others appear not to be?
- why is it that ideas of 'ability' have such a strangle hold on education?
- what is it like to live a life labelled as 'less able' or having 'special needs' as an adult learner of mathematics?
- how does the whole business of mathematical learning and activity, or more recently, numeracy, figure in the life of someone who has been labelled as struggling with this throughout her life?

These questions have provided the backdrop as we have worked together on this chapter. Along the way we have had to make decisions about what we identify as settings of experience of mathematics. We think that Sandra helping Lewis at home is such a setting but here there are clear links to the formal mathematical practices of school, for example, Sandra's description of the way she uses a vertical layout algorithm, managing to cope with her fear of zeros by missing them out. We have written briefly about settings in which Sandra has helped her daughter to cope with the mathematical demands of her life and about settings in which Sandra has supported others as they have worked together in adult numeracy classes, but we have said little about settings of Sandra's more general experience of mathematics, for example, her competence in saving money on a small income in order to give sizeable presents; her experience as a council worker in a highly responsible office job, that required, for example organising her time, responding to councillors' requests and using IT.

We have said little, too, about Sandra's experience of what is it like to live a life labelled as 'less able' or having 'special needs'. There is a story that could start like this:

Alison: you have been labelled in the past as handicapped, disabled, dunce and so on. So, Peter and I are interested in what it's like to live your life with labels like that.

Sandra: Oh, I could do pages on that!

Well, we don't have pages that look at this. We know that we could never tell the whole story and we believe that the very idea that we might tell the whole story, provide a complete description of any setting, is, to say the least, unrealistic.

Theories of situated cognition, in our understanding, support us as we work to make sense of Sandra's experiences of mathematics. We set out to work together to identify mathematical practices within which we could 'see' the constitution of mathematical identity. We have looked together at locations for these practices that include both adult numeracy classes and, more broadly, aspects of the life of one of us outside of formal educational arenas and settings. The mathematical identities that emerge suggest that we cannot explain Sandra's participation in formal mathematical settings unless we see her as responsible adult, parent, grandparent, friend, teacher and worker. But a broad thread in all of these identities is actually something that is not there; there is no bitterness or defeat, there is resolution, determination, persistence.

If we take this persistence as a 'given', we think we can explain some of what we see of Sandra's participation in the mathematical practices we have identified. But this doesn't answer what has become a central question for us: where does this persistence come from? Can we say that Sandra learned it? How was it learned? Where should we be looking for an appropriate context for that learning? Where are the practices within which we can locate this learning?

We don't doubt that the ways Sandra uses and learns mathematics are grown from and anchored in particular settings and histories. Theories of situated cognition are useful and will help us understand both what's happened in the past, and how Sandra sees herself in relation to mathematics now. But the situations in which she uses and studies mathematics, both for herself and in activities with others, and the situations in memory, or in other contexts, evoked through anything that is mathematics-related, are more complex than teachers or researchers can understand. There's no reason why that shouldn't be true for other people, and perhaps everyone. We can understand the notion of situatedness, and we can seek some understanding of the setting of students' learning, in order to help us understand students' mathematics and work out how to support further mathematics learning. What we (teachers and researchers) mustn't do is imply 'I know you better than you know yourself'.

Teachers' activity is situated; they can hope to plan in order to meet up with students' experiences and interests; from the perspective of situated cognition, they could conceptualise this in terms of setting up practices within which their students feel comfortable and legitimate participants and where it's reasonable to look forward to positive learning trajectories. The risk is that all this may turn out to be a form of patronage. Teachers should know that students are trying to collaborate in setting up situations in which the teacher's interest is addressed within what we might call a 'practice of politeness' (cf. Harris, 2001; Mullany, 2004), but only once some sort of ground level of understanding has been established. Teachers can make it possible for mathematical practices and relationships to be set up in their classrooms; they can make this impossible; having set up something that is possible, they can't know what is happening. The kind of 'practice of politeness' that we think we are talking about may resemble, superficially, what Lave (1997) calls 'faulty practice'. In her case, children are developing practices which keep them safe; in this case, adult students may be keeping the teacher safe.

8. CONCLUSION

So, are we close to the limits of what can be expected from theories of situated cognition, at least when it comes to trying to explain the origins of Sandra's sense of self? Maybe all we can say at this point is that the setting for what we might want to call Sandra's learned persistence is actually her life (or if we want to maintain some connection to aspects of our earlier discussion, the setting could be her life, seen as hugely complex collection of identities in unknowable arrays of practices). Perhaps a simpler way of saying this is that whilst educators have to know the learning is situated, they cannot hope to know all the situation.

We set out to paint rich pictures of the mathematical experiences of an adult student of numeracy, to situate these within practices within adult numeracy classes and to probe parallel mathematical practices that may be situated outside of formal classes. We wanted to address issues about the extent to which Sandra had been 'acquired' by a discourse of adult numeracy which is framed in terms of standards and targets in ways that, at best, are blind to, and at worst deny the identity of the student as responsible adult, parent, grandparent, friend and worker. In our activity we have been aware how the subjects of even the best-intentioned academic writing most often appear as ciphers that float underneath theory-laden words, how struggles to show the whole person too often succeed only in hiding them.

Our response to descriptions of students as handicapped, having a learning disability or having special needs has been, like Mehan (1993) and McDermott (1993), to see these as created by discursive and organisational practices. Our writing is informed by real personal experience of such descriptions being applied to ourselves, our friends, our students, our families, and we want it to be read as such.

From our perspective research activity is a creative one. As the researchers, we decide upon and define what will be taken to be the evidence (or text) on the basis of which we construct our narrative (Clough 2002). The research activity in which we are engaged fits with what Van Manen (1990) calls the:

> phenomenological and hermeneutical study of human existence: phenomenological because it is the descriptive study of lived experience (phenomena) in the attempt to enrich lived experience by mining its

> meaning; hermeneutics because it is the interpretive study of the expressions and objectifications (texts) of lived experience in the attempt to determine the meaning embodied in them. (p. 38)

We believe that the shift towards a discourse in which research is acknowledged to be a hermeneutic process of textual interpretation encourages the reconceptualisation of 'learners' and learning (Brown, 1997). Within this discourse students are rounded people the richness of whose lived experience is the central concern of the researcher and, to begin to understand that experience, it helps at least to try to escape some of the confines of classrooms as the site of learning.

For these reasons this chapter has acknowledged and combined the stories of each of us, with each in turn making use of the stories of the other two authors further to develop understandings of personal/mathematical experiences and the social contexts that have produced those experiences and constituted our mathematical and related identities. The final text is the result of negotiations designed to implement the hermeneutic process through questioning each other, responding to those questions and further questioning the narrative text that emerges from this process. It has been of central importance to us that all must not only sanction but also understand what is written.

There have been times in Sandra's life when she has had to be, as Peter put it, heroic. When she takes on life, she takes on the whole of it, including education and mathematics:

> It's a battlefield and I won't let go. If I lost my fighting will, we would be looking at a complete dunce. I know that's true. I must never leave grasp of that power.

REFERENCES

Boaler, J. (2000). Exploring situated insights into research and learning. *Journal for Research in Mathematics Education, 39*(1), 113-119.

Brown, A. (1997). *Mathematics Education and language: Interpreting hermeneutics and post-structuralism*. Dordrecht: Kluwer.

Clough, P. (2002). *Narratives and fictions in educational research*. Buckingham: Open University Press.

Gerofsky, S. (1996). A linguistic and narrative view of word problems in mathematics education. *For the Learning of Mathematics, 16*(2), 36-45.

Gray, J., Kattah, V., Lesley, Sandra, Tomlin, A., & Tracy. (1999). Maths – our ideas all came into one. *RaPAL Bulletin, 38*, 14-18.

Haacke, F. (1999). *Fun with mathematics. Adults Learning Mathematics*. Workshop given at the Sixth International Conference of Adults Learning Mathematics – a Research Forum (ALM-6). Sheffield Hallam University, UK..

Harris, S. (2001). Being politically impolite: Extending politeness theory to adversarial political discourse. *Discourse & Society, 12*(4), 451-472.

Lave, J. (1992). Word problems: A microcosm of theories of learning. In P. Light & G. Butterworth (Eds.), *Context and cognition: Ways of learning and knowing* (pp. 74-92). New York: Harvester Wheatsheaf.

Lave, J. (1993). The Practice of Learning. In S. Chaiklin & J. Lave (Eds.), *Understanding practice: Perspectives on activity and context* (pp. 3-32). New York: Cambridge University Press.

Lave, J. (1996). Teaching, as learning, in practice. *Mind, Culture and Activity, 3*(3), 149-164.

Lave, J. (1997). The Culture of Acquisition and the Practice of Understanding. In D. Kirshner & J. A. Whitson (Eds.), *Situated Cognition: Social, Semiotic and Psychological Perspectives* (pp. 17-36). London: Lawrence Erlbaum Associates.

McDermott, R. P. (1993). The acquisition of a child by a learning disability. In S. Chaiklin & J. Lave (Eds.), *Understanding practice: Perspectives on activity and context* (pp. 269-305). New York: Cambridge University Press.

Mehan, H. (1993). Beneath the skin and between the ears: A case study in the politics of representation. In S. Chaiklin & J. Lave (Eds.), *Understanding practice: Perspectives on activity and context* (pp. 241-267). New York: Cambridge University Press.

Mullany, L. (2004). Gender, politeness and institutional power roles: Humour as a tactic to gain compliance in workplace business meetings. *Multilingua, 23*(1-2), 13-37.

Tomlin, A. (2001). *Participatory approaches to work with adult basic mathematics students.*, Unpublished PHD thesis, King's College, London.

Tomlin, A. (2002). Literacy approaches in the numeracy classroom. *Literacy and Numeracy Studies, 11*(2), 9-24.

Tomlin, A. (2005). Who asks the questions. In M. Herrington & A. Kendall (Eds.), *Insights from research and practice: A handbook for adult literacy, numeracy and ESOL practitioners* (pp. 369-377). Leicester: NIACE.

Van Manen, M. (1990). *Researching lived experience: Human science for an action sensitive pedagogy*. New York: State University of New York Press.

Wilson, S., & Tomlin, A. (1999). How do you do maths? [Electronic Version]. *Adults Learning Maths Newsletter* 8, http://www.alm-online.org/Newsletters/newsletter8.htm#maths

Acknowledgements

We are particularly grateful to Andri Marcou, who did an immense amount of detailed work preparing the manuscript for publication, while she was a doctoral student at London South Bank University. Alan Bishop has been supportive of this project from the outset. The anonymous reviewers provided useful insights, encouragement and challenges which have been gratefully taken into account. The editorial team at Springer nurtured us from afar and tolerated unavoidable delays and changes. Alaster Douglas and Jingjing Zhang, while doctoral students at the University of Oxford, helped with the technical aspects of the video conference, making it all work.

Finally, our partners, Monique and John, kept us going.

Index of Authors

Adler, J., 52, 270, 314, 329
Almeida, A., 271
Andrade, F., 213
Andrews, J., 152
Araújo, C. R., 213
Artigue, M., 240
Asiala, M., 238
Askew, M., 60
Assude, T., 240
Axel, E., 186
Baker, D., 133
Bakhurst, D., 186
Barab, S. A., 260
Barton, B., 96, 157
Barton, D., 134
Becher, T., 260
Béguin, P., 27
Behr, M., 166
Bereiter, C., 33, 34
Bernstein, B., 326, 327
Berry, J., 126, 240, 241
Bezuidenhout, J., 238
Biglan, A., 260
Bingolbali, E., 241
Bishop, A. J., 228
Bkouche, R., 15
Blunk, L., 324
Boaler, J., 33, 34, 82, 83, 87, 308, 343
Boero, P., 110
Borges, O., 209, 212, 219, 220, 229
Bourdieu, P., 21
Bowers, J., 158
Boylan, M., 11
Braga, S. M., 216
Brito Lima, A. P., 218, 222
Brossard, M., 22, 23, 24, 25
Brown, A., 354
Brown, J. S., 157
Brown, M., 60
Bruner, J. S., 177
Burman, E., 327
Burton, L., 223, 328, 329
Cain, C., 82, 237
Cardoso, R., 271
Carey, D., 167
Carlgren, I., 2, 303
Carpenter, T. P., 167, 168
Carr, M., 62
Carraher, D. W., 210
Castela, C., 239, 240, 260
Chaiklin, S., 157, 185
Charlot, B., 15, 16
Chevallard, Y., 125, 237, 239
Chi, M. T. H., 126
Clark, M., 96
Clot, Y., 14, 26, 27, 214
Clough, P., 354
Cobb, P., 99, 158
Cockcroft, W., 303
Cognition and Technology Group at Vanderbilt, 71

Cole, M., 18, 34, 108, 109, 157, 184, 191, 198
Collins, A., 132, 157
Cooney, T., 303
Cooper, B., 326
Cottrill, J., 238
Crawford, K., 292, 301
Crook, C., 243, 260
Culpepper, S., 320
Curt, B. C., 11
Da Rocha Falcão, J. T., 209, 210, 211, 213, 218, 220, 222
Daniels, H., 239, 260
David, M. M., 35
Davydov, V. V., 108, 110, 111, 112, 119, 186, 292
Department for Education and Employment, 136, 137
Desanti, J. T., 15
Dienes, Z. P., 107
Dirks, M. K., 160
Dreyfus, T., 110, 112
Dubinsky, E., 238
Duffy, T. M., 260
Duguid, P., 132, 157
Dunne, M., 326
Eisenman, T., 155
Ellis, S., 126
Engestrom, Y., 34, 157, 177, 184, 187, 188, 199, 205
Engeström, Y., 324
Engeström, Y., 108, 237, 239
Ernest, P., 211, 212, 214, 217, 220, 222, 223, 224
Evans, J. T., 328
Faïta, D., 214
Fennema, E., 167
Fernandez, G., 214
Fevre, R., 62
Fischbein, E., 159, 160, 166
Fontinhas, F., 327
Forman, E. A., 123
Frade, C., 209, 212, 214, 216, 219, 220, 223, 224, 227, 228, 229
Freudenthal, H., 156, 159
Fuller, A., 325
Furlong, J., 62
Fuson, K. C., 167
Gardner, W., 126
Gay, J., 18
Gee, J. P., 109
Gergen, K., 63, 69, 77
Gerofsky, S., 347
Gibson, J. J., 32
Giddens, A., 185, 189, 190, 199, 324
Gillborn, D., 85
Gipps, C., 302
Glick, J. A., 18
Goldin-Meadows, S., 175
Goldstein, L. S., 62, 123
Gravemeier, K., 156
Gravemeijer, K., 110, 158
Graven, M., 308, 312
Gray, E., 110
Gray, E. M., 156
Gray, J., 336
Greenhough, P., 132, 134, 148, 149, 151, 152
Greeno, J., 32, 64
Griesemer, J., 132
Grigorenko, E., 219
Grugeon, G., 240
Guba, E. G., 241
Haacke, F., 339
Halliday, M. A. K., 173
Hamilton, M., 134
Harris, S., 353
Hasan, R., 173
Hausmann, R. G., 126
Hazin, I., 213
Heath, S. B., 134
Heckman, P. E., 157
Hedegaard, M., 185
Hemmi, K., 315, 329
Herrington, J., 328
Hershkowitz, R., 110, 112, 113, 126
Hiebert, J., 168
Hodgkin, R. A., 219
Hodkinson, H., 325
Hodkinson, P., 325
Holland, D., 82, 191, 237, 261
Holm, C., 322, 328
Holzman, L., 65, 66, 69
Houssart, H., 34
Hoyles, C., 34, 52, 55, 71, 101, 107, 110, 111, 125, 126, 127, 295
Hudson, B., 301, 303

Hughes, M., 132, 134, 136, 148, 149, 151, 152
Hutchins, E., 19, 20, 177
Imsen, G., 303
Jensen, U. J., 185
Jeong, H., 126
Kanes, C., 33
Kansanen, P., 303
Kattah, V., 336
Koukouffis, A., 155, 156, 160, 166, 174
Kutscher, B., 155, 167, 168
Lachicotte Jr., W., 82
Lachicotte, W., 237
Lagrange, J-b., 239
Lampert, M., 324
Latour, B., 213
Lave, J., 4, 18, 19, 20, 33, 34, 81, 82, 83, 84, 85, 96, 99, 102, 108, 109, 132, 135, 157, 177, 184, 185, 186, 188, 189, 190, 191, 199, 205, 215, 216, 231, 265, 267, 268, 269, 270, 292, 293, 294, 295, 301, 307, 308, 311, 313, 314, 315, 317, 318, 319, 320, 324, 325, 327, 329, 331, 341, 342, 347, 351, 353
Lehtinen, E., 108, 112
Lenfant, A., 240
Leontiev, A. N., 14, 18, 19, 20, 25, 28, 112, 157, 158, 186, 189, 211, 222, 292
Lerman, S., 33, 76, 82, 83, 232, 292, 293, 307, 308
Lesh, R., 166
Lewis, C., 149
Liebeck, P., 159, 162
Light, P., 243, 260
Linchevski, L., 155, 156, 157, 159, 161, 162, 167, 169
Lincoln, Y. S., 241
Lins Lessa, M. M., 213, 220
Lipka, J., 219
Locke, J., 106
Lopes, M. P., 35
Lytle, P. A., 160
Magajna, Z., 286, 287
Markova, A. K., 292
Matos, J. F., 186, 215, 216, 222, 223
Maull, W., 240, 241
McDermott, R., 65, 322
McDermott, R. P., 342, 354
McGinn, M., 330
McNeill, D., 175
Mehan, H., 354
Meier, E., 219
Middle-school Mathematics through Applications Program (MMAP), 64
Miettien, R., 184
Millett, A., 60
Mohatt, G., 219
Monaghan, J., 113, 124, 126, 287
Morais, A., 327
Moreira, V. G., 266, 267, 271, 272
Moro, C., 19
Mullany, L., 353
Murtaught, M., 188
Nascimento, J. C., 213
Neves, I., 327
Newman, F., 66
Noddings, N., 65
Noss, R., 71, 101, 107, 110, 111, 126, 127, 295
Nunes, T., 33, 210, 215
Ohlsson, S., 108, 112
Oliver, R., 328
Olson, D. R., 21, 22, 25
Ongstad, S., 303
Ozmantar, M. F., 113, 124, 126
Pallas, A., 312
Pepin, B., 303
Petersen, P. L., 167
Piaget, J., 107, 108, 111, 210, 211, 220, 221, 222
Pinto, M. M. F., 266, 271
Polanyi, M., 211, 212, 214, 215, 216, 217, 219, 222, 230, 231
Ponte, J. P., 216, 222, 223
Post, T., 166
Pozzi, S., 295
Prado, H., 271
Praslon, F., 240
Radford, L., 175
Rees, G., 62
Reeves, T. C., 328
Resnick, L. B., 19
Restivo, S., 190
Rocha, O., 188
Rogoff, B., 18, 19, 20, 33, 126
Romberg, T. A., 223
Roth, W. M., 330

Rouche, N., 15
Ryan, J. T., 156
Sahlberg, P., 126
Salway, L., 136, 152
Sampaio, V., 271
Santos, M. P., 186
Saxe, G. B., 33
Sazhin, S. S., 240
Scheller, L., 214
Schliemann, A. D., 210
Schoenfeld, A. H., 223
School Mathematcs Project, 89
Schwartz, B., 157
Schwarz, B. B., 110, 112
Schwingendorf, K., 238
Scribner, S., 18
Semadeni, Z., 159
Sfard, A., 156
Sharp, D. W., 18
Shorter, G., 96
Sierpinska, A., 110
Siler, S. A., 126
Silver, E., 166
Skemp, R., 107
Skemp, R. R., 16
Skinner, D., 82, 237
Smith, A., 303, 304
Snyder, W., 322
Solomon, Y., 149, 313
St. Julien, J., 37
Star, S. L., 132
Sternberg, R., 219
Street, B., 133, 134, 135, 143, 148
Tall, D., 110, 266
Tall, D. O., 156
Thom, R., 15, 303
Thompson, A. G., 303
Tirosh, D., 220
Tizard, B., 132
Tomlin, A., 133, 336, 341, 347
Torrance, N., 21
Toulmin, S., 185
Treffers, A., 156
Trowler, P., 260
Unwin, L., 325
Van Manen, M., 87, 354
van Oers, B., 110, 111
Varenne, H., 65
Vergnaud, G., 211, 213, 220, 221, 222
Vinner, S., 266, 271
Volosinov, V. N., 25
Vygotsky, L. S., 15, 18, 19, 20, 21, 22, 23, 24, 25, 64, 157, 220, 221, 222, 292, 322, 324, 327
Wake, G. D., 156, 177, 178
Walkerdine, V., 132, 158, 327
Warin, J., 149
Watson, A., 1, 7, 35, 37, 71, 81, 82, 84, 86, 87, 99, 132, 215, 216, 222, 229, 268, 269, 270, 283, 287, 311
Wedege, T., 311, 318
Weil, S. W., 62
Weissglass, J., 157
Wenger E., 307, 308, 310, 311, 314, 315, 317, 318, 319, 320, 321, 322, 324, 325, 327, 329, 331
Wenger, E., 3, 7, 19, 32, 35, 131, 132, 133, 149, 150, 152, 157, 177, 184, 185, 186, 191, 199, 205, 206, 215, 216, 231, 237, 265, 268, 292, 293, 294, 301
Wenger, E., 81, 82, 83, 99
Wertsch, J. V., 63, 120, 156, 157, 239
Whitenack, J., 158
Wigner, E. P., 219
Wiliam, D., 83
Williams, J. S., 155, 156, 157, 159, 160, 161, 162, 166, 167, 169, 174, 175, 177, 178
Wilson, S., 336
Winbourne, P., 7, 33, 35, 37, 71, 81, 82, 84, 86, 87, 96, 99, 215, 216, 222, 224, 229, 268, 269, 270, 283, 287, 311
Winter, J., 136, 152
Woolgar, S., 213
Yamauchi, T., 126
Yanez, E., 219
Yee, W. C., 136, 152
Yin, R. K., 173
Youdell, D., 85
Young-Loveridge, J., 69
Zevenbergen, R., 83
Zumpano, A., 271

Index

abstraction 20, 23, 105, 161, 188, 225, 305, 321, 326
activity 13, 35, 52, 66, 82, 96, 101, 105, 121, 128, 157, 169, 177, 180, 186, 188, 190, 201, 207, 213, 222, 236, 243, 267, 274, 293, 299, 308, 314, 319, 324, 326, 351, 361
adult 342, 344, 362
affect 215, 234, 262, 265, 309, 316, 335
affordance 4, 34, 46, 52, 56, 58, 63, 66, 177, 209
alignment 84, 103, 319
apprenticeship 19, 33, 35, 39, 82, 83, 219, 236, 268, 299, 314, 323, 335
artefact 20, 54, 106, 109, 120, 127, 186, 187, 189, 193, 197, 201, 207, 320, 331

becoming 84, 236, 276, 290, 326
boundary 6, 109, 111, 132, 135, 143, 150, 153, 158, 181, 219, 274, 320, 322, 327, 337

classroom 9, 34, 39, 51, 63, 68, 70, 85, 95, 100, 159, 176, 179, 228, 275, 291, 294, 319, 333, 350
cognition 3, 14, 17, 21, 34, 37, 71, 79, 112, 158, 162, 180, 189, 221, 249, 254, 262, 265, 319, 326, 343
communication 21, 124, 173, 178, 181, 204, 208, 238, 285, 305, 309, 332
communities of mathematical practice 7, 37, 51, 83, 86, 111, 238, 277, 295, 351

discourse 8, 15, 24, 51, 66, 110, 159, 176, 218, 221, 287, 291, 295, 298, 332, 334, 335, 362, 363
discussion 2, 3, 72, 94, 121, 205, 209, 245, 287, 321, 344

enculturation 36, 222
everyday 5, 14, 25, 30, 111, 132, 153, 158, 181, 185, 207, 300, 315, 333, 349

generalisation 53, 108, 178
goals 7, 109, 160, 179, 189, 191, 217, 264, 267, 277, 293, 298, 314, 335

home 98, 132, 143, 145, 329, 333, 348, 360

identity 2, 5, 7, 29, 35, 62, 64, 71, 81, 83, 85, 92, 100, 101, 102, 133, 143, 145, 151, 180, 219, 243, 313, 319, 322, 326, 335, 360, 362
institutions 10, 64, 109, 127, 244, 246, 262, 265, 267, 275, 319, 329
intentional teaching 6, 33, 39, 274

knowing 2, 132, 214, 216, 221, 224, 228, 238, 310, 336

language 12, 21, 34, 51, 111, 148, 159, 179, 181, 218, 221, 224, 227, 231, 277, 291, 298
legitimate peripheral participation 35, 57, 83, 100, 103, 219, 237, 299, 317, 330, 361
local communities of practice 7, 37, 46, 51, 58, 84, 220, 235, 274, 276, 290

mediation 4, 20, 30, 36, 99, 106, 109, 120, 160, 178, 193, 205, 215, 320, 331

narrative 11, 93, 322, 325, 328, 363
norms 21, 34, 43, 46, 71, 123, 246, 267, 319
novices *See* apprenticeship
numeracy 69, 134, 136, 146, 298, 351, 360, 362

out-of-school 8, 30, 93, 96, 98, 134, 145, 152, 153, 158, 170, 172, 177, 181, 214, 217, 226, 237, 300

parents 135, 144, 150, 152, 309, 361
pedagogy 16, 66, 78, 125, 310, 333, 338

school 5, 13, 17, 19, 20, 25, 30, 36, 38, 64, 68, 70, 78, 81, 84, 87, 93, 99, 100, 124, 132, 134, 144, 148, 152, 160, 177, 179, 181, 214, 218, 226, 229, 237, 245, 268, 273, 276, 278, 290, 293, 306, 309, 314, 318, 334, 338, 347, 351, 360
semiotic 158, 178, 215, 225
social practice 34, 57, 96, 123, 132, 180, 186, 191, 194, 205, 207, 218, 221, 238, 245, 274, 299, 308, 314, 317, 325, 331, 333, 337

talk *See* discussion
teaching 6, 17, 25, 34, 53, 65, 69, 78, 84, 87, 88, 96, 100, 102, 125, 137, 153, 161, 163, 170, 181, 216, 222, 228, 249, 260, 263, 267, 274, 293, 309, 314, 317, 321, 329, 334, 338, 358
tool 3, 14, 19, 27, 29, 36, 46, 49, 54, 66, 110, 120, 127, 159, 179, 187, 200, 203, 220, 292, 298, 309, 331, 337
trajectories 6, 83, 317, 320, 325, 330, 361
transfer 5, 36, 101, 133, 159, 179, 314

workplace 5, 195, 202, 218, 280, 288, 293, 297, 300, 303, 307, 332, 335, 344

Mathematics Education Library

Managing Editor: A.J. Bishop, Melbourne, Australia

H. Freudenthal: *Didactical Phenomenology of Mathematical Structures.* 1983. ISBN 90-277-1535-1 HB; 90-277-2261-7 PB

B. Christiansen, A. G. Howson and M. Otte (eds.): Perspectives on Mathematics Education. Papers submitted by Members of the Bacomet Group. 1986 90-277-1929-2 HB; 90-277-2118-1 PB

A. Treffers: Three Dimensions. A Model of Goal and Theory Description in Mathematics Instruction. TheWiskobas Project. 1987. ISBN 90-277-2165-3

S. Mellin-Olsen: The Politics of Mathematics Education. 1987. ISBN 90-277-2350-8

E. Fischbein: Intuition in Science and Mathematics. An Educational Approach. 1987. ISBN 90-277-2506-3

A.J. Bishop: Mathematical Enculturation. A Cultural Perspective on Mathematics Education.1988 ISBN 90-277-2646-9 HB; 0-7923-1270-8 PB

E. von Glasersfeld (ed.): Radical Constructivism in Mathematics Education. 1991. ISBN 0-7923-1257-0

L. Streefland: Fractions in Realistic Mathematics Education. A Paradigm of Developmental Research. 1991. ISBN 0-7923-1282-1

H. Freudenthal: Revisiting Mathematics Education. China Lectures. 1991. ISBN 0-7923-1299-6

A.J. Bishop, S. Mellin-Olsen and J. van Dormolen (eds.): Mathematical Knowledge: Its Growth Through Teaching. 1991. ISBN 0-7923-1344-5

D. Tall (ed.): Advanced Mathematical Thinking. 1991. ISBN 0-7923-1456-5

R. Kapadia and M. Borovcnik (eds.): Chance Encounters: Probability in Education. 1991. ISBN 0-7923-1474-3

R. Biehler, R.W. Scholz, R. Sträßer and B. Winkelmann (eds.): Didactics of Mathematics as a Scientific Discipline. 1994. ISBN 0-7923-2613-X

S. Lerman (ed.): Cultural Perspectives on the Mathematics Classroom. 1994. ISBN 0-7923-2931-7

O. Skovsmose: Towards a Philosophy of Critical Mathematics Education. 1994. ISBN 0-7923-2932-5

H. Mansfield, N.A. Pateman and N. Bednarz (eds.): Mathematics for Tomorrow's Young Children. International Perspectives on Curriculum.1996. ISBN 0-7923-3998-3

R. Noss and C. Hoyles: Windows on Mathematical Meanings. Learning Cultures and Computers.1996. ISBN 0-7923-4073-6 HB; 0-7923-4074-4 PB

N. Bednarz, C. Kieran and L. Lee (eds.): Approaches to Algebra. Perspectives for Research and Teaching.1996. ISBN 0-7923-4145-7 HB; 0-7923-4168-6, PB

G. Brousseau: Theory of Didactical Situations in Mathematics. Didactique des Mathématiques 1970-1990. **Edited and translated by N. Balacheff, M. Cooper, R. Sutherland and V. Warfield**. 1997. ISBN 0-7923-4526-6

T. Brown: Mathematics Education and Language. Interpreting Hermeneutics and Post-Structuralism. 1997 ISBN 0-7923-4554-1 HB. Second Revised Edition. 2001. ISBN 0-7923-6969-6 PB

D. Coben, J. O'Donoghue and G.E. FitzSimons (eds.): Perspectives on Adults Learning Mathematics. Research and Practice. 2000. ISBN 0-7923-6415-5

R. Sutherland, T. Rojano, A. Bell and R. Lins (eds.): Perspectives on School Algebra. 2000. ISBN 0-7923-6462-7

J.-L. Dorier (ed.): On the Teaching of Linear Algebra. 2000. ISBN 0-7923-6539-9

A. Bessot and J. Ridgway (eds.): Education for Mathematics in the Workplace. 2000. ISBN 0-7923-6663-8

D. Clarke (ed.): Perspectives on Practice and Meaning in Mathematics and Science Classrooms. 2001. ISBN 0-7923-6938-6 HB; 0-7923-6939-4 PB

J. Adler: Teaching Mathematics in Multilingual Classrooms. 2001. ISBN 0-7923-7079-1 HB; 0-7923-7080-5 PB

G. de Abreu, A.J. Bishop and N.C. Presmeg (eds.): Transitions Between Contexts of Mathematical Practices. 2001. ISBN 0-7923-7185-2

G.E. FitzSimons: What Counts as Mathematics? Technologies of Power in Adult and Vocational Education. 2002. ISBN 1-4020-0668-3

H.Alrø andO. Skovsmose: Dialogue and Learning in Mathematics Education. Intention, Reflection,Critique.2002. ISBN 1-4020-0998-4 HB;1-4020-1927-0 PB

K. Gravemeijer, R. Lehrer, B. van Oers and L. Verschaffel (eds.): Symbolizing, Modeling and Tool Use in Mathematics Education. 2002. ISBN 1-4020-1032-X

G.C. Leder, E. Pehkonen and G. T rner (eds.): Beliefs: A Hidden Variable in Mathematics Education? 2002. ISBN 1-4020-1057-5 HB; 1-4020-1058-3 PB

R. Vithal: In Search of a Pedagogy of Conflict and Dialogue for Mathematics Education. 2003. ISBN 1-4020-1504-6

H.W. Heymann: Why Teach Mathematics? A Focus on General Education. 2003. **Translated by T. LaPresti** ISBN 1-4020-1786-3

L. Burton: Mathematicians as Enquirers: Learning about Learning Mathematics. 2004. ISBN 1-4020-7853-6 HB; 1-4020-7859-5 PB

P. Valero, R. Zevenbergen (eds.): Researching the Socio-Political Dimensions of Mathematics Education: Issues of Power in Theory and Methodology. 2004. ISBN 1-4020-7906-0

D. Guin, K. Ruthven, L. Trouche (eds.): The Didactical Challenge of Symbolic Calculators: Turning a Computational Device into a Mathematical Instrument. 2005. ISBN 0-387-23158-7

J. Kilpatrick, C. Hoyles, O. Skovsmose (eds. in collaboration with Paola Valero): Meaning in Mathematics Education. 2005. ISBN 0-387-24039-X

H. Steinbring: The Construction of New Mathematical Knowledge in Classroom Interaction: An Epistemological Perspective. 2005. ISBN 0-387-24251-1

M.Borba, M. Villarreal: Humans-with-Media and the Reorganization of Mathematical Thinking: Information and Communication Technologies, Modeling, Visualization and Experimentation. 2005. ISBN 0-387-24263-5 HB; ISBN 0-387-32821-1 PB

G. Jones (ed): Exploring Probability in School: Challenges for Teaching and Learning. 2005. ISBN 0-387-24529-4

D. DeBock, W. Van Dooren, D. Janssens, and L. Verschaffel: The Illusion of Linearity: From Analysis to Improvement. 2007. ISBN 978-0-387-71082-2

K. Francois and J. P. Van Bendegem: Philosophical Dimensions in Mathematics Education.2007. ISBN 978-0-387-71571-1

E. Filloy, L. Puig, and T. Rojano: Educational Algebra: A Theoretical and Empirical Approach. 2007. ISBN 978-0-387-71253-6

B. Barton: The Language of Mathematics. 2007. ISBN 978-0-387-72858-2

P. Winbourne, A. Watson: New Directions for Situated Cognition in Mathematics Education. 2007. ISBN 978-0-387-71577-3

E. DeFreitas, K. Nolan: Opening the Research Text. 2007. ISBN 978-0-387-75463-5

A. Watson, P. Winbourne: New Directions for Situated Cognition in Mathematics Education. 2007. ISBN 978-0-387-71577-3

Printed in the United States of America